Lakes of
New York State

VOLUME I: Ecology of the Finger Lakes

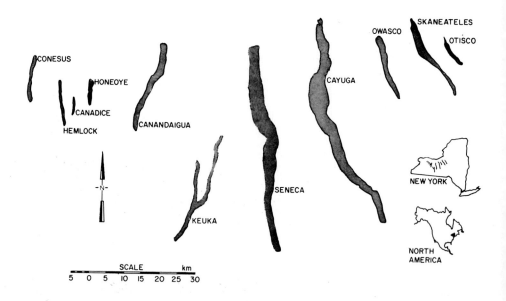

The Finger Lakes

Lakes of New York State

VOLUME I
ECOLOGY OF THE FINGER LAKES

EDITED BY

Jay A. Bloomfield

New York State Department of Environmental Conservation
Albany, New York

ACADEMIC PRESS   New York   San Francisco   London   1978
A Subsidiary of Harcourt Brace Jovanovich, Publishers

ACADEMIC PRESS, INC.
111 Fifth Avenue, New York, New York 10003

United Kingdom Edition published by
ACADEMIC PRESS, INC. (LONDON) LTD.
24/28 Oval Road, London NW1

Library of Congress Cataloging in Publication Date

Main entry under title:

Lakes of New York State.

    Includes indexes.
    CONTENTS: v.1 Ecology of the Finger Lakes
    1. Lakes—New York (State) 2. Limnology—
New York (State)   3. Lake ecology—New York (State)
I. Bloomfield, Jay A.
GB1625.N7L34   574.5'2632'09747   77-6588
ISBN 0-12-107301-7

Contents

THE LIMNOLOGY OF CAYUGA LAKE
Ray T. Oglesby

THE LIMNOLOGY OF CONESUS LAKE
Herman S. Forest, Jean Q. Wade, and Tracy F. Maxwell

v

THE LIMNOLOGY OF CANANDAIGUA LAKE
Stephen W. Eaton and Larry P. Kardos

LIMNOLOGY OF EIGHT FINGER LAKES: HEMLOCK, CANADICE, HONEOYE, KEUKA, SENECA, OWASCO, SKANEATELES, AND OTISCO
W. R. Schaffner and R. T. Oglesby

List of Contributors

Numbers in parentheses indicate the pages on which the authors' contributions begin.

STEPHEN W. EATON (225), Department of Biology, St. Bonaventure University, St. Bonaventure, New York

HERMAN S. FOREST (121), Environmental Resource Center, State University College of Arts and Science, Geneseo, New York

LARRY P. KARDOS (225), Department of Biology, St. Bonaventure University, St. Bonaventure, New York

TRACY F. MAXWELL (121), Environmental Resource Center, State University College of Arts and Science, Geneseo, New York

RAY T. OGLESBY (1, 313), Department of Natural Resources, New York State College of Agriculture and Life Sciences, Cornell University, Ithaca, New York

W. R. SCHAFFNER (313), Department of Natural Resources, New York State College of Agriculture and Life Sciences, Cornell University, Ithaca, New York

JEAN Q. WADE (121), Environmental Resource Center, State University College of Arts and Science, Geneseo, New York

Foreword

Between 1926 and 1938 the New York State Department of Conservation published a series of biological surveys of the river drainage basins in the State. The studies carried out and the resulting reports must be considered classic biological inventories and, even today, represent the best set of biological information on the surface waters of New York State.

Anyone working in the State quickly realizes their value. Their influence on management of these waters and on the scope and direction of subsequent studies has been and continues to be so prominent that one wonders what would have been done had they not existed. Rarely does a study begin or is a report published on any lake or stream in the State that does not utilize these as basic references.

However, even as the last of the surveys was completed in 1938, the "desirability of up-to-date revisions of former surveys" was recognized. Unfortunately, this has not been done, and an examination of the original surveys reveals this need for updating. Since the original surveys, conditions in some systems have changed, the state of the art of measuring and interpreting data has advanced, and numerous studies provide a wealth of additional information.

Because of the demonstrated values of these biological inventories and the increased interest in lake management, the Department of Environmental Conservation commissioned the preparation of this treatise on lakes in New York State. The contributions were to be inventory reports modeled after the original biological surveys, but were to be expanded to cover the physical, chemical, and biological state of the lake and its drainage basins. Preparation of the contribution was not to include carrying out of any special field studies but was to be based on existing data and information. Fortunately, in New York State, for most of the important lakes, a great amount of data has been collected and reported as part of studies directed at one or another aspect of the lake. Additionally, the State has a reservoir of resident limnologists who by virtue of their location and interest were uniquely capable of authoring these articles.

The purpose of these articles may be explicitly stated as follows:

1.   To provide in a unified manner an authoritative set of current data from which lake management decisions can be made.

2.   To begin to meet the requirements of the 1970 Federal water pollution control legislation which requires "that the State shall prepare an identification and classification according to the trophic condition of all publicly-owned freshwater lakes in such State."

3.   To provide a uniform set of data from which comparative lake studies could be made.

4.   To provide a set of uniform baseline data on lakes from which future changes can be measured.

5.   To provide for students a set of real data that will bring to life their classroom experience.

6.   To determine where gaps in our knowledge of New York State lakes exist, and provide a basis for investment in future studies and research.

In an age when there is an increased awareness of the value of the environment, when environmental laws and regulations are multiplying at an unprecedented rate, when environmental legal proceedings are commonplace, and when environmental impact statements have become a way of life, these volumes should prove useful.

The idea of this treatise series was originally conceived in 1972 and, after many setbacks, the project commenced in 1974. The leadership, assistance, and patience of so many people were instrumental in their completion that space prevents a listing of their names and contributions. However, I feel compelled to mention the commitment of Mr. Eugene Seebald, Director of the Division of Pure Waters, who provided steadfast support for the effort, even when short-term needs suggested utilization of the required resources in other areas.

Leo J. Hetling
Albany, New York

Preface

There are over 4000 lakes in New York State and the most prominent are the Finger Lakes, a group of eleven elongated bodies of water of glacial origin in the west-central portion of the State. Seven of the Finger Lakes lie within the Oswego River drainage basin; the remainder are within the Genesee River basin.

The Finger Lakes have a rich legacy of ecological research beginning with the classic investigations of Birge and Juday (1912).[1] The next major publications dealing with the Lakes were the Biological Surveys of the Oswego River System (1928)[2] and the Genesee River System (1927).[3] These documents describe the biota, water chemistry, and physical nature of the Lakes in the context of establishing a scientific framework for management of the Finger Lakes sport fishery. Research done on the Lakes subsequent to the Biological Surveys has involved many investigators from institutions of higher learning in New York State.

The intent of this volume is to describe the state of each Finger Lake and its respective watershed. The information presented should be considered as "anatomy" rather than "physiology," for many of the finer points of the dynamics of each Lake have been glossed over. The content is presented in order to document existing knowledge rather than to present data interpretation, laboratory methodology, or ecological theory. The volume represents a framework on which future ecological studies of the Finger Lakes can be based. Much of the material included herein is of a general nature and hence should be of interest to most ecologists working on temperate zone freshwater lakes. The individual contributions could also be

[1] E. A. Birge and C. Juday, A limnological study of the Finger Lakes of New York. *Bull. Bur. Fish. Doc. 791* (1912) **32,** 525–609 (1914).

[2] New York State Conservation Department, "A Biological Survey of the Oswego River System," Suppl. 17th Annu. Rep., 1927, 248 pp. Albany, New York, 1928.

[3] New York State Conservation Department, "A Biological Survey of the Genesee River System," Suppl. 16th Annu. Rep., 1926, 86 pp. Albany, New York, 1927.

used to complement a textbook on limnology. Oglesby discusses the most frequently studied Finger Lake, Cayuga, with the aim of documenting the Lake's trophic status. Mass budgets for the important ions are presented to emphasize the dominance of activities and processes in the atmosphere and the terrestrial ecosystem which determines the status of the Lake.

Forest takes a more classic approach to the ecology of Conesus Lake, and assesses the structure of the Lake's plant and animal communities and how these communities interact with the abiotic components of the environment. Eaton and Kardos assess the condition of Canandaigua Lake from the standpoint of fish population dynamics, examining the various physical, chemical, and biological aspects of the Lake's ecosystem. Last, Schaffner presents a summarized comparative account of the eight less well-studied Lakes, including Hemlock, Canadice, Honeoye, Keuka, Seneca, Owasco, Skaneateles, and Otisco. His task was unenviable because of the limited nature of data available on these Lakes, particularly information related to the higher trophic level organisms such as fish. Yet definite comparisons and conclusions are made as to each lake's productivity.

As a summary, the tabulation below presents current trophic information available for all eleven Finger Lakes culled from these contributions and an excellent review article by Oglesby and Schaffner (1975).[4] It lists three commonly used indicator measurements: average summer Secchi disc depth, average summer chlorophyll a concentration, and average winter

Lake	Average summer Secchi disc depth		Average summer chlorophyll a concentration		Average winter total phosphorus level	
	Meters	Rank[a]	mg/m^3	Rank[a]	mg/m^3	Rank[a]
Conesus	4.9	7	6.3	5	17.6	3
Hemlock	3.5	4	6.4	4	10.9	7
Canadice	5.2	8	4.4	7	9.2	9
Honeoye	3.0	3	13.2	1	16.2	4
Canandaigua	4.1	6	2.6	9	10.1	8
Keuka	7.0	10	3.3	8	11.5	6
Seneca	3.6	5	7.1	3	17.8	2
Cayuga	2.1	1	8.7	2	21.1	1
Owasco	3.0	2	5.6	6	14.7	5
Skaneateles	7.0	10	1.4	11	7.7	11
Otisco	5.7	9	2.2	10	8.4	10

[a] Ranked 1, most eutrophic; 11, least eutrophic.

[4] R. T. Oglesby and W. R. Schaffner, The response of lakes to phosphorus. *In* "Nitrogen and Phosphorus: Food Production, Waste and Environment" (K. S. Porter, ed.), pp. 25–60. Ann Arbor Sci. Publ., Ann Arbor, Michigan, 1975.

total phosphorus level. By evaluating these three crude measures, one can rank the eleven Lakes into a unilateral trophic list. If this technique seems arbitrary to the reader, one is encouraged to read on and make one's own personal judgments at both the ecosystem and regional level.

This publication was supported under a United States Environmental Protection Agency Annual Water Pollution Control Program Grant M002060-03-3 to the New York State Department of Environmental Conservation for Federal Fiscal Year 1975.

The contributions have been reviewed by the New York State Department of Environmental Conservation and approved for publication. Approval does not signify that the contents necessarily reflect the views and policies of the Department of Environmental Conservation.

Jay A. Bloomfield

Contents of Other Volumes

The Limnology of Cayuga Lake

Ray T. Oglesby

INTRODUCTION

Cayuga Lake is a superb limnological laboratory, a fact recognized by E. A. Birge and C. Juday when they carried out the first scientific studies of it in 1910. It is also a resource of great natural beauty and considerable economic importance to the people of New York State. Despite these facts and the added stimulus which the presence of a major university near its shores might have been expected to exert, Cayuga was the site of only sporadic and largely specialized investigations by limnologists until the mid-1950's. Beginning at that time and continuing at intervals to the present, a series of research projects by graduate students and faculty from Cornell University have both explored discrete limnological facets of Cayuga and provided the data base for a more comprehensive view of the lake and its basin. A significant impetus to research was provided in the late 1960's by a proposal to site a nuclear power plant (since modified in permit applications to one burning coal) on the shore of Cayuga Lake. Extensive investigations relating to this proposal were carried out in 1968–1969 (Cornell University and Cornell Aeronautical Laboratory) and from 1973 onward (Nuclear Utilities Service Corporation).

During the past few years, several investigations have concentrated on nutrient–phytoplankton relations. It was this work which led the author to prepare an initial report to the North American Lakes Project (part of the Comparative Projects for Monitoring Inland Water Programme of the Organization for Economic Cooperation and Development, Paris) that has served as the nucleus for this contribution. The information on nutrient budgets, especially, is designed to fit into a larger motif formulated by R. A. Vollenweider.

This contribution should provide a definitive summary of the existing literature concerned with Cayuga Lake. As will be apparent to the reader, there are many aspects of the limnology as yet incompletely worked out and others for which investigations have yet to be initiated. Among current research which, when published, should fill some of these gaps is an intensive study of phytoplankton standing crops and the factors which act as determinants on them (Godfrey, 1977), a paleolimnological investigation of the upper 3 m of sediment (McKenna, 1977), and research on the rooted macrophyte beds of the north end of the lake by Professor J. Peverly (Cornell) and Professor G. Miller (Eisenhower College).

In addition to summarizing extensive data, I have attempted to synthesize as many limnological aspects as possible into a rational framework that defines temporal (historical and seasonal) and spatial patterns. In some few instances, suggestions are offered concerning cause and effect relations.

Despite the many gaps in our knowledge and understanding of Cayuga, the information summarized in the following pages, when viewed as a whole, qualifies this lake as one of the best studied in the world. It is hoped that this compendium and discussion of information on Cayuga will benefit the limnologist in his study of lake processes and will also serve as a basis and stimulant for enlightened management of this important resource.

GEOGRAPHIC DESCRIPTION OF WATER BODY

The center of Cayuga Lake is located at latitude 42° 41′ 30″ N and 76° 41′ 20″ W. Its altitude is 116.4 m (382 ft) above sea level (Greeson and Williams, 1970), and the catchment area (including the lake's surface) is 2033 km² (785 square miles) according to the United States Department of the Interior (1971).

General Climatic Data

Birge and Juday (1914) cite nine occasions from 1795 to 1912 when Cayuga Lake was completely frozen over. No subsequent records could be found, but D. A. Webster (personal communication, 1973) states that the ice cover was virtually complete during one winter in the late 1940's or early 1950's. In a typical year sheets of ice extend out from the north and south ends only to about the 5- to 10-m depth contours with maximum coverage in February.

The nearest long-term climatological data collection stations are at Ithaca and Geneva. Exact locations have changed slightly and there are informational gaps, but data are available from the year 1828 for Ithaca and from 1850 for the latter site. Current locations are, for Ithaca, Caldwell Field located on the Cornell campus at an elevation of 290 m (950 ft) about

4 km east of the southern end of the lake, and for Geneva, the New York Agricultural Experiment Station at an elevation of 187 m (615 ft) some 19 km west–northwest of Cayuga Lake's north end.

The general climatic conditions of the Cayuga Lake area and summaries of temperature and precipitation data have been described by Dethier and Pack (1963a,b). The climate is of the humid continental type with warm summers and long, cold winters. The area lies on or near the major west to east track of cyclonic storms and hence is characterized by variety and frequent periods of stormy weather, particularly in winter.

Air temperatures at Ithaca of 32°C (90°F) or higher usually occur on 7 to 10 days a year, and winter temperatures of −18°C (0°F) or colder are normally recorded on 6 to 10 days from early December through early March. The freeze-free season for exposed areas near 1000 ft elevation averages about 145 days.

Ithaca's annual precipitation ranges from 71 cm (28 in.) to 117 cm (46 in.), but in most years is between 76 cm (30 in.) and 102 cm (40 in.). About half of this normally falls during May through September. Very heavy [5 cm (2 in.) or greater] precipitation occasionally falls from a single storm during the warmer months.

One of the more persistent climatic features of the Cayuga Lake area is cloudiness, especially during the winter months. Ithaca averages about 175 cloudy days a year and Geneva 170. The percentage of possible sunshine at Ithaca is less than 30% in November and December and increases to a maximum of 60% in June and July.

Prevailing winds in the area are from the southwest during the summer and the northwest in winter. Velocities at Ithaca average 11–16 km/hr (7–10 mph) from May through October and 18–19 km/hr (11–12 mph) during the colder months. Evaporation from a free surface (pan) is measured at Ithaca from May through October. Average monthly evaporation data are summarized in Table 1 and Fig. 1 together with mean values for other climatic parameters.

Cayuga Lake is located in a basin that opens into rather flat terrain at its north end but becomes progressively steeper toward the south (Fig. 2). On the east side of the lake this rise becomes an obvious feature about one-third of the lake's length from its northern terminus, and a similar rise occurs on the west side slightly farther to the south. As Birge and Juday (1914) described the basins of both Cayuga and Seneca ". . . the lakes are bounded by a steep slope, often precipitous at the bottom, which reaches in places 100 meters or more in height. Above this steep slope there is for much of the way a more or less definitely marked shelf, and above this there is another rise to more considerable isolated heights." The upland plateau or

TABLE 1

Mean Climatological Values at Ithaca, New York with Number of Years of Record Given in Parentheses[a]

Month	Temperature (°C), 1942–1962			Precipitation (cm) (21)	Wind vel. (km/hr) (30)	Evaporation (cm) (18)	Sunshine (% possible) (38)	Solar radiation (10^{-2} gm cal/cm^2) (10)
	Daily max. (21)	Daily min. (21)	Monthly (21)					
Jan	0	−8.9	−4.4	5.16	19.3	—	30	35.4
Feb	1.1	−8.9	−3.9	5.61	18.0	—	40	54.5
Mar	5.6	−5.0	0.6	6.83	18.9	—	42	84.6
Apr	13.3	1.1	7.2	7.67	18.0	—	46	108.0
May	19.4	6.1	12.8	9.37	14.6	10.4	54	149.1
Jun	25.0	11.7	18.3	8.91	13.2	13.5	60	164.8
Jul	27.2	13.9	20.5	9.55	12.2	15.0	62	172.4
Aug	26.7	13.3	20.0	9.78	12.2	12.7	59	147.0
Sep	22.2	9.4	16.1	7.87	13.7	8.9	55	108.3
Oct	16.7	4.4	10.6	8.38	15.3	5.8	44	78.0
Nov	8.3	−0.6	3.9	7.39	17.9	—	29	37.9
Dec	1.7	−7.2	−2.8	5.84	17.9	—	24	30.6
Annual Avg.	13.9	2.2	8.3	7.70	15.9		45	97.6
Total				92.36				1170.6

[a] From Dethier and Pack (1963a).

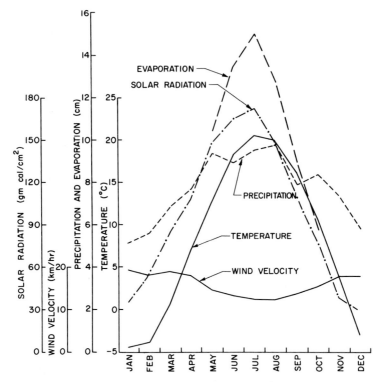

Fig. 1. Monthly mean values for solar radiation, wind velocity, precipitation, evaporation, and temperature at Ithaca, New York.

shelf is at an elevation of 250–300 m with hills beyond occasionally extending to about 600 m.

The effects of the surrounding topography on climatic conditions immediately over the lake's surface is largely a matter of casual observation and conjecture. Wright (1969a) has reviewed the available evidence and concluded that the shape of the basin probably exerts considerable influence on wind speed and direction in two ways. First, the generally north–south orientation of the basin channels the prevailing winds along the long axis of the lake; and second, downslope winds pouring cooler air into the basin occur during the warmer months at times when nighttime cooling is appreciable. In the autumn the passage of this cool air over the warmer water surface often results in dense, local fogs. On calm, sunny mornings the entire fog bank will sometimes rise intact as a fingerlike cloud that mirrors the shape of the lake surface. The large-scale culture of grapes and orchard crops on the slopes surrounding all of the Finger Lakes attests to the

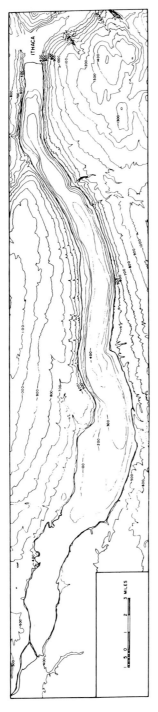

Fig. 2. The bathymetry of Cayuga Lake and the topography of its basin. Modified from Birge and Juday (1914). Ithaca is at the southern end of the lake. All contours are in feet. Conversion to metric units was not carried out since the object of this figure is to provide an overall view of the lake's bathymetry and the land form in its immediate vicinity.

ameliorating effects of the water surfaces on air temperatures, but quantitative data are not available.

North–south differences in the local climate may be estimated by comparing data from Ithaca with that taken at Aurora, New York. The latter climatological station was established in 1957. It is located about 3.5 km (2.2 miles) east of the lake and about one-third the distance from the north to the south end at an elevation of 253 km (830 ft). Comparing Ithaca and Aurora in terms of annual means through 1966 (Anonymous, 1972), temperatures were, respectively 7.7°C (46°F) and 6.7°C (44°F), total precipitation 89.5 cm (35.22 in.) and 78.7 cm (31.00 in.), and total evaporation for May through October 62.4 cm (24.55 in.) and 80.2 cm (31.59 in.). Annual average solar radiation for the period 1965–1972 was significantly different (t test, $\alpha = 0.05$) between Aurora and Ithaca with the latter being 8% greater.

General Geological Characteristics

Von Engeln (1961) has provided the authoritative description of the geomorphic processes that shaped the Finger Lakes and their surrounding landscapes, and the following summation is abstracted from his description.

For a period of 325 million years the Finger Lakes region was part of a vast inland sea. During this time evaporative processes and precipitation of particles from suspension resulted in the accumulation of bed after bed of sand, mud, lime, and salt. Eventually these materials were compressed into rocks with a total depth of some 2400 m (8000 ft).

About 200 million years ago the land was uplifted. At this time drainage was to the south through the Susquehanna system. Over the next 100 million years the uplifted land was gradually eroded into a peneplain which was then disrupted by additional uplifting. Postpeneplain headwater erosion produced deep north–south gaps in the landscape.

About 1 million years ago this pattern of fluvial cutting was interrupted by the advent of the Ice Age. Great masses of ice ground and gouged their way into a large funnel created by the low relief of land to the north of Cayuga Lake giving way to steep-sided valleys to the south. Twice glaciers advanced in this fashion with a long interglacial period between. The first glacial invasion caused major modifications in the rock topography while the second brought about great changes in the aspect of the countryside.

Postglacial time in the Finger Lake's region extends over a period of about the last 9000–10,000 years. During this period, sculpturing by running water has brought about notable landscape changes with the cutting of the many scenically famous rock gorges that are such a characteristic feature of the region.

During the last period of uplift, rock formations in the Finger Lakes region were generally given a moderate inclination. The resultant cuestas, with their escarpment profiles accentuated by the downcutting of streams through the less resistant rock strata, are characteristic landscape features of the area. Three major rock formations were of special importance in determining postglacial topography. From north to south these were the Onondaga limestone, 30–45 m (100 to 150 ft) thick; the shallow Tully limestone, 5–6 m (15–20 ft) in thickness; and the Portage escarpment composed of sedimentary shales and sandstones. The latter, though not delineated by a sharp crest, is a dominant relief element in the topography of the Finger Lakes region, and its front characteristically consists of a steep slope 275–300 m (900–1000 ft) high.

The drainage divide of the preglacial peneplain was originally near the Tully escarpment, but, by the time of the first ice advance, headwater erosion by northward flowing streams had captured the flow of water as far southward as Fall Creek. The present hooking of the Salmon Creek basin near its point of entry into Cayuga Lake and the pattern of forking by its two main branches constitute a classic illustration of a captured tributary with reversed direction of drainage.

The most dramatic landform features in the Cayuga basin are the result of glacial action, and this is particularly true of the depression in which the lake itself lies. Von Engeln describes this as a true rock-scoured basin, citing its shape and the fact that its bottom is below sea level as evidence. The east–west tributaries, lying athwart the direction of major ice flows, were excavated to lesser depths. As a result, they were left "hanging" far above the main valley and now cascade into it through series of scenically spectacular, although individually not very high, waterfalls. Taughannock Falls, 67 m (215 ft) high, is the only major gorge with an enduring cap rock (limestone) capable of resisting erosion and is therefore an exception to this general pattern.

The first of the two ice advances was responsible for most of the great glacial erosion phenomena of the region. While glacial erosion has played a part in the formation of the major gorges, the principal sculpturing agent has been downcutting by their streams during the interglacial and postglacial periods. The second glaciation filled much of the interglacial gorges with debris, and this is still being downcut in places.

As the glaciers retreated northward their meltwater formed a series of high-level proglacial lakes bordered by the ice front on the north and the highlands that surround the head of the Cayuga basin to the south. A series of 11 such lakes was formed during the 1000 or so years required for the last ice sheet to retreat from the basin. Their emphemeral history is marked by successive lines of lake sediment deposits and even occasional wave-cut cliffs.

The retreat of the second glaciation was retarded for several hundred years a short distance south of the head of Cayuga Lake. The massive end-moraine formed by the deposition of glacial debris insured that, once northern routes to the Mohawk and St. Lawrence drainage were passed by the retreating ice, the direction of flow of Cayuga Lake and its precursors would be to the north rather than south to the Susquehanna.

Details of bedrock composition (Fig. 3) show a dominant shale–limestone mixture extending along most of the immediate border of the lake. The major portion of the drainage basin (located to the south and east) is, however, composed mostly of shales and siltstones. Bedrock geology in the northernmost 8 km of the basin is more complex and varied with narrow east–west bands consisting respectively of the Marcellus formations, the Onondaga limestone escarpment, the Helderberg group, and the Akron Dolostone.

In addition to occurring as the massive deposits of end-moraines, debris from the last glaciation was generally spread over the basin and is manifested in both the structure of the soils and in aspects of the landscape. There is regularly a basal layer of ground moraine, sometimes referred to as till, boulder clay, or hardpan, which is normally covered by other types of deposited material some of which have maintained their integrity. Others have served as the material from which postglacial soils were formed.

The northern two-thirds of the lake basin is dominated by moderately coarse-textured soils with calcareous substrata. Those of the major tributaries and highlands surrounding the southern part of the basin present a diverse and complex assemblage and, in general, are less well drained and more acid. Soil distributions are shown in Fig. 4. Dominant soil types are Honeoye–Lima, 23.3%; Erie–Langford, 21.5%; and Howard–Langford–Valois, 15.7%.

Vegetation

The natural flora of the Cayuga Lake basin was first described in an authoritative and comprehensive fashion by Dudley (1886). He also provided a scholarly review of material pertaining to the "primitive," i.e., prior to settlement by the white man, flora of the region. A picture emerges of a basin densely forested in its southern portion with hilltops covered with white pine and hemlock and the sides and valleys between with oak, birch, beech, ash, hemlock, basswood, and elm. That the forest cover was dense is attested to both by the writings of early travelers and by references in the decrees of Hiawatha to the Cayugas as the people whose "habitation was the dark forest."

The landscape of the Cayuga basin as a whole was, however, given variety by the influence of the Indian and by the presence of marshes at the

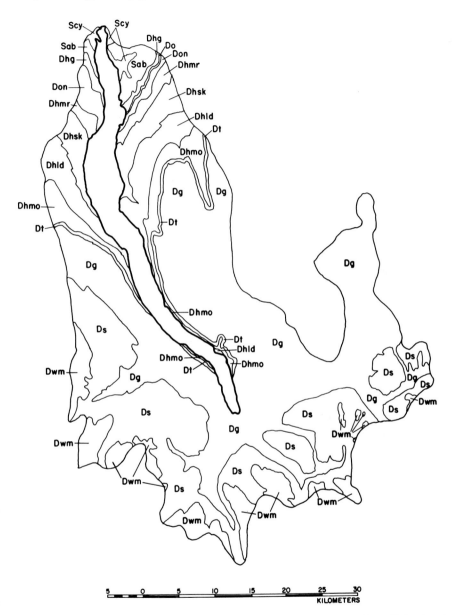

Fig. 3. Geological formations in the Cayuga Lake basin. Redrawn from Rickard and Fisher (1970).

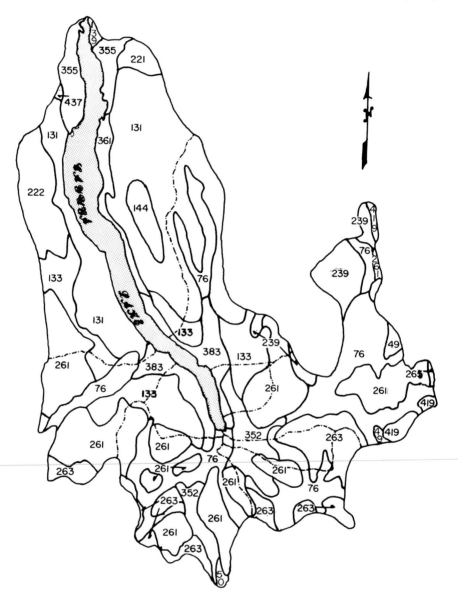

Fig. 4. Soils of the Cayuga Lake basin. Key to map symbols (from Cline and Arnold, 1970) is given in the tabulation that follows.

Map symbol	Soil association	Other associated soils

I. Areas dominated by medium and moderately coarse-textured mesic soils with sand or gravel substrata
a. Dominantly well- and moderately well-drained soils not subject to flooding

49	Howard	Phelps, Fredin, Halsey
50	Howard–Chanango	Phelps, Castile, Halsey

b. Dominantly well-drained soils in complex pattern with soils having compact substrata

76	Howard–Langford–Valois	Erie, Halsey, Madrid

II. Areas dominated by medium and moderately coarse-textured soils with compact loamy substrata
a. Dominantly well- and moderately well-drained soils with calcareous substrata

131	Honeoye–Lima	Kendaia, Lyons
133	Lansing–Conesus	Appaton, Kendaia, Lyons
139	Ontario–Hilton	Camillus, Madrid, Newstead, Sun
144	Lima–Kendaia	Honeoye, Lyons, Nallis

III. Areas dominated by moderately fine and fine-textured soils with compact, clayey substrata
a. Dominantly moderately well- and somewhat poorly drained soils in complex pattern with moderately deep soils over bedrock

221	Cazenovia–Wassaic	Ovid, Romulus, Aurora
222	Darien–Angola	Aurora, Danley, Varick

IV. Areas dominated by medium and moderately coarse-textured mesic soils with fragipans and compact substrata
a. Dominantly well- and moderately well-drained soils

239	Langford–Erie	Canaseraga, Ellery, Alden

b. Dominantly somewhat poorly drained soils

261	Erie–Langford	Ellery, Aldan, Papakating
263	Volusia–Lordstown	Chippawa, Mardin, Arnot
265	Volusia–Mardin–Lordstown	Chippawa, Arnot, Tuller

V. Areas dominated by moderately coarse to fine-textured stone-free soils with silty or clayey substrata
a. Dominantly well- to moderately well-drained soils with clayey substrata

352	Hudson–Rhineback	Madalin, Fonda, Collamer
355	Schoharie–Odessa	Cayuga, Churchville, Lakemont

b. Dominantly somewhat poorly and poorly drained soils

361	Canandaigua–Collamer	Canadice, Niagara, Alden

c. Dominantly moderately well- to poorly drained soils in complex pattern with soils having compact substrata or rockland

383	Hudson–Cayuga	Churchville, Rhineback, Madalin

VI. Areas dominated by moderately deep and shallow soil with bedrock substrata
a. Dominantly well-drained, moderately deep soils in complex pattern with deep soils having compact substrata

419	Lordstown–Valousia–Mardin	Arnot, Chippawa, Bath

b. Dominantly well-drained shallow soils

437	Benson–Wassaic	Hilton, Appleton, Lyons

north end. Writing in 1671, a Jesuit priest (Raffieux, 1671–1672) stated that in regard to the northeastern part of the basin "Cayuga is the most beautiful country I have seen in America. It is a country situated between two lakes (Cayuga and Owasco) and is no more than four leagues wide, with almost continuous plains bordered by beautiful forests." The plains he referred to were undoubtedly the oak openings kept free of brush and other trees by annual burnings, a practice that provided browse for deer (Thompson, 1972). The Cayugas were agriculturists as well as hunters, and they maintained fields of corn and extensive orchards. The latter were mainly peach and apple, both probably introduced by the Jesuits.

Cayuga Lake is the lowest in elevation of the Finger Lakes, and this has resulted in the formation of extensive marshes around the northern end. Dudley (1886) points out that this unique characteristic was "recognized by the 'Six Nations' or ancient Kanonsionni, who called what we now know as Cayuga Lake, 'Tiohero',* the *lake of flags or rushes, or lake of the marsh*." This area extends northward from about Canoga and Farley's Point and contains areas of typical marsh plants such as *Typha* and *Scirpus*.

The orchards and fields of the Indians were laid waste by the punitive expedition of Sullivan during the War of the American Revolution. Ten years later the first white settlement was started at Ithaca. Dudley concluded from interviews with older residents of the town that at about 1830–1835 the Indian oak clearings were still characteristics of the hillsides about Ithaca and that the hilltops retained their forests of white pine and hemlock. With the connection of Cayuga Lake to the Erie Barge Canal in 1821 and the completion of a lock in 1829 (Whitford, 1906) came a large influx of settlers, and the softwood forests were probably cut over most of their extent within a few decades.

In "The Cayuga Flora" Dudley (1886) considered the plants of the area according to their habitat (the Montezuma and Cayuga marshes, Cayuga Lake shore, Cayuga Lake including its inlets and outlets, the alluvial floodplains and creek bottoms, ravines, the woodland, sphagnum swamps, and open peat bogs, and the watershed marshes and ponds) and their geographical associations. He concluded that the hardwood forests were little changed in their composition from primitive conditions. His catalog of plants contained a total of 1160 species of which 963 were native to the Cayuga flora.

Dudley concluded that the plants of the Cayuga basin represented the southern and southeastern boundary of "that vast area including the Ohio valley, the Great Lakes and a part of Canada" including some subarctic

* In early writings also spelled Thiohero, Tichero, or Choharo and also written Deyohero in the Canienga or Mohawk dialect.

TABLE 2

Composition of Major Forest Types in the Cayuga Basin

Area covered	Forest type	Common associated trees
Northern one-fourth of basin	Elm–ash–red maple	Willow, sycamore, beech, cottonwood
The remainder of the lake border and valleys	Oak–hickory	Yellow poplar, elm, red maple, hard maple, hemlock, beech
Uplands	Beech–hard maple	Hemlock, basswood, white pine, white ash, black cherry, red maple, birch

forms driven southward by the ice sheets. The flora also shows traces of a connection with the Alleghenian plants to the south.

A composite description of the present forests has been compiled from Ferguson and Mayer (1970), Stout (1958), and interviews with foresters on the Cornell faculty (Table 2). The extent of cover by trees less than or equal to 30 ft in height, natural forest and plantation forest is shown in Table 3 (Child *et al.*, 1971).

The nonforest vegetative cover in the Cayuga basin falls mainly into various agricultural categories or is in a transition state between agriculture

TABLE 3

Forest Cover in the Cayuga Lake Basin

Drainage basin	% of total land use in basin			
	Forest ≤ 30 ft	Natural forest	Plantation forest	Total forest
Cascadilla	36.0	21.5	2.3	59.8
Fall	20.5	9.9	2.1	32.5
Inlet	27.3	17.7	2.3	47.3
Six Mile	27.7	24.8	3.7	56.2
Taughannock	17.2	13.8	1.1	32.1
Salmon	17.0	7.1	0.2	24.3
Cayuga 1[a]	17.0	12.2	0.9	30.1
Cayuga 2	12.0	12.1	0.4	24.5
Cayuga 3	13.7	4.7	0.1	18.5
Cayuga 4	17.1	5.6	0.1	22.8
Cayuga 5	11.8	2.7	0.2	14.7
Cayuga 6	9.8	3.2	0.6	13.6

[a] Cayuga 1–6. Six arbitrary, approximately 10-km segments (with 1 to 6 extending south to north) taken at right angles to the long axis of the basin.

and forest brushland. The latter is especially common in the southern portion of the basin and represents a gradual abandonment of marginal agricultural lands that began with the Farm Resettlement Program in the 1920's and 1930's. Existing agriculture is principally in the form of field and forage crops, grain, and rotated tillable pasture. Despite ameliorating effects of the lake on the immediately surrounding climate, orchards and vineyards are not in much evidence in the Cayuga basin although vine culture is presently on the increase. Agricultural land use is summarized in Table 4.

Population

Data from the 1970 census (United States Bureau of the Census, 1970a,b) are given in Table 5. An estimated factor has been applied to correct census boundaries to those corresponding to the watershed. Students at Cornell University and Ithaca and Wells Colleges, about 20,000 in total, are included so that the actual population averaged over the period of a year is probably less than the number shown.

Land Usage

The most recent and most quantitative land use description for the Cayuga Lake basin is that of Child et al. (1971). Data from New York State's Land Use and Natural Resources Inventory (Shelton et al., 1968), collected in 1968, were analyzed by major sub-watersheds and for segmented portions of the rest of the basin. The original data resolved land use according to 1-hectare units and contained some 130 use classifications.

Tables 3 and 4 above describe in detail the extent and nature of forest and agricultural coverage within the major subdivisions of the basin. Other land uses and a summary of the extent of agriculture and forests are shown in Table 6. The dominant land use category in the Cascadilla, Six Mile, and Inlet watersheds is "Forest" while in the remainder of the basin it is "Active Agriculture." Definitions of the use category terms are as follows:

1. Active agriculture—sum of the following seven types:
    (a) Orchard—intensively managed commercial orchards
    (b) Vineyards—intensively managed commercial vineyards
    (c) Horticulture or floriculture—sod and seed farms, nurseries
    (d) High intensity cropland—intensive production of vegetables (fresh and processed market vegetables) and small fruits (berries); all muckland truck crops
    (e) Cropland/cropland pasture—cultivated; field crops, forage crops, grains, dry beans, and rotated tillable pasture
    (f) Permanent pasture, generally unimproved, not tillable
    (g) Specialty farms—mink, pheasant and game, aquatic agriculture, and horse farms

TABLE 4

Land in the Cayuga Lake Basin Used for Agriculture and That in Inactive Agriculture

Drainage basin	Orchard	Vineyard	Horticulture	High intensity crop	Cropland pasture	Permanent pasture	Inactive agriculture	Total
				% of total land use in basin				
Cascadilla	1.0	0.0	0.0	0.0	21.9	0.4	7.0	30.3
Fall	0.1	0.0	0.0	0.3	37.9	5.9	17.8	62.0
Inlet	0.0	0.0	0.1	0.1	24.4	2.2	19.5	46.3
Six Mile	0.1	0.0	0.0	0.0	22.4	1.1	13.1	36.7
Taughannock	0.0	0.0	0.1	0.0	33.9	3.5	26.6	64.1
Salmon	0.0	0.0	0.1	0.0	57.0	3.7	13.0	73.8
Cayuga 1[a]	0.3	0.1	0.0	0.1	28.6	2.1	22.2	53.4
Cayuga 2	0.1	0.0	0.1	0.0	43.8	1.3	26.4	71.7
Cayuga 3	0.0	0.0	0.1	0.1	70.4	1.1	7.6	79.3
Cayuga 4	0.0	0.1	0.0	0.0	63.9	1.5	7.7	73.2
Cayuga 5	0.1	0.3	0.0	0.0	76.8	1.8	5.0	84.0
Cayuga 6	0.4	0.0	0.0	0.0	75.2	1.3	6.7	83.6

[a] Cayuga 1–6. Six arbitrary, approximately 10-km segments (with 1 to 6 extending from south to north) taken at right angles to the long axis of the basin.

TABLE 5

Population in the Cayuga Lake Basin, 1970

Census unit	Population	Estimated proportion living in basin	Total
Caroline	2,536	0.9	2,282
Danby	2,141	0.9	1,927
Dryden	9,572	1.0	9,572
Enfield	2,028	1.0	2,028
Groton	4,881	0.3	1,384
Ithaca (city)	26,226	1.0	26,226
Ithaca (town)	15,620	1.0	15,620
Lansing (town)	5,972	1.0	5,972
Newfield	3,390	0.9	3,051
Ulysses (town)	4,315	1.0	4,315
Total for Tompkins County			72,377
Hector (town)	3,671	0.5	1,835
Total for Schuyler County			1,835
Covert	2,027	1.0	2,027
Fayette (town)	2,997	0.2	599
Lodi	1,287	0.2	257
Ovid	3,107	0.3	982
Romulus	4,284	0.2	880
Varick	1,700	0.3	510
Total for Seneca County			5,854
Genoa	1,720	1.0	1,720
Ledyard	1,886	1.0	1,886
Scipio	1,290	0.9	1,161
Springfort	1,911	1.0	1,911
Venice	1,261	0.9	1,135
Summerhill	670	1.0	670
Total for Cayuga County			8,463
Virgil	1,692	1.0	1,692
Total for Cortland County			1,692
Total for Cayuga Lake basin			90,221

2. Inactive agriculture—tillable cropland permanently or temporarily removed from agricultural use, in a transition stage between agricultural land and forest brush-land
3. Forest—sum of the following three types:
   (a) Forest brushland and brush pasture—regenerating lands with brush cover and pole stands to 30 ft in height; 40–50 years of age
   (b) Forest land—natural stands with 50% or more in excess of 30 ft in height, and greater than 50 years of age
   (c) Forest plantation—artificially, stocked; any species, age, class, or size
4. Residential—sum of the following seven types:
   (a) High density—50-ft frontage
   (b) Medium density—50- to 100-ft frontage
   (c) Low density—over 100-ft frontage
   (d) Strip development—4 or more residences per 1000 ft population
   (e) Rural hamlet—all communities of less than 1000 population
   (f) Farm labor camps
   (g) Shoreline developed
5. Commercial—sum of the following three types:
   (a) Urban—downtown
   (b) Shopping centers
   (c) Commercial strips
6. Industrial—sum of light and heavy manufacturing
7. Outdoor recreation—all types when they predominate land use
8. Extractive—stone, gravel, metal, underground mines
9. Public—institutions, bases, disposal areas, equipment centers, etc.
10. Transportation—sum of highway, railroad, airport, canal, port, shipyard, locks, service facilities

Use of Water

Water authorities using Cayuga Lake for a potable supply are Seneca Falls, Hibiscus Harbor, Aurora, Lansing (via wells on the lake shore), and Trumansburg. The towns of Ithaca, Dryden, and Lansing began operation in 1976 of a pumping and treatment facility at Bolton Point, located on the east shore about 2 km north of the south end. Many lakeshore cottages and residences draw their water directly from the lake. The City of Ithaca receives its water from impoundments on Six Mile Creek.

Cayuga Lake is extensively used for recreation. Three state, one city, and one town park are located on its shore. Swimming and other water contact sports are generally confined to June through mid-September while boating and fishing extend over a longer period. Stevens and Kalter (1970) estimated that the boating and fishing demands for the summer quarter in 1970 were, respectively, 106,360–143,720 and 117,880–137,110 activity days. Their projected demands for 1985 were for fishing, 187,020–212,570 activity days and for boating, 174,090–206,320 activity days. The direct and indirect economic impact of recreational expenditures by nonresident boaters and

TABLE 6

Cayuga Lake Basin	Drainage basin (km²)	Active agriculture (km²)	Active agriculture (%)	Forest (km²)	Forest (%)	Residential (km²)	Residential (%)
Cascadilla	34.66	8.50	24.5	20.72	59.8	1.77	5.1
Cayuga 1[a]	104.53	32.89	31.5	31.45	30.1	6.93	6.6
Cayuga 2	110.60	50.04	45.2	27.14	24.5	2.82	2.5
Cayuga 3	133.82	94.53	70.6	24.70	18.5	2.12	1.6
Cayuga 4	125.09	82.02	65.6	28.38	22.7	1.41	1.1
Cayuga 5	154.02	121.58	78.9	22.63	14.7	1.45	0.9
Cayuga 6	69.99	53.96	77.1	9.51	13.6	0.84	1.2
Fall	378.18	167.11	42.8	122.91	32.5	7.40	1.9
Inlet	242.80	65.16	26.8	114.82	47.3	4.35	1.8
Salmon	221.69	134.77	60.8	53.66	24.2	1.49	0.7
Six Mile	122.73	28.87	23.5	69.03	56.2	5.42	0.4
Taughannock	172.36	64.53	37.4	57.49	33.3	0.85	0.5
	1870.47	903.96	48.3	582.44	31.1	36.85	2.0

[a] Difference between total land area in a watershed and the subtotal of the uses shown is accounted for by land in inactive agriculture (Table 5).

[b] Cayuga 1–6. Six arbitrary (approximately 10-km) segments with 1 being southern-most, 6 being northernmost.

fishermen during the summer quarter of 1970 was calculated to be in the range of $569,331 to $658,358 and projected to be $921,119–$1,077,041 by 1985. Allee (1969) has described recreational activities associated with the lake as shown in Table 7.

Youngs and Oglesby (1972) have described the present and past sport fisheries in Cayuga Lake. The principal limnetic species harvested by anglers is the lake trout. Rainbow trout and salmon are occasionally taken from the open waters of the lake as well. Both the lake trout and rainbow trout are also fished from shore during the spring and fall. An annual harvest of at least 5675 kg of lake trout is realized in an average year. Smallmouth bass, yellow perch, bullheads, largemouth bass, northern pike, and chain pickerel are among the principal species taken by fishermen from the littoral zone. An intensive tributary fishery exists for smelt and rainbow trout during the time of their spawning migrations.

Sewage and Effluent Discharge

A consulting engineering report (O'Brien and Gere Consulting Engineers, 1965) has described the principal industrial effluents generated in the Ithaca

Land Use in the Cayuga Lake Basin[a]

Commercial		Industrial		Outdoor recreation		Extractive		Public		Transportation	
(km²)	(%)	(km²)	(%)	(km²)	(%)	(km²)	(%)	(km²)	(%)	(km²)	(%)
0.20	0.6	0.05	0.1	0.00	0.0	0.18	0.5	0.66	1.9	0.11	0.3
0.51	0.5	0.18	0.2	2.53	2.4	1.07	1.0	0.92	0.9	2.82	2.7
0.06	0.1	0.00	0.0	0.84	0.8	0.00	0.0	0.34	0.3	0.09	0.1
0.12	0.1	0.00	0.0	0.01	<0.1	0.00	0.0	0.53	0.4	0.10	0.1
0.01	<0.1	0.00	0.0	0.22	0.2	0.00	0.0	3.37	2.7	0.00	0.0
0.02	<0.1	0.02	<0.1	0.01	<0.1	0.04	<0.1	0.46	0.3	0.05	<0.1
0.08	0.1	0.01	<0.1	0.25	0.4	0.18	0.3	0.19	0.3	0.22	0.3
0.97	0.3	0.62	0.2	1.40	0.4	0.70	0.2	5.31	1.4	0.66	0.2
0.87	0.4	0.10	<0.1	5.88	2.4	0.67	0.3	1.53	0.6	0.44	0.2
0.08	<0.1	0.00	0.0	0.02	<0.1	0.51	0.2	0.79	0.4	0.09	<0.1
0.54	0.4	0.58	0.5	0.14	0.1	0.37	0.3	1.21	1.0	0.31	0.3
0.09	0.1	0.00	0.0	1.84	1.1	0.13	0.1	0.21	0.1	0.02	<0.1
3.55	0.2	1.56	0.1	13.14	0.7	3.85	0.2	15.52	0.8	4.91	0.3

area (Table 8). The only other sources of industrial discharge of potential significance to the main body of the lake are from the Cargill, Inc. rock salt mining and storage operations at Portland Point and the Milliken Power Station operated by the New York State Electric and Gas Corporation (Fig. 5). Cargill mines rock salt from deep shafts that extend out under the lake at a depth of 1200 ft or more below the bottom of the lake. Large quantities are stored, pending shipment, in sheds and in open piles covered with plastic sheeting. In the past very high concentrations of chloride were

TABLE 7

Recreation Uses in the Late 1960's

Use category	Amount
Parks	500,000 visitor days
Fishing	100,000 fishermen days
Boating	
Daily launchings	13,000 boat days
Seasonal moorings	180,000 boat days
Cottages	350,000 cottage days

TABLE 8

Waste Discharges from Ithaca Industries

Morse Chain Company

Quantity: 4731 m^3/day (1,250,000 gal/day)

Waste characteristics: Water is used for cleaning, pickling, and plating. Waste contains trace amounts of associated plating chemicals and process oils

Waste treatment: Acid solution from cleaning tanks is neutralized with caustic soda before discharge to sewers. Oils are separated and reclaimed. Water from separation tank is disposed of on land

Waste disposal: Sanitary to city sanitary sewers and industrial to sanitary sewers and land

National Cash Register

Quantity: 378 m^3/day (100,000 gal/day)

Waste characteristics: Approximately 43% is cooling water. Remaining 57% is process waste containing cyanides, acids, or alkalies

Waste treatment: Two 8500-gal capacity holding tanks receive plating wastes before their discharge to municipal sewers. The concentration of cyanide is reduced to 0.2 ppm or less before discharge

Waste disposal: Sanitary and industrial to city sanitary sewer

Ithaca Gun Company

Quantity: Not given

Waste characteristics: Dragout from quenching operations, dragout from rinsing blued parts, overflow from alkaline soda bath, and cooling water

Waste treatment: Holding tank is provided for cooling water before return to Fall Creek

Waste disposal: Sanitary to city sanitary sewers and industrial to sanitary sewers and Fall Creek

---

detected at times in a stream (Gulf Creek) draining the Cargill property (Likens, 1974b). Discharges at the Milliken Station involve the introduction of heat and chlorine to the lake. An ecological study of the lake that included the immediate environment of the plant (Oglesby and Allee, 1969) did not reveal any significant adverse effects of the discharge, although at times very low pH's were encountered in the discharge and possible pollutants in drainage from coal piles were not investigated. The New York State Electric and Gas Corporation is presently proposing the construction of a second, larger fossil-fueled electric generating plant adjacent to the Milliken Station.

Effluents of possible localized significance enter the lake at the north end from industries located on or near the Seneca River. The principal industrial concentration is located upstream from the lake in Seneca Falls.

Discharges of treated municipal wastes are described in Table 9. Approximately 52% of the basin's total population is served by sewage

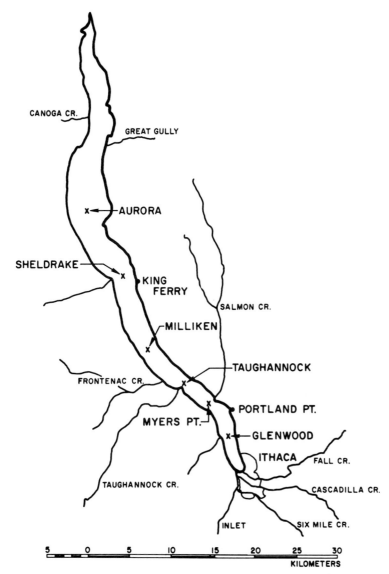

Fig. 5. Location map of Cayuga Lake showing principal tributaries and sites of sampling stations (denoted by ×) used in the various limnological studies.

TABLE 9

Annual Average Discharges from Sewage Treatment Facilities to Cayuga Lake

Source	Year of record	Population served	Treatment	Flow m³/day	Flow mg/day	Source of information
Interlaken	1969	800	Secondary	151	0.04	Daniell and Long Engineers (1969)
Aurora	1973	< 1,000	Tertiary	530	0.14	D. Brooks, personal communication
Cayuga	1970	693	Holding ponds	262	0.07	Cayuga County Health Dept.
Cayuga Heights	1973	< 8,000	Tertiary (P removal	2,839	0.75	G. Gleason, personal communication
Dryden	1973	1,400	Secondary	568	0.15	E. Bell, personal communication
Ithaca	1973	33,000	Secondary	18,925	5.0	F. Liguori, personal communication
Trumansburg	1972–1973	< 2,000	Secondary	606	0.16	G. Gleason, personal communication
Approx. totals		46,893		23,881	6.31	

treatment plants. The remainder include lakeside cottages and numerous small villages many of which are located along the banks of tributary streams.

MORPHOMETRIC AND HYDROLOGIC DESCRIPTION

Surface Area and Volume

According to Birge and Juday (1914) the surface area of Cayuga Lake is 172.1 km² (66.4 square miles); its length 61.4 km (38.1 miles); the maximum width 5.60 km (3.50 miles), and the average 2.80 km (1.74 miles); shore length is 153.4 km (95.3 miles) and shoreline development 3.35. They gave the maximum depth of Cayuga Lake as being 132.6 m (435 ft) and the mean depth as 54.5 m (172.3 ft).

Birge and Juday state that the volume, computed for a lake elevation of 116 m (381 ft), is 9379.4 × 10⁻⁶ m³ (331,080 × 10⁻⁶ ft³) and the volume development is 1.23. Their calculations indicate that a plane at the 40.0-m depth would divide the lake into two equal volumes.

The water level in Cayuga Lake is regulated by Mud Lock at the north end. Since this is part of the Erie Barge Canal system, the lock is operated by the New York State Department of Transportation. Outflow management is complicated by the interconnection of Cayuga with Seneca Lake through the Seneca River and by operation functions that at times are in conflict. The general strategy is to draw the lake down in mid-December to minimize ice damage to shoreline structures and for maximizing storage during the period of heavy spring runoff. At minimum level during January or February, Cayuga is still 0.67–0.75 m (2.2–2.5 ft) above the United States Geological Survey's base line elevation above sea level of 114.8 (376.6 ft). A maximum recorded lake elevation of 117.7 m (386.3 ft) occurred after Tropical Storm Agnes (June 26, 1972). A lake elevation of 116 m (381 ft) is given by Birge and Juday (1914), and the New York Barge Canal Authority uses a value of 115.2 (378 ft) as a datum above sea level.

Liu and Tedrow (1973) have used systems analysis techniques to construct a hydraulic optimization model for Cayuga Lake and the other bodies of water with which it is interconnected. In their model lake levels would be controlled according to the operational function of storage (flood control and recreation), diversion (water supply diversion and irrigation), and release (navigation and flood control).

Details of Hydrography

Maximum depths are associated with a trough extending from about Myer's Point north to Long Point (Fig. 2). The relationships of depth to area and volume are given in Table 10 (from Birge and Juday, 1914).

Relative Volumes of Stratified Layers

Not infrequently, summer temperature profiles for Cayuga Lake show a marked gradient from the surface to a depth of 25–60 m (depending on the time of year), but the discontinuity characteristic of a thermocline is lacking. At other times secondary thermoclines persist for considerable periods so that it is again difficult to define the lake in terms of the idealized three layer model. Internal seiches introduce a final complication which can be dealt with only by considering sequences of temperature profiles.

Recognizing these factors, Singley (1973) has defined the relative volumes of the various thermal strata based on extensive temperature data from 1950–1952 and 1968–1972 (see below) and has set arbitrary boundaries for the epilimnion and hypolimnion. He defines the latter as the portion of the lake found below the 6.1°C (43°F) isotherm and the epilimnion as being comprised of that water above the depth where temperature has dropped by 15% of the difference between the surface and 4.4°C (40°F). The 15% criterion was chosen because it corresponded well with those observed profiles that were ideally stratified. Singley's computations are shown, after conversion to metric units, in Table 11 and volume ratios have been added. The ratios of epilimnion to other strata are also shown graphically in Fig. 6.

Nature of Lake Sediments

Observations of bottom material in the limnetic portion of the lake indicate that the sediments in the central trough are predominantly a fine-textured mixture of clay and silt. Numerous fathometric profiles have failed to reveal any protruding rock formations. While the littoral zone at the ends of the lake and in gently sloping areas around the deltas have soft bottoms, that around much of the rest of the lake is rock, often with loose shale scattered on it, or, occasionally, sand and gravel. Ludlam (1967) has studied the banding in an extensive series of corings taken throughout the lake south of Sheldrake. He attempted to correlate patterns with both annual runoff cycles and periods of abnormally intense flooding. Vogel's study of rooted macrophytes at the south end of the lake (see below) provides us with the only detailed description of the physical–chemical nature of the surficial sediments in Cayuga. Additional inferences are possible from studies which have been conducted on the bottom deposits of Seneca Lake.

TABLE 10

Hydrographic Details for Cayuga Lake

Depth (m)	Area km²	Area %	Length of contours (km)	Depth (m)	Area km²	Area %	Volume[a] m³ × 10⁻⁶	Volume %
0	172.1	100.0	153.8	0–10	47.4	27.5	1435.7	15.4
5	138.7	78.9	116.5	10–20	12.7	7.4	1183.4	12.6
10	124.8	72.5	100.9	20–30	8.0	4.6	1080.2	11.5
20	112.0	65.1	93.3	30–40	10.1	5.9	989.6	10.6
30	104.0	60.5	90.8	40–50	12.0	7.0	878.7	9.4
40	93.9	54.6	86.4	50–60	8.3	4.8	777.3	8.3
50	81.9	47.6	83.9	60–70	7.8	4.8	696.4	7.4
60	73.6	42.8	80.4	70–80	7.1	4.1	622.0	6.6
70	65.8	38.2	76.3	80–90	8.0	4.6	546.8	5.8
80	58.7	34.1	72.4	90–100	8.4	4.9	464.8	5.0
90	50.7	29.5	68.9	100–110	8.4	4.9	380.5	4.1
100	42.3	24.6	58.2	110–120	17.9	10.4	244.2	2.6
110	33.9	19.7	53.9	120–130	15.3	8.9	79.1	0.8
120	16.0	9.3	39.4	130–133	0.8	0.4	0.7	
125	7.4	4.3	30.3					
130	0.8	0.4	6.2					
132.6	0.0	—	—					

[a] Erroneously given by Birge and Juday as thousands of cubic meters.

TABLE 11

Depth, Volume, and Volume Ratios of Thermal Strata in Cayuga Lake

Dates	Depth of layer boundaries (m)		Layer volumes (m³ × 10⁻⁹)			Epilimnion: metalimnion plus hypolimnion	Epilimnion: hypolimnion
	Epilimnion	Hypolimnion	Epilimnion	Metalimnion	Hypolimnion		
May 21–Jun 3	1.2	25.0	0.1755	2.9715	6.2260	0.019	0.028
Jun 4–Jun 17	3.0	28.6	0.4386	3.1130	5.8298	0.049	0.075
Jun 18–Jul 1	4.9	32.0	0.6990	3.2262	5.4336	0.081	0.129
Jul 2–Jul 15	6.4	36.3	0.9169	3.3960	5.0657	0.108	0.181
Jul 16–Jul 29	7.3	36.3	1.0499	3.2545	5.0657	0.126	0.207
Jul 30–Aug 12	7.6	36.3	1.0924	3.2262	5.0657	0.132	0.216
Aug 13–Aug 26	9.8	40.5	1.3980	3.3394	4.6412	0.175	0.301
Aug 27–Sep 9	11.9	40.5	1.6584	3.0847	4.6412	0.215	0.357
Sep 10–Sep 23	13.7	41.8	1.8735	2.9715	4.5280	0.250	0.414
Sep 24–Oct 7	13.7	41.8	1.8735	2.9715	4.4528	0.250	0.414
Oct 8–Oct 21	15.5	44.5	2.0914	2.9998	4.3016	0.286	0.486
Oct 22–Nov 4	19.8	53.6	2.5951	3.2545	3.5375	0.382	0.734
Nov 5–Nov 18	25.3	< 57.9	3.1979	2.7451	3.4243	0.518	0.934
Nov 19–Dec 2	27.7	< 57.9	3.4526	2.4904	3.4243	0.584	1.008

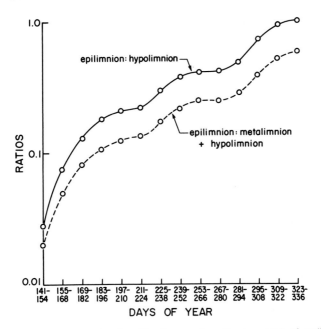

DAYS OF YEAR

Fig. 6. Ratios (by volume) of epilimnion to hypolimnion and of epilimnion to metalimnion plus hypolimnion as a function of season.

Over 80 cores, averaging 3 m in length, were collected throughout Cayuga Lake by Ludlam (1967). Despite the inherent distortions in many of these due to density currents, slumping and other disturbances, he was able to develop a rational correlation between visually apparent bandings and major runoff events. In water depths exceeding 45 m, cores consistently exhibited a pattern of couplets. Through deductive processes and a study of the allochthonous material transported by the major tributary, Fall Creek, he concluded that approximately one couplet per year was deposited with spring runoff contributing a dark band (rich in organic matter near the mouths of streams) and the lower portion of an overlying light band, while the remainder of the light band was laid down during the rest of the year. Ludlam estimated that between October 1, 1962 and October 1963, 3.2×10^{10} gm dry weight of clastic material entered the lake via Fall Creek. This would provide an average sedimentation rate of 0.1 g/cm^2, and this is probably conservative since precipitation was only 82% of normal in that year and the spring runoff occurred gradually. He also estimated that during the spring and winter thaws of 1964, 2×10^8 gm dry weight of organic detritus was transported by Fall Creek. Extraneous major bands were correlated with periods of known flooding. With these serving as approximate dating layers it appears that some 1.2–1.4 m of sediment have been deposited in the profundal zone of Cayuga Lake during the past 100 years.

Vogel (1973) collected 33 samples of the top 5–10 cm of the bottom materials along transects located in the southernmost kilometer of Cayuga Lake. The greatest water depth sampled was 5.5 m and the average was 3.2 m. In physical composition the sediments were about 10–20% clay at all locations. The remainder was made up of a mixture of sand and silt with the former being more abundant (up to 80%) near the mouths of tributaries and in the southeastern corner. Silt predominated (45–85%) elsewhere. Organic content of the bottom materials ranged from less than 1 to 5%. Vogel also reported values for a number of chemical components (total nitrogen, total phosphorus, K, Mg, Ca, Mn, Fe, Al, and Zn) and a pH range of 6.7–8.2. No obvious correlations can be seen between the various physical and chemical parameters. Most of the latter ranged over an order of magnitude in concentration. Total phosphorus, an item of particular interest, varied between 3.5 and 27.5 ppm with a mean of 11.9 ppm.

In a follow-up study to assess the effects of Tropical Storm Agnes (June 21–22, 1972), Oglesby et al. (1976) took cores at the 11 stations Vogel (1973) had established for quantitative plant collections. It appeared that the material imported by this storm was distributed unevenly over the study area. The sand fraction increased at stations where sand was the dominant component before the storm. In the southwestern corner of the lake, a layer of fine clay had been added to the sediments. Elsewhere silt still predominated and as much as several centimeters may have been deposited in some areas.

Comparable geologies, fairly similar land use patterns in the basins, and morphometric similarities between Cayuga and Seneca lakes suggest that studies done on bottom sediments from the latter may be applicable to Cayuga. Sass (1972) has reported the work of a large group of researchers in mapping the characteristics of the upper 15–20 cm of sediments in the southern 5 km of Seneca Lake. Depth of water overlying sampling sites ranged from 6.1 to more than 128 m. Thirty-two samples were taken along transects crossing the study area. Median values for the various size fractions were silt, 71.8%; sand and larger (including marked boulder and gravel delta aprons), 25.2%; and clay, 3.2%. The highest concentrations of silt occurred in the southeastern part of the lake and also in the deepest part of the study area. Sand and larger particles were concentrated along the west shore where there are greater number of tributaries. The most common mineral components of the clay were illite and chlorite which occurred in a 2:1 ratio. Kaolinite was found in trace amounts at two locations.

Professor Donald Woodrow (Hobart and William Smith College, Geneva, New York) has been collecting deep cores from Seneca Lake during the past several years. Although his mapping and characterization of these are as yet incomplete, he has indicated (personnel communication) that cores from the limnetic zone in the central and northern portions of the basin are comprised of an upper layer of black clays followed by a layer of varved clays containing much shell material, and finally a layer of pink, varved clays.

Seasonal Variation of Monthly Precipitation and Maximum and Minimum Conditions

Long-term monthly averages for precipitation were previously presented in the section entitled General Climatic Data, and annual hydrologic extremes are noted in the following section entitled Water Inflow, Outflow, and Renewal Time. There remains the question of short-term events.

The heaviest single storm rainfall of record occurred on July 7–8, 1935 when 20.89 cm (8.22 in.) fell with 20.07 cm (7.90 in.) in a 24-hr period and 3.05 cm (1.20 in.) during a single hour (Johnson, 1936). Professor D. A. Webster (personal communication) recalls that Cayuga Lake remained visibly turbid for the entire summer of 1935 as a result of suspended material introduced by this series of severe thunderstorms.

The second most severe precipitation event on record was associated with the passage of Tropical Storm Agnes through the region on June 21–25, 1972. Rainfall amounted to 17.91 cm (7.05 in.) with 90% of this being the result of steady rains on the twenty-first and twenty-second. Some of the effects of this storm on phosphorus concentrations and on the rooted macrophyte growth at the south end of the lake are described below.

A review of weather records indicates that precipitation which produced substantial flooding also occurred in 1916 (May 15–17), 1914 (June 7 and 28), 1913 (heavy rains associated with snowmelt in late March), 1906 (June 15 and 17), and 1896 (June 6–10 and the third week of June). Ludlam (1967) states that earlier records of high lake levels and/or recorded storms indicate incidents of significantly higher than normal rainfall also occurred in 1878, 1865, and 1863.

Weekly mean precipitation together with probabilities of achieving certain levels are shown in Table 12 (Dethier, 1966). From this data and the known dates of major storms and maximum lake levels, it is apparent that severe rainfall with heavy runoff is most likely to occur in June and July. In a typical year, however, maximum runoff is in late March or early April associated with a combination of spring rains and thawing. This pattern is readily apparent when hydrographs representing mean flow of the tributaries are examined.

TABLE 12

Percent Probability of Receiving at Least the Amounts of Precipitation Indicated during 1-week periods[a] (Dethier, 1966).

Week beginning	Precipitation (cm)				
	Trace[b] or none	0.51	1.52	2.54	5.08
Jan 3	1	67	24	9	1
Jan 10	1	68	18	4	0
Jan 17	3	63	19	5	b
Jan 24	2	65	19	5	b
Jan 31	3	67	21	6	b
Feb 7	1	72	25	9	1
Feb 14	1	79	32	11	1
Feb 21	0	82	38	16	2
Mar 1	0	81	37	15	1
Mar 8	1	80	43	21	3
Mar 15	2	76	40	21	4

Week beginning	Precipitation (cm)				
	Trace[b] or none	0.51	1.52	2.54	5.08
Jul 5	8	77	47	28	8
Jul 12	5	82	51	30	8
Jul 19	2	84	51	28	6
Jul 26	2	82	51	31	8
Aug 2	2	80	50	30	8
Aug 9	6	74	45	28	8
Aug 16	7	71	45	29	10
Aug 23	10	68	43	28	10
Aug 30	13	65	40	26	9
Sep 6	12	71	44	27	8
Sep 13	12	70	42	24	6

Date						Date					
Mar 22	2	77	40	20	3	Sep 20	10	72	42	24	6
Mar 29	1	82	43	20	2	Sep 27	11	66	36	19	4
Apr 5	2	84	45	20	2	Oct 4	10	66	37	22	6
Apr 12	2	86	49	24	3	Oct 11	8	66	38	22	6
Apr 19	3	80	44	21	3	Oct 18	6	70	42	25	7
Apr 26	2	79	46	25	6	Oct 25	3	75	45	27	7
May 3	4	74	44	25	7	Nov 1	2	78	42	22	5
May 10	7	76	48	30	9	Nov 8	3	76	41	21	4
May 17	6	77	49	30	8	Nov 15	4	71	34	16	3
May 24	4	80	51	30	8	Nov 22	7	71	38	20	4
May 31	2	81	48	27	6	Nov 29	5	70	33	15	2
Jun 7	3	79	48	28	7	Dec 6	4	69	35	17	3
Jun 14	3	82	51	30	7	Dec 13	2	66	28	12	2
Jun 21	6	78	49	29	8	Dec 20	1	69	32	15	3
Jun 28	7	78	51	32	10	Dec 27	1	72	29	11	1

[a] From Dethier (1966).
[b] Percent less than 0.5 but greater than zero.

Water Inflow, Outflow, and Renewal Time

Henson *et al.* (1961) estimated the renewal times of water in Cayuga Lake using two methods of calculation. In reviewing their values of 8.75 and 9.1 years mean retention time, Wright (1969b) pointed out that the former was arrived at by an invalid methodology and proceeded to recalculate new estimates based on more extensive stream-flow and precipitation data and on gauged outflow. He argued that the latter parameter could be used with more justification, and from 36 years of outflow measurements he determined that 7.8% of the lake's volume is replaced in an average year, giving a water renewal time of 12.8 years.

Wright (1969b) correctly emphasizes the need for calculating water renewal time on a year-by-year basis due to cyclic changes in evaporation and precipitation. Thus for the 36 year period of outflow data which he cites annual flow-through ranged from 389.7×10^{-6} m^3 to 1157.3×10^{-6} m^3, giving a variation of from 8.1 to 24.1 years in the water renewal rate.

Singley (1973) has addressed the problem of extrapolating the limited data on gauged flow to the basin as a whole. As can be seen in Table 13, only the flow from 34% of the drainage basin has been measured, and the gauging of Salmon Creek was discontinued after a 10-year period. Using the presumed relationship of watershed area to tributary flow, Singley has calculated a hydrologic budget for 1965 with minimum and maximum estimates of ungauged flow (Table 14). The range of water renewal times (17.3–

TABLE 13

Major Tributaries to Cayuga Lake and Their Drainage Areas

Name	Land drainage (km²)	Percent of total land drainage
Cayuga Inlet at mouth	409.2	22
Buttermilk Creek	31.3	
Cascadilla Creek	35.5	
Enfield Creek	79.0	
Six Mile Creek	128.5	
Cayuga Inlet at USGS gauge	91.2	5
Fall Creek at mouth	331.5	18
Fall Creek at USGS gauge	326.3	18
Salmon Creek at mouth	231.0	12
Salmon Creek at USGS gauge	211.6	11
Taughannock Creek	173.5	9
Total gauged	629.1	34
Total area drained	1145.2	61

TABLE 14

Hydrologic Budget for 1965

	Area (km²)	Mean flow (m³/sec)	Mean flow/km²	Total flow (m³ × 10⁶)	Percent lake renewed	Water renewal time
Cayuga Inlet, gauged	95.1	0.45	4.73	14.4		
Cayuga Inlet, ungauged	314.2	1.97	6.27[a]	62.3		
Fall Creek, gauged	321.2	2.68	8.34	84.9		
Fall Creek, ungauged	10.4	0.08	7.69[a]	2.5		
Salmon Creek, gauged	212.4	1.21	5.70	37.9		
Salmon Creek, ungauged	18.1	0.10	1.81[a]	3.1		
Canoga Creek, gauged	8.3	0.06	0.05	2.0		
Canoga Creek, ungauged	6.7	0.05	7.46[a]	1.4		
Precipitation into lake	173.3	3.50	0.02	110.4		
Subtotal	1159.7	10.10		318.9		
Ungauged tributaries (min. estimate total inflow)	872.8	4.22	4.84[a]	133.0		
Subtotal	2033.2	14.32		452.8	4.8	20.8
Ungauged tributaries (max. estimate total inflow)	872.8	7.16	8.20[a]	226.4		
Subtotal	2033.2	17.26		543.4	5.8	17.3
Outflow estimate		11.39		359.4		
Evaporation estimate				155.6		
Total outflow				515.1	5.5	18.2

[a] Flow rate per area of ungauged portion of the tributary watershed assumed to be the same as for gauged portion.

20.8 years) arrived at by this method agrees well with the 18.2-year estimate derived from outflow data. The year 1965 was exceptionally dry (precipitation was 27.61 cm below and evaporation 5.82 cm above the mean for 1942–1962), and hence the long renewal time.

The preceding estimates have all omitted flow from the Seneca River. In a technical sense this river is a tributary, and no studies have been done to prove that materials carried by it do not mix significantly with the water in the lake. However, its west to east pattern of flow and the fact that it intercepts the lake at its most northern point have been taken as strong presumptive evidence that the Seneca River does not substantially affect the main body of the lake. The above determinations of water renewal time have also ignored seasonal fluctuations in lake level [the lake volume of 9379.4×10^{-6} m³ given by Birge and Juday (1914) seems to have been used in all calculations].

Groundwater inflow and outflow into Cayuga Lake are completely unknown. Substantial springs are found in the basin, for example in the villages of Union Springs and Slaterville Springs, as are potentially porous rock formations and filled valleys. These factors suggest that significant groundwater flow remains a distinct possibility.

Currents

Currents in Cayuga Lake have never been definitively studied. From the size and shape of the basin, high mean water depth, and low degree of shoreline development, it would be logical to postulate that the general current pattern might conform closely to classic theoretical models. Sundaram et al. (1969) have used such assumptions, together with limited measurements, to assess some aspects of the current regime. They concluded that a surface wind drift is present which amounts to 2–3% of the wind velocity above the lake. Winds are light during the stratified period resulting in correspondingly low velocities in surface drift. Citing measurements made on September 17–24, 1968 as typical of this period, Sundaram and his co-workers found that wind-induced surface currents were less than 3 cm/sec. At other times during the stratified season, major currents, which could be as great as 50 cm/sec, were associated with seiche motions. Significant hypolimnetic currents, shown in an example to be as high as 10 cm/sec, were only found in association with seiches.

A series of surface temperature measurements made with an airborne sensor revealed cross-lake temperature differences indicative of geostrophic effects (e.g., deflection of surface waters away from the east shore on a northerly wind with consequent upwelling of the deeper waters). Circulation of this nature appears to be further enhanced by topographic features. Thus, on June 20, 1968, during a period of strong northwest wind, surface temperatures on the south side of a small point (Lake Ridge Point located 0.7 km south of the Milliken power plant and extending about 90 m out into the lake) were as low as 10°C when, at the same time, they were 13.5°C at the west side of the lake.

Henson *et al.* (1961) and Wright (1969d) have also cited east–west isothermal distortions that could be taken as evidence of geostrophic effects. The former were likewise the first to publish gross observations on currents based on the movement of turbidity plumes. They state

> The fact that more than half of the surface drainage enters the lake in the southern third of the basin . . . presupposes a predominently northerly current, and this current is expected to be deflected along the eastern shore due to Coriolis effect.
>
> That such a current exists is well known. It can be observed from the elevated area around Ithaca by the difference in the appearance of the water, and is especially noticeable after heavy rains when the inflow water is turbid. This current has been observed to extend as far north as Taughannock Point, a distance of more than 10 miles.

An aerial photograph of such a turbidity plume is shown in Plate 1. Casual observations over the last 5 years indicate that the shape of this plume varies in response to both the amount of tributary discharge and the prevailing wind. Suspended material discharged from Salmon Creek typically spreads out into a plume that often has a southwesterly trend. That from Taughannock Creek seems to be transported downward into deeper water with perhaps a component of flow to the south. Density currents undoubtedly play an important role in the distribution of allochthonous materials but these have never been measured.

LIMNOLOGICAL CHARACTERIZATION

Temperature, Heat Content, and Internal Waves

Beginning with winter isothermy, a generalized temperature regime for the water column at mid-lake would show minimum homothermy at a temperature of 1.3°–3.3°C sometime between late February and early April. Gradual warming but continued homothermous conditions are characteristic of April and early May. Between mid-May and early June the bottom waters become effectively isolated from the surface although some slight warming of the deepest waters continues to occur. A definite division

Plate 1. Turbidity plume at the south end of Cayuga Lake.

into a persistent three-layered system is present by mid-June or early July. Maximum summer bottom temperatures are largely a function of mixing in May and early June and vary from year to year between 4.1° and 5.5°C. Maximum summer surface temperatures range from about 20° to 27°C. During some years, periods occur when no definite thermocline is present despite a substantial thermal gradient in the water column, and at other times auxiliary thermoclines complicate the pattern.

Maximum bottom temperatures of 6.6°–9.6°C are associated with fall homothermy which occurs between early November and December. The water column continues to mix until minimum homothermy is reached. Inverse stratification is a transient phenomenon during the colder months and, except in the rare years when an extensive ice cover forms, Cayuga Lake is essentially a completely mixed system from December into May. Temperature values and dates of occurrence for some of the cardinal points in the annual temperature cycle are given in Table 15. The references cited refer to those presented in Table 16.

Temperature data have been collected by almost all scientists who have conducted investigations on the limnetic portion of Cayuga Lake. The important references are cited in Table 16 with information which indicates their content. Additional but unpublished profiles have been recorded for 1940 and 1942–1947 (D. A. Webster), 1965 (Federal Water Quality Administration), 1971 (R. T. Oglesby), and 1972–1973 (P. J. Godfrey, D. G. Hennick, and R. T. Oglesby). The earliest studies (Birge and Juday, 1914, 1921) on the Finger Lakes involved only isolated summer and winter measurements during 1910–1911, 1914, and 1917–1918, but from these few data Birge carried out a classic analysis of heat distribution and flux. Of the subsequent workers, Hess (1940), Howard (1963), and Henson et al. (1961) have developed interesting discussions on the temperature regime. Wright (1969c), Sundaram et al. (1969), and Singley (1973) have all addressed the question of what effects a proposed nuclear power plant,* located adjacent to the current Milliken Station electric generating plant (Fig. 5), would have on the thermal regime of the lake. Both Sundaram et al. and Singley approached this as a basic heat exchange problem and provide quantitative descriptions of interchanges at the air–water interface, heat transfer processes within the lake, and internal waves and seiches. The latter has also used extensive temperature data from 1950–1952 and 1968–1972 to develop an average pattern of temperature and heat content for the water column at biweekly intervals. In considering the published temperature

---

* In 1973 the New York State Electric and Gas Corporation announced their decision to construct a fossil fuel rather than a nuclear power plant.

TABLE 15

Date(s) of Occurrence (in Parentheses) and Temperatures (°C) for Important Events in the Annual Temperature Cycle of Cayuga Lake

Year	Ref. no.[a]	Max. spring homothermy	Isolation of bottom water	Permanent summer stratification	Max. summer bottom	Max. summer surface	Max. fall homothermy	Min. winter homothermy
1927	4	—	(14V)	(20VI)	5.5 (29VII)	19.3 (26VIII)	7.4 (4XII)	3.2 (16IV)
1928	4	—	(20V)	—	4.9 (29IX)	21.9 (12VIII)	—	2.3 (25III)
1939	5	—	—	—	—	24.0 (20VIII)	8.1 (14XI)	—
1940	5	3.8 (24V)	(12VI)	—	—	—	—	1.3 (3V)
1950	6	—	—	—	5.4 (26VII)	22.8 (11VIII)	6.7 (5XII)	—
1951	6	4.4 (28IV)	(1V)	(28VI)	5.5 (4VIII)	25.6 (29VIII)	—	3.3 (28III)
1952	6	4.4 (28IX)	(16V)	(1VII)	5.0 (27VII)	26.7 (22VII)	9.6 (4XI)	2.8 (18III)
1957	7	—	—	(27VI)	5.0 (—VIII)	23.0 (15VIII)	6.1 (13XII)	—
1958	7	3.7 (24IV)	(15V)	(19VI)	4.5 (19–20VIII)[c]	22.2 (17VIII)	—	1.8 (27II)
1968	8	—	—	—	—	23.1 (15–16VII)	—	1.5 (18III)
1969	8	—	—	—	—	—	—	—
1968[b]	11	—	(—V)	(18VI–1VII)	4.7 (—VIII)[c]	22.2 (4VIII)	6.6 (—XI)[c]	1.7 (22III)
1969[b]	11	—	(—V)	(18VI–1VII)	4.3 (—VIII)[c]	—	—	1.4 (13III)
1970[b]	11	—	(—V)	(18VI–1VII)	5.3 (—VIII)[c]	24.3 (4VIII)	6.6 (—XI)[c]	1.5 (5III)
1971[b]	11	—	(—V)	(18VI–1VII)	4.1 (—VIII)[c]	21.8 (4VIII)	6.8 (—XI)[c]	2.0 (5III)

[a] Numbers refer to references in Table 16.
[b] Based on biweekly averages.
[c] 60-m depth rather than bottom.

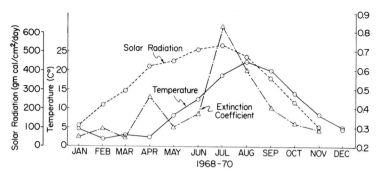

Fig. 7. Mean monthly solar radiation, light extinction coefficients, and euphotic zone temperatures, 1968–70 (Peterson, 1971).

records for Cayuga Lake, it should be borne in mind that sampling frequencies have generally been lower for the colder months, that stations for which data have been collected vary, and that internal waves and seiches may produce temperature profile anomalies in specific instances.

Peterson (1971) has estimated mean monthly euphotic zone temperatures (Fig. 7) for a 3-year period. The annual average is 10°C.

From both observed and calculated values Sundaram *et al.* (1969) concluded that internal seiches in Cayuga Lake have a typical period of 65–70 hr, a maximum amplitude (which would occur at the two ends of the lake) of about 18.5 m, and maximum seiche-induced current velocities (which would be found at the center of the lake) of 44 cm/sec in the epilimnion and 10 cm/sec in the hypolimnion. They also identified the presence of progressive, short period (5 min) internal waves in the lake.

The first heat budgets for Cayuga Lake were calculated by Birge (Birge and Juday, 1914, 1921), and Henson *et al.* (1961), Sundaram *et al.* (1969), and Singley (1973) have prepared them for subsequent years. The latter reviewed earlier computations and, together with data for 1968–1970, has calculated the average biweekly heat flux of the lake. Minimum heat content (9.1×10^5 kcal/m²) is in late February through mid-March, and the maximum (45×10^5 kcal/m²) is bimodal with the first peak in mid- to late August and a second in mid-September.

Specific Conductance

An isolated value of 596 μmhos/cm² was reported by Berg (1966) for the mid-1950's, but the first data of any comprehensiveness were for July 1968 through April 1969 (Wright, 1969e), when variations with depth and between stations were determined. Wright found that specific conductance

TABLE 16

Important Sources of Information on Temperature Distribution and Heat Flux in Cayuga Lake

Reference	Location(s)[a] in lake	Date(s) of measurement	Comments
(1) Birge and Juday (1914)	Sheldrake	Aug 11, 1910; Feb 13, 1911; Sep 2, 1911	Annual heat budgets, distribution of heat with depth, and mean summer and winter temperatures computed
(2) Birge and Juday (1921)	Sheldrake	Sep 4, 1914; Aug 16, 1917; Aug 30, 1918	Summer heat income and distribution of heat computed
(3) Wagner (1927)	3 km and 39 km from south end	Jun–Sep 1927	Only high and low temperatures given
(4) Burkholder (1931)	2.5 km from south end	2-week intervals Apr–Sep 1927, monthly Oct 1927–Feb 1929	5-m intervals to 50 m and bottom (about 60 m)
(5) Hess (1940)	Taughannock Point	Jul 1939–Jul 1940	Temperatures taken at variable intervals in order to accurately define periods of transition to and from homothermous conditions
(6) Henson et al. (1961)	Taughannock Point	Jun 1950–Nov 1953 at weekly intervals in summer and at least once each month at other times	Seasonal temperature cycles and anomalies in vertical profiles described, heat budgets calculated

Reference	Location/stations	Period	Description
(7) Howard (1963)	Taughannock Point	Apr 1957–Aug 1958 at weekly intervals	Isotherm plots of annual temperature cycle, discussion of thermal classification, computation of internal seiche
(8) Wright (1969c)	Seven stations on long axis of lake and two additional cross-lake samples at Milliken	Jul 1968–Apr 1969 at 2 week intervals during warmer months and monthly in winter	Isotherm plots by station and average for lake
(9) Sundaram et al. (1969)	Four fixed buoys with thermistor-bearing cables 1000–2500 ft off shore from Milliken Station	Aug–Dec 1968	Thorough mathematical analysis of heat exchanges between lake and air and within lake, observations and calculations on internal waves and internal seiches, and airborne infrared radiometer studies of "surface" temperatures in vicinity of Milliken Station
(10) Peterson (1971)	Not specified but data probably for Portland Point	All months	Mean euphotic zone temperatures by months
(11) Singley (1973)	1950–1952 for Taughannock Point, 1968–1972 Portland Point and Milliken Station combined	Calculations based on 1950–1952 and 1968–1972 data	Average temperature and heat content profiles by 2-week periods with standard deviations, quantitative model of stratified layers, heat budgets

[a] See Fig. 5 for locations.

in the epilimnion increased steadily during the summer from a July 1 value of 475 μmhos/cm² (25°C) to nearly 600 μmhos/cm² in October. Values increased with depth, being greatest near the bottom. Conductivity was generally 600 μmhos/cm² or higher at depths greater than 75 m from early August onward. In mid-November a slight vertical gradient still persisted. From January to the middle of April specific conductance was the same throughout the water column at about 600 μmhos/cm².

Although an annual cycle was not completed by Wright, it seems likely that the lower epilimnetic conductivities of early summer are the result of electrolyte dilution by heavy surface runoff in the spring and early summer. Conversely, it is postulated that the relatively high specific conductances characteristic of the deep waters result from ion solubilization or groundwater input within the lake basin proper. Both mechanisms receive credence from the observed low electrolyte content of tributary streams (Wright, 1969e) and the relatively high sodium chloride content of the lake compared with that of surface drainage (R. T. Oglesby, unpublished data; Likens, 1974b).

Dahlberg (1973) determined specific conductances as a function of location and depth on five occasions (May through October) in 1972. The mean value was 566 μmhos/cm² and the minimum and maximum were, respectively, 445 and 703 μmhos/cm². During the stratified period statistically significant increases with depth were observed. Seasonally, specific conductance was about 570 μmhos/cm² throughout the water column in May. The epilimnetic average had decreased to 510 μmhos/cm² on June 13–14 while electrolyte content at mid-depth and near the bottom increased slightly. A substantial decrease was apparent throughout the water column in samples taken July 25–August 3 with surface values averaging around 465 μmhos/cm². From this point on specific conductance increased markedly for the August 29–30 samples and those taken on October 3–4. By mid-November conductivity at the surface (the only depth sampled) had decreased to an average of 559 μmhos/cm².

The decreased specific conductance observed by Dahlberg (1973) between the June and July samples was almost certainly the result of electrolyte dilution from surface runoff produced by Tropical Storm Agnes. Some 17 cm of precipitation was dumped on the basin June 21–22, 1972 by this single storm.

Light and Solar Radiation

Edward Birge (Birge and Juday, 1921) measured light absorption in the water column at a station off Sheldrake Point on July 29, 1918 (Fig. 8). Only relative values were given.

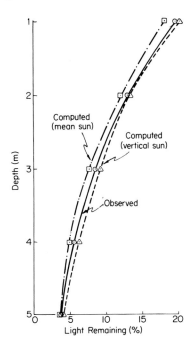

Fig. 8.   Attenuation of light with depth during mid-summer of 1918 off Sheldrake.

Howard (1963) computed extinction coefficients for 1957–1958 and used these data in obtaining his productivity estimates. However, no values are given nor does he include any information on the light profiles upon which his extinction coefficients were based. In commenting on the relation of light penetration to the depth of the euphotic zone Howard states "The frequent observation of net production at 20 meters shows that the estimation of the trophogenic zone depth as being the depth where one percent of the surface light penetrates is not valid for Cayuga Lake. To give an example, when the extinction coefficient is 0.40 there is only 0.04 percent of the surface light remaining at 20 meters. To estimate the depth of the euphotic zone on the one percent rule, for a coefficient of 0.40 the real euphotic zone would be about two times deeper than the estimate."

Data on solar radiation and extinction coefficients have been summarized for 1968–1970 (Table 17) by Peterson (1971). Additional data on solar radiation were presented previously in the section on climatology.

A somewhat subjective but empirically significant indication of light penetration in the water column is given by the Secchi disc transparency. Wright (1969f) has compared all previous Secchi disc measurements and those which he made during 1968. After sorting the data into 2 week

TABLE 17

Monthly Means of Solar Radiation and Extinction Coefficients[a]

Month	Solar radiation (gm cal/cm^2/day)	Extinction coefficient (k)
Jan	112 (3)	0.250 (1)
Feb	221 (3)	0.292 (1)
Mar	293 (3)	0.250 (1)
Apr	421 (3)	0.463 (2)
May	450 (3)	0.301 (4)
Jun	511 (3)	0.370 (3)
Jul	529 (3)	0.854 (4)
Aug	472 (3)	0.598 (10)
Sep	358 (3)	0.403 (4)
Oct	234 (3)	0.321 (2)
Nov	103 (3)	0.286 (3)
Dec	93 (3)	—

[a] Number of values averaged shown in parentheses.

periods and by location, he convincingly argued that quantifiable trends could not be established from the available information. Nevertheless, where reasonably comparable data exist for the summer, transparencies were generally lower for the 1950–1968 period than for 1910–1927. Limited transparency measurements for the 1940's indicated intermediate values between the two periods for which most data are available. Minimum Secchi disc readings (about 2–4.5 m) have been found by all workers to occur between mid-June and mid-September, but occasionally, high transparencies (ca. 6 m) do occur during the summer. By October–November transparency has typically increased to 6 m or greater.

Color

Dahlberg (1973) is the only researcher to publish values for color. For his six samplings between May and November 1972 he reported mean, minimum, and maximum values of 6, 0, and 20 mg/liter (presumably he meant color units). Highest values occurred in July (about 15 color units averaged for the surface, metalimnetic, and hypolimnetic samples) and were also relatively high on June 13–14. Color levels were low during the rest of the sampling period with none being detectable at the surface (the only depth sampled) on November 19–27.

pH

In reviewing the published data on hydrogen ion concentration in Cayuga Lake a general picture emerges of minimum water column averages (pH

7.7–8.0) during the winter months. An increase to a pH of about 8 occurs prior to thermal stratification. During the summer, hypolimnetic pH decreases fairly rapidly to a low of 7.7–7.8 with occasional values to 7.5 prior to autumnal mixing. At the same time pH in the epilimnion reaches a maximum (as high as 9.0) averaging 8.2–8.4. The hydrogen ion concentration then drops during the mixing period and reaches its winter minimum in January or February.

Sufficient data exist for 1927 (Wagner, 1927; Burkholder, 1931), 1950–1952 (Henson *et al.*, 1961), and 1968 (Wright, 1969j) to permit a reasonable comparison of pH to be made for the summer-stratified periods of these years. The data in Table 18 represent similar depths and dates on which pH was measured. Burkholder's values were determined at a station near Glenwood with a maximum depth of 65 m, and all but one of the samples composited for the 1927 description are his. The exception is a single profile from a station several kilometers north of Taughannock taken by Wagner on August 10, and values are in good agreement with those from Glenwood for August 12. All of Henson's data are from Taughannock. Hydrogen ion concentrations at both stations are available for 1968. The data indicate that a slight decrease in pH may have taken place between 1927 and the 1950–1968 period in both epilimnion and hypolimnion. If this is real, a proportionate decrease throughout the water column would seem to reflect a change in ionic makeup.

Although data for the winter period are very limited, they tend to confirm this possibility. Burkholder (1931) reported a pH of 8.3 ($n = 5$), averaged for the water column, on February 27, 1929, Henson *et al.* (1961) a pH of 8.0 ($n = 2$) on March 18, 1952, and Wright (1969j) pH's of 7.8 ($n = 3$) at Taughannock and 7.7 ($n = 3$) at Glenwood on February 17, 1969.

Dahlberg (1973) presents a graphical plot of hydrogen ion concentration, averaged for all stations and given for the five occasions in 1972 at the three depths he sampled, which agrees well with the pH's and their seasonal pattern of change as described by Henson *et al.* (1961) and Wright (1969j). His minimum, maximum, and mean values for all depths were, respectively, 7.3, 8.4, and 8.0.

Dissolved Oxygen

The general features of dissolved oxygen distribution in Cayuga Lake are as follows. At fall homothermy, about mid-November to early December, the entire water column is 80–90% saturated. Dissolved oxygen increases during the course of the winter isothermal period, but at the time of initial thermal isolation of bottom waters, dissolved oxygen is still only 90–95% of saturation. Over the summer-stratified season, daytime concentrations in the epilimnion decrease from about 10 mg/liter early in the summer to 9

TABLE 18

Comparable pH Values, Obtained during the Summer Months, for Cayuga Lake[a]

	1927 (G)[b]	1950(T)	1951(T)	1952(T)	1968(T)	1968(G)
Epilimnion						
Min.	7.9	8.1	8.1	8.1	8.0	7.5
Max.	9.0	8.3	8.5	8.4	8.8	8.7
Avg.	8.5 ($n = 44$)	8.2 ($n = 13$)	8.3 ($n = 10$)	8.2 ($n = 9$)	8.4 ($n = 8$)	8.2 ($n = 9$)
Hypolimnion						
Min.	7.8	7.5	7.6	7.8	7.5	7.7
Max.	8.8	8.1	8.1	8.1	8.2	8.2
Avg.	8.2 ($n = 45$)	7.9 ($n = 22$)	7.9 ($n = 14$)	7.9 ($n = 14$)	8.1 ($n = 19$)	8.0 ($n = 20$)

[a] G, Glenwood; T, Taughannock; and n, number of measurements averaged. See Fig. 5 for locations.

[b] Includes a single profile (5 measurements) taken several kilometers north of Taughannock.

mg/liter in September as solubilities become lower due to increasing temperature. However, supersaturation is fairly common in the upper 5 m during periods of high primary productivity. As fall cooling occurs, concentrations increase but percent saturation in the upper portion of the water column tends to be the lowest of the year during autumnal mixing. During stratification, percent saturation decreases with depth in the epilimnion, usually increases slightly in the metalimnion, and then decreases again proceeding downward through the hypolimnion, with the rate of change being greatest in the stratum between 100 m and the bottom. Principal sources of detailed information on dissolved oxygen in Cayuga Lake are presented in Table 19.

Wright (1969g) has provided both the greatest amount of information for a single year and the most comprehensive discussion of oxygen in the lake. During 1968, the maximum epilimnetic oxygen averaged for all stations ($n = 13$) was 119% of saturation in July. During the two samplings carried out in this month, there was a definite south to north cline of decreasing oxygen just as there was in chlorophyll a, with by far the highest levels of both being in the southern 5 km. The highest discrete saturation attained in 1968–1969 was 169% at the surface on July 15 off Glenwood. Hypolimnetic oxygen was at 95% of saturation (12 mg/liter) during late May and early June and decreased steadily through September. The lowest recorded value in 1968 was 6.2 mg/liter (49% saturated) September 26 in 116 m of water at Sheldrake. By the end of October, when the lake was still thermally stratified, partial reoxygenation of hypolimnetic waters had occurred almost to the bottom. Over the 6 months of stratification, oxygen decreased at an average rate of about 0.3 mg/liter/month in the upper epilimnion and about 1.0 mg/liter/month at 120 m.

Wright (1969g) compared all available information on dissolved oxygen in Cayuga Lake in an attempt to define historical trends. Because of the temporal and spatial scatter and the questionable nature of some of the early data (e.g., Henson et al., 1961), it was impossible to say whether any small scale trends toward changed summer saturation values in the epilimnion or in hypolimnetic oxygen depletion had taken place. However, it is obvious from Wright's comparisons that no major changes of this kind had occurred over the 6 decades since the first dissolved oxygen measurements.

Dahlberg (1973) comments that in spite of the potentially large input of allochthonous organic material by Tropical Storm Agnes in June of 1972, dissolved oxygen in the hypolimnion remained relatively high throughout the stratified season. He reports concentrations of 6.8 and 7.9 mg/liter near the bottom off the Milliken Station and Sheldrake, respectively, on October 3–4, 1968. By November 19 bottom dissolved oxygen had increased to 9.6 mg/liter at Milliken Station.

TABLE 19

Important Sources of Published Data on Dissolved Oxygen in Cayuga Lake[a]

Reference	Location(s) in lake	Date(s) of measurement	Comments
Birge and Juday (1914)	Sheldrake	Aug 11, 1910	Single profile to 122 m
Wagner (1927)	3 km and 39 km from south end	Jun–Aug 1927	High, low, and avg. in profiles to 60 m
Burkholder (1931)	2.5 km from south end	Jun 20–Sep 14, 1927 biweekly	Profiles to 50 m and bottom (55–65 m)
Henson et al. (1961)	Taughannock	Jun 21–Nov 14, 1950; May 15–Nov 13, 1951; and Mar 18–Sep 24, 1952, weekly in summer and at least once each month at other times	Profiles at 10–20 m intervals to 60–100 m, accuracy of data questionable (Wright, 1969h)
Wright (1969g)	Seven stations on long axis of lake and two cross-lake stations at Milliken	Jul 1, 1968–Apr 15, 1969, 2-week intervals during warmer months and monthly at other times	Isopleth plots, station and lake averages, raw data, historical comparisons, and discussion on long-term and seasonal trends
Dahlberg (1973)	Same as Wright (1969g) with added stations off Milliken	May 17–Nov 11, 1972	Only selected values given

[a] See Fig. 5 for locations.

Gates (1969) and Dahlberg (1973) have both reported the results of standard (5-day, 20°C) biochemical oxygen demand (BOD) tests on Cayuga Lake water. The former conducted limited testing in 1968 and obtained BOD's of 1–2.5 mg/liter for epilimnetic samples and about half this for waters from the hypolimnion. Dahlberg (1973) recorded average (lakewide at three depths on five occasions in 1972) BOD's as being greater than 2 mg/liter in May, less than 1 mg/liter in mid-June, and about 1 mg/liter for the other three samplings. There was a tendency for biochemical oxygen demand to decrease with depth.

Phosphorus

The earliest published observations on phosphorus in Cayuga Lake were for 1968–1969 (Barlow, 1969; Wright, 1969i). Samples were taken at 10 m-intervals from five to seven south–north and two cross-lake stations beginning July 1 and continuing at about biweekly intervals through the end of November 1968. Additional samples were taken on January 21 and February 17, 1969. Wright found little consistent difference in soluble reactive phosphorus between stations. However, there were several occasions in midsummer when concentrations were higher at the two closest to the head end of the lake. Values averaged for all stations are given in Table 20.

Barlow (1969) discussed the relationship of phosphorus to phytoplankton production in Cayuga Lake and also compared the distribution of total phosphorus (TP) to that determined as soluble reactive (SRP). During the stratified season SRP was always a small fraction (5–15%) of the total in the epilimnion but constituted a major portion throughout the water column after turnover. Reactive phosphorus was nearly always at least 50% and sometimes as much as 90% of the total in the hypolimnion. Rather surprisingly, Barlow found an almost uniform vertical distribution (about 20 μg/liter) of total phosphorus over the period the lake was sampled.

The role of zooplankton in the recycling of phosphorus in Cayuga Lake was studied by Peterson (1971). He concluded that during April–June metabolic requirements of phytoplankton for phosphorus were met by mechanical recycling within the well-mixed water column. During July–November, however, the estimated rate of primary production could not have been maintained on the basis of this supply mechanism but would had to have been met through recycling by zooplankton (and presumably other heterotrophs).

Soluble reactive, soluble unreactive (SUP), and particulate phosphorus (PP) were all measured by Peterson (1971) during the 1968–1970 field work related to his study. Averaged over the entire period (n = 133) total phos-

TABLE 20

Soluble Reactive Phosphorus (μg/liter) during 1968–1969

Depth (m)	1968									1969	
	Jul		Aug		Sep	Oct		Nov		Jan	Feb
	1–2	15–16	1–2	19–20	19–20	16–17	29–30	11–14	26	21	17
00–09	2.1	3.7	2.1	1.3	1.0	1.0	1.4	1.8	4.7	12.2	11.4
10–19	3.5	3.3	2.0	3.2	2.6	1.2	—	2.6	—	—	—
20–29	3.2	4.5	3.5	4.0	1.4	1.3	2.1	5.7	—	—	—
30–39	4.5	5.4	5.8	6.0	5.7	3.1	2.0	7.3	6.1	16.0	12.7
40–49		6.2	7.2	6.2	6.0	6.6	3.6	4.4	6.4	12.5	10.3
50–59			8.3	—	—	9.6	8.8	8.1	—	—	12.1
60–69			10.7	11.4	10.8	13.3	—	10.0	11.7	13.1	13.0
70–79			13.1	12.9	10.5	—	13.1	17.7	9.8	12.2	12.6
80–89					10.2	14.0	13.7	20.1	—	—	—
90–99					9.7		17.9	21.9	13.4	12.7	11.6
100–9									15.9	12.9	12.7

phorus (TP) concentration was 18 μg/liter and ranged from 9.1 to 56.7 μg/liter. In comparing TP for June–August for the 3 years, means were 20.9, 15.8, and 14.5 μg/liter, respectively, for 1968, 1969, and 1970. The difference between the first 2 years was statistically significant. Total phosphorus on an areal basis for 1969–1970 is given in Table 21. Commenting on the atypically high value in April of 1969 and the decrease back to a more normal level in May, Peterson speculated that phosphorus may have been removed from the water column as an adsorbant on soil particles distributed in the lake due to heavy runoff in April. Expressed as a percentage of TP, the various phosphorus fractions measured by Peterson showed little change in the hypolimnion. Epilimnetic SRP decreased markedly during the stratified season with concomitant increases in SUP (≡ dissolved organic phosphorus) and PP. Expressed another way, SRP:TP increased, SUP:TP decreased, and PP:TP decreased (but not at as rapid a rate as SUP:TP) with depth during the stratified season. During the mixed period SRP was still relatively depleted in the upper 35 m, but other forms varied only slightly with depth.

Seasonal changes in SRP concentration in the upper portion of Cayuga Lake during 1972 are shown in Fig. 9. Values represent unpublished data obtained independently by two groups of workers at Cornell University. The very large peak in late June and early July represents input associated with Tropical Storm Agnes, which dumped some 17 cm of precipitation on the basin June 21–22. Observations indicated that as the allochthonous inorganic suspended material settled a substantial amount of the phosphorus fraction measured as SRP went with it. The same phenomenon was concurrently observed in four of the other Finger Lakes which were under study and agrees with the earlier observation by Peterson (1971) in his study on Cayuga Lake.

TABLE 21

Total Phosphorus in a 100-m Water Column

1969					
Sampling date:	Winter	VII-7	VII-30	VII-14	XI-11
No. depths sampled:	5	6	9	6	8
Tot. P (gm/m²)	1.74	1.64	1.69	1.50	1.86

1970						
Sampling date:	IV-7	V-9	VI-2	VI-22	VII-16	VII-19
No. depths sampled:	8	8	8	8	8	8
Tot. P (gm/m²):	2.21	1.67	1.50	1.64	1.46	1.56

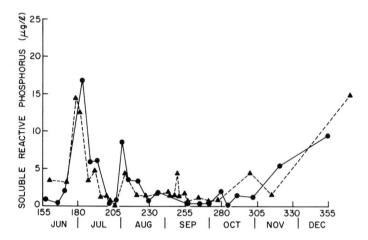

Fig. 9. Soluble reactive phosphorus in the upper waters of Cayuga Lake during 1972. Circles represent concentrations for the epilimnion at five stations as determined by P. Godfrey, W. Schaffner, and R. Oglesby, Department of Natural Resources. The triangles show values determined from integrated (0–10 m) samples taken off Portland Point by J. Barlow and B. Peterson, Section of Ecology and Systematics.

Nitrogen

Nitrogen in the form of nitrate appears to be present at almost all times and all places in Cayuga Lake in excess of the metabolic needs of phytoplankton. However, summer concentrations do vary from year to year, seasonally, and with depth. Unpublished data on the chemistry of tributaries into Cayuga Lake indicate the major seasonal input of nitrate is during late March or early April when spring snowmelt typically occurs. Lake concentrations are still about 800–900 μg/liter NO_3^-–N in mid-May. Until the summer of 1973, when P. Godfrey (personal communication) found occasional nitrate–N concentrations of less than 50 μg/liter, the residual was always considerably in excess of published minimum requirements for phytoplankton growth. A small but distinct vertical cline of increasing nitrate is present during the summer. In this same period organic nitrogen increases in the epilimnion but remains at low levels in the profundal zone.

The first published datum point for nitrogen in Cayuga Lake was given by Berg (1966), who reported a NO_3^-–N concentration of 632 μg/liter for the mid-1950's. A survey by the Federal Water Quality Administration (unpublished notes, 1965) in 1965 gives the first comprehensive information. Vertical profiles, with depth intervals dictated by thermal structure, were sampled at a series of stations along the south–north axis of the lake on three occasions (Table 22). Their ammonia–N values appear to be

TABLE 22

Concentrations (µg/liter) of Various Forms of Nitrogen, Averaged for All Depths and
Stations, during 1965[a]

Sampling dates	NO$_3$⁻-N	NH$_3$-N	Organic N	NO$_3$⁻-N:NH$_3$-N:Organic N
May 18–21	809 (44)	328 (43)	96 (15)	8.4:3.4:1
Jul 9–10	839 (37)	145 (38)	148 (25)	5.7:1:1
Oct 11–12	721 (30)	84 (30)	232 (28)	3.1:0.4:1

[a] Number of samples shown in parentheses after the values. Weight ratios for the
different forms are given.

somewhat high for a lake with well-oxygenated waters. For the lake as a
whole there was a definite seasonal shift from the dissolved inorganic to the
organic form of nitrogen. Individual values for all forms are highly variable,
making it difficult to assess trends in spatial distribution.

Barlow (1969) and Wright (1969i) defined the variations of NO$_3$⁻–N with
season and depth for the latter half of 1968 and early 1969. The pattern of
concentrations obtained by averaging data from the seven to nine stations
routinely sampled is shown in Fig. 10. The action of vertical mixing in
redistributing nitrate to the upper waters of the lake is strikingly apparent in
the data for November 26. Prior to this, a progressive decline of NO$_3$⁻–N in
the euphotic zone is apparent. Both Wright and Barlow, emphasize,
however, that levels were never low enough to limit primary production.

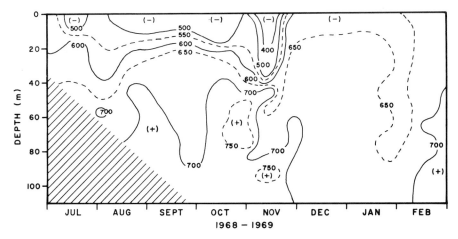

Fig. 10. Isopleths of nitrate nitrogen (µg/liter) in the water column during 1968–
1969. No data were available for depths of 80 m or greater until mid-September.

Barlow reinforces this argument in several ways. From vertical gradients of nitrate during the stratified season, he estimated its downward transport and from this calculated the "renewal ratio" of nitrogen to phosphorus to be 88:1 compared with an estimated assimilation ratio of 25:1. He also examined the relationship between changes in NO_3^--N concentration and the ratio of reactive to total phosphorus and found this to be nonlinear ($Y = 30.8 + 5.4 \ln X$, where $X = RP/TP$), another indication that nitrate was nonlimiting.

Dahlberg (1973) measured the nitrogen content of samples taken at three depths (epilimnion, metalimnion, and hypolimnion) on five occasions in 1972 from a lakewide network of stations. His data are shown, averaged for all depths and stations and as station averages for surface samples, in Table 23. An increase of nitrate with depth during the stratified season is indicated. Kjeldahl nitrogen decreased markedly between the surface and hypolimnion but was sometimes maximum in the metalimnion. Note the very low NH_3-N values compared with those reported by the Federal Water Quality Administration (unpublished notes, 1965).

Alkalinity and Acidity

Cayuga Lake has a well-developed carbonate–bicarbonate buffer system with a total alkalinity on the order of 100 mg/liter as $CaCO_3$. Winter values are typically higher than this and an annual minimum occurs in July–September. The variation within a year is 10–15 mg/liter. During the stratified season, there is a slight increase in alkalinity with depth.

Although the principal limestone formation in the basin is the Tully escarpment (see below), shallow outcroppings and deposits of glacially carried limestone are common throughout resulting in generally high calcium carbonate–bicarbonate concentrations in tributaries. However, Salmon Creek and the smaller northern tributaries do reflect the greater mass of calcareous materials in their watersheds with higher alkalinities. No obvious south–north gradient is evident in the lake.

The earliest published data on the alkalinity of water in Cayuga Lake are those of Wagner (1927), who recorded concentrations determined for a single profile taken off Frontenac Point on August 10, 1927, and Burkholder (1931), who measured the methyl orange alkalinity at 5-m depth intervals for a station south of Glenwood during the summer of 1927, on September 29, 1928, and on February 27, 1929. Phenolpthalein alkalinity was determined on the last two occasions only with none being found in February, and small concentrations (maximum 1.5 mg/liter as $CaCO_3$ at the surface) extending to 30 m being present in September. Average methyl orange alkalinities obtained by Burkholder and by Wagner are shown

TABLE 23

Average Nitrogen Concentrations (μg/liter) for 1972

Sampling date	All depths				Surface			
	No. samples	NO_3^--N	NH_3-N	Kjeldahl N	No. samples	NO_3^--N	NH_3-N	Kjeldahl N
May 10–17	19	881	0	236	7	788	0	309
Jun 13–14	20	871	—	321	7	733	—	402
Jul 25–Aug 3	19	991	36	496	7	901	43	524
Aug 29–30	19	927	—	286	7	781	—	402
Oct 3–4	19	1013	—	398	7	933	—	429
Nov 19–27	—	—	0	—	7	607	—	250

in Table 24. Neither found any substantial differences in the vertical distribution.

Henson et al. (1961) described the distribution of total alkalinity as a function of depth and season for 1950–1952 at Taughannock. Depths sampled were usually 0, 10, 30, 60 and 100 m. A tendency for values to increase slightly with depth during the summer-stratified period is apparent in their data. Means representing averages of all depths were, for 1950, 100 mg/liter as $CaCO_3$ ($n = 14$, SE = 1.4); for 1951, 93 mg/liter as $CaCO_3$ ($n = 11$, SE = 2.0); and for 1952, 96 mg/liter as $CaCO_3$ ($n = 8$, SE = 1.83). They stated that "Some phenolpthalein alkalinity was commonly found at the surface and at 10 meters, and occasionally extended to a depth of 30 meters. It was found at the surface from June to November in 1950, from June to September in 1951, and beginning in July, 1952."

The total alkalinities given by Henson et al. show considerable variation between years and are consistently lower than the methyl orange alkalinity concentrations reported for the summer of 1927. A trend to summer minima is apparent, but the extent of seasonal decrease and the dates within this period when concentrations are lowest also differ from year to year.

Wright (1969j) analyzed samples taken as vertical profiles throughout the lake for alkalinity between July 1968 and April 1969. He states that "The methyl orange alkalinity was usually between 100 and 110 mg/liter (as $CaCO_3$). . . It was higher in the epilimnion than in the hypolimnion and decreased in both as stratification progressed." Wright believed that an

TABLE 24

Methyl Orange Alkalinity during 1927–1929 Averaged for Entire Water Column[a]

Sampling date	Concentration (mg/liter as $CaCO_3$)
Jun 20, 1927	107 (10)
Jun 30, 1927	105 (11)
Jul 12, 1927	107 (11)
Jul 29, 1927	103 (11)
Aug 10, 1927	104 (12)
Aug 12, 1927	103 (11)
Aug 26, 1927	103 (11)
Sep 14, 1927	101 (11)
Sep 29, 1928	99 (6)
Feb 27, 1929	103 (5)

[a] Number of samples given in parentheses.

extremely high (282 mg/liter as $CaCO_3$) methyl orange alkalinity value obtained at the surface off Sheldrake on March 18, 1969 was the result of waste products from large numbers of Canada geese found at this location.

Phenolpthalein alkalinity was found intermittently during the summer of 1968 and into mid-October with a maximum value of 10.1 mg/liter as $CaCO_3$. It was generally confined to the upper 20 m of the water column.

Dahlberg (1973), sampling at three depths and stations throughout the lake during 1972, reported a methyl orange alkalinity range of 84–112 mg/liter as $CaCO_3$ with a mean of 99.7 mg/liter for his five May to November samplings. Concentrations decreased with depth and with time during the stratified season. Phenolpthalein alkalinity was found only during June–August.

The only published values of acidity are those of Dahlberg (1973). Mean, minimum, and maximum concentrations were, respectively, 2.6, 0, and 9.3 mg/liter. Increases during the stratified season were noted for metalimnetic and hypolimnetic samples.

Major Ions

Cayuga Lake has a well-developed calcium carbonate buffer system, and concentrations of sodium and chloride are unusually high for an inland lake in the northeastern United States. Some of the earlier limnological studies on the lake included routine measurement of alkalinity (see above), but the first complete listing of major ionic constituents was that given by Berg (1966). Values were based on limited and unpublished data collected in the 1950's by the United States Geological Survey (C. O. Berg, personal communication). During 1965 approximately 100 samples taken during May, July, and October from stations throughout the lake and at up to five depths were chemically analyzed (Federal Water Pollution Control Administration, unpublished notes, 1965). Dahlberg (1973) has reported on a similar series of analyses for samples taken in 1973. He gave the following mean concentrations (mg/liter): calcium, 44; magnesium, 10; sodium, 51; potassium, 2.6; bicarbonate, 121.6; sulfate, 36.3; and chloride, 81.

Data from the three investigations are summarized as milliequivalants of anions and cations in Table 25. Positive and negative ions appear to be reasonably well balanced, although anions are lower by about 0.3 mEq for the 1965 and 1973 data. Fluctuations in total ionic composition between years are fairly large and may well relate to varying ratios between evaporation and precipitation. Data are as yet insufficient to assess this with any confidence. Sulfate appears to have decreased by some 30% between the mid-1950's and 1965.

TABLE 25

Major Anions and Cations in Cayuga Lake from Berg for the Mid-1950's, the Federal Water Pollution Control Administration for 1965, and Dahlberg for 1973[a]

	1950's	1965	1973		1950's	1965	1973
Ca^{2+}	2.35	2.19	2.20	HCO_3^-	2.11	2.14	1.99
Mg^{2+}	0.79	0.90	0.82	SO_4^{2-}	1.14	0.80	0.76
Na^+	2.65[b]	2.94	2.22	Cl^-	2.48	2.83	2.28
K^+	—	0.05	0.07				
	5.79	6.08	5.31		5.73	5.77	5.03

[a] The latter two represent lakewide averages.
[b] Determined as sodium plus potassium.

The sodium chloride in Cayuga Lake does not come entirely from tributary inflow (unpublished analyses of streams in the basin, and Likens, 1974b). Berg (1966) has postulated that the lake basin proper may be intersected by salt strata. These are known to underlie the central part of the lake at a depth of over 350 m below the bottom of the lake, but saline surface springs are found a few kilometers to the north of the lake (Dudley, 1886).

Trace Metals and Micronutrients

Wright (1969h) presents a table of trace element concentrations in Cayuga Lake for a single set of samples taken in January 1969. Although analytical methods were not specified, it is obvious, from both the wide ranges of values given for each element and the extremely high levels represented by these ranges, that techniques were not appropriate to the problem.

The concentrations of zinc, copper, lead, cadmium, and cobalt in the euphotic zone were studied by Mills and Oglesby (1971) during June, July, and August of 1970. The trace metals were concentrated from solution by extraction with dithizone in carbon tetrachloride and were then measured by atomic absorption spectrophotometry. Mean values in soluble form were zinc, 2.7 μg/liter; copper, 0.60 μg/liter; lead, 0.12 μg/liter; cadmium, 0.54 μg/liter; and cobalt, 0.005 μg/liter. Concentration of all decreased substantially (an order of magnitude, except cobalt, which decreased by about 50%) during the latter half of the study period. Levels for all were among the lowest so far described for surface waters anywhere in the world.

Dahlberg (1973) has reported the results of analyses for trace metals in Cayuga Lake during May–November 1973. No attempt was made to

TABLE 26

Trace Element Concentrations during 1973 as Reported by Dahlberg (1973)[a]

	May 10–17	Jun 13–14	Jul 25– Aug 3	Aug 29–30	Oct 3–4	Nov 19–27
Fe	0.08	0.12	0.03	0.04	0.08	0.06
Al	0.05	—	0.02	—	—	0
Cu	0.02	0.02	0.06	0	0.03	0.01
Zn	0.04	—	0.02	—	—	0
Hg	4.0	1.2	2.1	14.1	0.61	1.7

[a] Data are averaged for three depths and all stations and are expressed in mg/liter except for mercury which is in μg-liter.

concentrate samples, and chromium manganese, cadmium, nickel, lead, and tin were all generally below the level of detection by direct atomic absorption analysis. His results for other trace metals are summarized in Table 26. Presumably, Dahlberg's values include particulate as well as soluble forms of these elements, and this may account for the concentrations of copper being higher than those reported by Mills and Oglesby (1971). Dahlberg states that iron was the only trace metal that consistently showed a gradient of increasing concentration with depth during the period of summer stratification.

Unpublished data on trace metals were also collected by the author during 1973 and are presented in Table 27. The analyses were done by Cornell's Soil Testing Laboratory (Department of Agronomy) using atomic absorption analysis. These represent the only determinations so far reported for manganese in Cayuga Lake.

The only data on organic micronutrients in Cayuga Lake are those reported for vitamin B_{12} by Mills and Oglesby (1971). They measured

TABLE 27

Manganese and Iron Concentrations (mg/liter) during 1973

Nature of samples	Sampling date	Location	Fe	Mn
Surface and bottom composite	IV-26	Sheldrake	0.03	0.03
Surface and bottom composite	IV-28	Myers Pt.	0.003	0.002
Surface and bottom composite	VI-16	Sheldrake	0.22	—
	VI-17	Myers Pt.	0.20	—
Surface and 50 m composite	VIII-29	Taughannock	0.002	0.001
Bottom	VIII-29	Myers Pt.	0.019	0.005

TABLE 28

Vitamin B_{12} Averaged for Two Representative Stations in the Euphotic Zone from Late June to Mid-August 1970

Date	Vitamin B_{12} (mg $\times$ $10^{-3}/m^3$)		
	0 m	5 m	10 m
Jun 12	1	1.5	—
	$n = 1$	$n = 2$	
Jul 16	7.5	5	3
	$n = 2$	$n = 2$	$n = 2$
Aug 6	8	7	3
	$n = 2$	$n = 2$	$n = 2$
Aug 17	8	7	6
	$n = 1$	$n = 1$	$n = 1$

concentrations in both tributary streams and the euphotic zone of the lake during the summer of 1970. Values showed considerable variation between tributaries but were spatially homogenous at a given depth in the lake. Mean concentrations of soluble vitamin B_{12} over the course of the study were 9 mg $\times$ $10^{-3}/m^3$ for streams and 5 mg $\times$ $10^{-3}/m^3$ for Cayuga.

Vitamin B_{12} in the euphotic zone of the lake exhibited patterns of increase during the summer and decrease with depth (Table 28). The ratio of soluble cobalt to cellular B_{12} in June ranged from 0.03 to 0.62% while August values were 3.2 to 11.0 [ratios comparable to those obtained by Benoit (1957) for Linsley Pond]. *In situ* nutrient enrichment studies during late summer did not indicate any increase in carbon uptake due to vitamin B_{12} additions.

Phytoplankton

Pigments

The first reported chlorophyll *a* data for Cayuga Lake were those of Hamilton (1969) for 1966 (Table 29). Maximum cell counts occurred in June but had decreased markedly before the chlorophyll peak on the twenty-eighth. The June maximum was associated with a phytoplankton community dominated by silicious forms. Hamilton did not specify the depths at which samples were taken, but it can be inferred from his article that these were from the surface waters.

Spatial as well as temporal variations in chlorophyll *a* and phaeophytin were first reported for the year 1968 and for early 1969 (Wright, 1969k), and these data were analyzed and discussed by Barlow (1969) in terms of both their intrinsic meaning and their relationships to other measurements. During the 1968–1969 period 12 samplings were made at from seven to nine stations located on the long axis of the lake and at two additional points extending laterally across the lake at the Milliken Station.

Chlorophyll *a* concentrations (Wright, 1969k) averaged for all stations are given in Table 30. Samples were taken at 10-m intervals, although the deeper strata were not sampled early in the summer and some depths were omitted later on. Chlorophyll *a* was usually greater than 1 μg/liter at depths less than 25 m through the middle of November, at which time the lake was mixed to a depth of 80 m with commensurate chlorophyll increases in some of the deeper strata. However, only on February 17, 1969 could chlorophyll *a* be considered uniform in its vertical distribution. Variation between stations was often considerable. On July 1 a decreasing gradient from south to north was very marked with a surface concentration of 16.0 μg/liter occurring at the most southerly station, and only 1.7 μg/liter chlorophyll *a* being present at the sampling location nearest the northern end of the lake. A similar, though less pronounced, longitudinal gradient occurred on September 5. The three sampling locations extending laterally across the lake at Milliken Station exhibited some gradation up to the middle of September, with the station farthest to the east usually having the highest concentration of chlorophyll *a*.

Phaeopigments in 1968–1969 exhibited differences from chlorophyll in their seasonal pattern of change. A slight maximum appeared to be associated with that of chlorophyll on July 1, but a more pronounced peaking of phaeopigments followed a slight increase of the former on September 5–6. Barlow's (1969) data on cell volumes (see below) indicate that the September maximum was preceded by a "bloom" of *Coelosphaerium* which was not reflected in chlorophyll increases. The vertical distribution of

TABLE 29

Summer Chlorophyll a Concentrations (μg/liter) during 1966

Date	Concentration	Date	Concentration
Jun 18	3.8	Jul 29	1.5
Jun 23	5.3	Aug 2	1.5
Jun 26	5.9	Aug 5	1.5
Jun 28	7.2	Aug 9	1.6
Jul 6	5.9	Aug 10	1.0
Jul 20	1.2	Aug 16	1.0
Jul 22	1.9	Aug 18	1.3
Jul 26	1.7		

TABLE 30

Phaeophytin and Chlorophyll a

Phaeophytin Concentrations (µg/liter) in Cayuga Lake during 1968 and Early 1969 Averaged for All Stations[a]

Depth (m)	1968										1969	
	VI 1-2	VI 15-16	VIII 1-2	VIII 19-20	IX 5-6	IX 19-20	X 16-17	X 29-30	XI 14	XI 26	I 21	II 17
00-09	2.8	2.2	1.1	0.8	2.4	4.2	2.8	2.5	1.8	0.7	0.6	0.5
10-19	1.3	1.6	1.1	0.4	1.2	1.2	1.4	—	1.2	—	—	—
20-29	1.3	1.4	0.7	0.6	0.4	1.2	1.5	1.0	1.9	—	—	—
30-39	0.9	1.0	0.5	0.2	0.4	0.7	1.0	1.8	0.6	1.0	0.6	0.7
40-49		0.9	0.4	0.4	0.6	0.4	0.6	1.4	0.4	0.5	1.1	0.5
50-59			0.3	—	—	—	0.4	0.7	1.0	—	—	0.8
60-69			0.6	0.2	0.4	0.3	0.4	—	1.2	0.6	0.5	0.5
70-79			1.8	0.6		0.2	—	0.3	0.5	1.0	1.3	0.6
80-89						0.2	0.4	0.4	0.2	—	—	—
90-99						0.5		0.8	0.8	0.7	0.3	0.7
100-9									1.3	1.2	1.8	0.6

Chlorophyll a Concentrations (µg/liter) in Cayuga Lake during 1968 and Early 1969 Averaged for All Stations[a]

Depth (m)	1968										1969	
	VI 1-2	VI 15-16	VIII 1-2	VIII 19-20	IX 5-6	IX 19-20	X 16-17	X 29-30	XI 14	XI 26	I 21	II 17
00-09	6.0	3.7	3.6	3.8	4.0	1.9	1.6	1.5	2.5	1.8	1.3	1.4
10-19	1.1	2.1	2.0	0.9	1.8	1.3	2.5	—	2.0	—	—	—
20-29	0.7	1.2	1.1	0.5	1.5	0.7	1.1	1.1	1.4	—	—	1.5
30-39	0.8	0.7	0.8	0.4	0.4	0.2	0.5	0.8	0.6	1.8	1.1	1.6
40-49		0.2	0.4	0.3	0.0	0.2	0.8	0.5	0.2	1.4	0.6	1.4
50-59			0.4	—	—	—	0.2	0.2	1.4	—	—	1.4
60-69			0.4	0.4	0.2	0.3	0.3	—	0.9	0.6	1.4	1.6
70-79			0.3	0.2		0.2	—	0.4	1.0	1.2	0.4	1.3
80-89						0.2	0.2	0.2	0.2	—	—	—
90-99						0.2		0.2	0.2	0.7	0.9	1.6
100-9								0.1	0.2	0.2	0.8	1.4

[a] The number of samples averaged ranged from 1–16 (Wright, 1969k).

phaeopigments followed a pattern similar to that of chlorophyll except that
increases near the bottom were apparent in autumn.

Barlow (1969) compared the chlorophyll *a* concentrations found in 1968
with cell volume measurements and cell counts. He states that "The correlation between chlorophyll and total cell volume is only $r = 0.11$ ($n = 48$),
but between chlorophyll and numbers is $r = 0.57$ a significant improvement
in the relation. Apparently the composition of the population has an
important effect on the relation between chlorophyll and cell volume, for
partial correlation coefficients between chlorophyll and diatoms was $r = 0.35$, whereas between chlorophyll and blue-greens it was $r = -0.34$. Both
coefficients are highly significant. Presumably chlorophyll is a poor indicator
of blue-greens because of the importance of the unique phycobilin pigments
in these algae."

Chlorophyll *a* was measured by Dahlberg (1973) in his 1972 studies on
Cayuga Lake but only for surface samples, except during July and October.
On these two occasions, and especially the former, there were substantially
higher levels at greater depths than those reported by Barlow (1969) for
1968. Average concentrations for the three depths sampled were 3.9, 4.0,
and 3.4 μg/liter on July 27 for the epilimnion, metalimnion, and hypolimnion, respectively, and 4.6, 2.3, and 1.9 μg/liter on October 3 for comparable strata. While differences between stations were often appreciable, no
persistent pattern of longitudinal gradation was apparent from Dahlberg's
data.

Dahlberg's (1973) investigation was carried out as a consulting project
relating to the proposed construction of a nuclear power plant on the shore

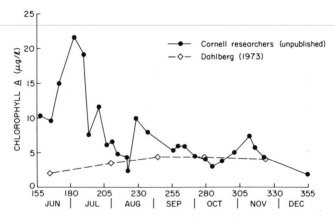

Fig. 11. A comparison of chlorophyll *a* in surface samples during 1972 from two
groups of researchers working independently.

TABLE 31

Mean Monthly Chlorophyll *a* (µg/liter) in the Euphotic Zone during 1968–1970

Month	Concentration	n	Month	Concentration	n
Jan	1.1	1	Jul	6.0	5
Feb	1.4	1	Aug	4.3	10
Mar	—		Sep	5.2	4
Apr	2.1	2	Oct	1.6	2
May	4.1	4	Nov	3.0	3
Jun	4.0	3	Dec	—	

of Cayuga Lake. In a critique of the resulting environmental feasibility report, Cornell scientists drew, for comparative purposes, on as yet unpublished chlorophyll data* which they had collected during 1972. The results, shown in Fig. 11, represent samples taken by both groups at mid-lake off the Milliken Station and are for comparable depths. The weekly samples of the Cornell group reveal considerable short-term variation in chlorophyll *a* concentrations and also indicate a much higher phytoplankton standing crop for 1972, especially in early summer, than do the five values taken from Dahlberg's report. The Cornell data also show a significantly higher chlorophyll *a* concentration for 1972 than that reported by Barlow (1969) for 1968, and generally higher summer and autumn values than those reported by Peterson (1971) for 1968–1970 (Table 31).

Primary Production

Howard's (1963) study of primary production in Cayuga Lake during 1957–1958 represents one of the first attempts to provide such estimates in fresh water using the ^{14}C uptake method. *In situ* measurements were made at 5-m intervals from the surface to 20 m at mid-lake off Taughannock Point, usually at 1-week intervals. Results were expressed on both a volumetric and an areal basis (Table 32). The latter are undoubtedly underestimates for most of 1957 since net assimilation still occurred at the greatest depths (20 m) for which productivity was determined.

Seasonally, areal primary productivity rose from May to August 1 of 1957. A decrease on August 8 was followed by a maximum on the fifteenth succeeded by another decrease, an increase beginning on September 11 which reached a maximum in mid-October, and a sharp decline followed by a rise to the highest value of 1957 on November 22. Productivity rates

* P. J. Godfrey, W. R. Schaffner, and R. T. Oglesby, Department of Natural Resources, Cornell University, Ithaca, New York.

TABLE 32

Primary Production Rates and Efficiencies for Cayuga Lake during 1957–1958[a]

	Seasonal means						
	mg CO_2/liter/hr						Cal CO_2/cal of light
Dates	0 m	5 m	10 m	15 m	20 m	gm/m²/hr	× 100
Spring (May 15–Jun 21, 1957)	0.0062	0.0159	0.0081	0.0086	0.0033	0.1671	0.22
Summer I (Jun 21–Aug 20, 1957)	0.0215	0.0251	0.0094	0.0091	0.0037	0.2800	0.28
Summer II (Jun 21–Sep 11, 1957)	0.0194	0.0235	0.0100	0.0084	0.0051	0.2695	0.32
Fall (Sep 27–Nov 22, 1957)	0.0262	0.0219	0.0106	0.0149	0.0198	0.3747	1.21
Winter (Jan 10–Mar 6, 1958)	0.0068	—	0.0035	—	0.0022	0.0837	0.720
Spring I (Apr 17–Jun 19, 1958)	0.0045	0.0033	0.0020	0.0027	0.0009	0.0529	0.198
Spring II (May 8–Jun 19, 1958)	0.0038	0.0049	0.0014	0.0010	00002	0.0456	0.092
Summer (Jun 30–Aug 21, 1958)	0.0019	0.0021	0.0022	0.0014	0.0013	0.0364	0.064

	Totals and means						
	mg CO_2/liter/hr						Cal CO_2/cal of light
Dates	0 m	5 m	10 m	15 m	20 m	gm/m²/hr	× 100
12 month (May 15, 1957–May 8, 1958)							
Sum	0.3046	0.3751	0.2097	0.2079	0.1892	5.7443	
Mean	0.0122	0.0170	0.0084	0.0090	0.0076	0.2297	0.532
16 month (May 15, 1957–Aug 21, 1958)							
Sum	0.4334	0.4106	0.2308	0.2221	0.1922	6.1819	
Mean	0.0120	0.0121	0.0064	0.0065	0.0055	0.1700	0.388

[a] From Howard (1963).

decreased during the winter and remained at low and relatively constant levels through the spring and summer of 1958.

There appeared to be little correlation between the short-term patterns of change in the various strata. Productivity at 5 m was generally slightly greater than at the surface except during the summer of 1957. Carbon fixation rates were similar for the 10- and 15-m strata and were generally lower than those for 0 and 5 m, and the temporal pattern of change was less erratic at this depth. Primary production was relatively higher in the deeper strata during the spring of 1957 and the summer of 1958.

Primary production rates were four to seven times lower in 1958 than for comparable periods in 1957. Cell count differences between the two years correlated well with productivity, being about five times lower for 1958. Howard (1963) found that the trophogenic zone in Cayuga Lake was about twice as deep as would have been estimated by taking this as the depth at which 1% of the surface light was available.

Production efficiency (as estimated by the ratio of the caloric value of daily production to that of the available light energy) was lower in spring and summer relative to other seasons and was lower in 1958 than for comparable periods in 1957. During the former year, extinction coefficients were higher and there was no net production at 15 and 20 m during July and August. Efficiency was inversely correlated with standing crop, and lowest efficiences were associated with populations dominated by diatoms while the highest occurred when blue–greens were the most abundant forms.

In a comparison with data from the literature for other bodies of water Howard (1963) found that, on an areal basis, Cayuga Lake during 1957 was about half as productive as Lake Erie, approximately equal to Lake Mendota and Canyon Ferry Reservoir, and twice as great as Linsley Pond. However, based on the 1958 data, areal production was among the lowest reported in the literature up to that time. On a volumetric basis (mg CO_2/liter) Cayuga was similar to Lake Maggiore, about an order of magnitude less than Lake Erie during 1957, and had the lowest production of the 23 bodies of water with which it was compared for 1958. Production efficiencies during 1957 were similar to those reported for Lakes Maggiore and Mendota and during 1958 for Lake Erken.

Barlow (1969) determined the net rate of ^{14}C fixation by Cayuga Lake phytoplankton on two occasions each during July and August and once in October 1968. Data, expressed only as radiation counts, indicated similar ^{14}C uptake rates during July and August, with those in October being about two orders of magnitude lower. There was evidence of a slight inhibition of photosynthesis at the surface. Maximum rate, with the exception of one set of the July measurements, was at 3 m. Below this stratum there was a regular, negative exponential decline with depth. Barlow (1969) estimated that the compensation depth, based on 1% of the surface radiation level, would have been between 5 and 8 m on the four occasions (once each month, July–October) when he measured total radiation in the water column.

Peterson (1971) measured the photosynthetic fixation of ^{14}C on three dates in August of 1969 and once each in May, June, July, and August 1970. His experiments were conducted at a mid-lake station located between Myers and Taughannock points. The effects of both temperature and light on the rate at which the phytoplankton fixed carbon were examined. Over a temperature range of about 7–25°C he found a Q_{10} of 2.5 (ln P_{max} = 0.0089T – 0.90; r = 0.75; n = 14) for the rate of photosynthesis per unit of chlorophyll. Using Howard's (1963) data he computed the regression of ^{14}C fixation rate on temperature and again obtained a significant correlation (ln P = 0.12T + 1.86; r = 0.67; n = 27). Since Howard did not measure chlorophyll concentration, the two regressions are not directly comparable.

Photosynthesis was also significantly affected by light (P = 0.29 ln %I_0 + 0.99; r = 0.48; n = 49). When these two environmental factors were considered together, 77% of the variation in photosynthetic rate could be accounted for.

Using different methods of estimating primary production from his 1969–1970 data, Peterson (1971) obtained mean values of 366, 409, and 418 mg C/m²/day. For a comparable time of year Howard (1963) estimated rates of 342 μg CO_2/cm²/day (= 1041 mg C/m²/day) for 1957 and 66 μg CO_2/cm²/day (= 201 mg C/m²/day) during 1958.

Calculating photosynthetic carbon uptake per day from chlorophyll, temperature, and light data, Peterson (1971) was able to obtain a bell-shaped curve for 1968–1970 comprised of 36 points with a June–August maximum averaging around 400 mg C/m²/day, but containing isolated single values as high as 1100–1200 mg C/m²/day on two occasions in late summer. Using unpublished data obtained by Barlow in his 1968 study, he was also able to investigate the differences in photosynthetic rates at nine stations located along the long axis of the lake and to further define seasonal patterns of change. An analysis of variance indicated significant differences between stations, but there was no clear-cut pattern of any one station being the most productive. Seasonally, production for most stations averaged about 250 mg C/m²/day during August and September. It was generally more variable between locations in October, averaging about 200 mg C/m²/day, but still remaining between 250 and 300 for several stations. A marked decrease to approximately 100 mg C/m²/day occurred in November, and winter values ranged between 50 and 200 with a tendency for the more northerly stations to be higher.

Like Howard (1963), Peterson (1971) measured ^{14}C uptake at 5-m intervals from the surface to 20 m. His data indicate a definite inhibition of photosynthesis at the surface with a maximum at 5 m. Based on curves shown for four dates during May and July of 1970, net carbon fixation was slight at 15 m, and there was virtually none in the 20-m stratum.

Algal Assays

Short-term (8-hr) batch culture bioassays on Cayuga Lake phytoplankton were performed by Hamilton (1969) on samples collected at mid-lake off Taughannock Point during June–August and in October 1966. Experiments were carried out in a light chamber with intensities approximating those characteristic of the epilimnion.

Results indicated that silicon (added as Na_2SiO_2) stimulated photosynthesis in 8 of 12 experiments conducted from July 20–August 18 and in October, and that Si was inhibitory in late October. Phosphorus (added as K_2HPO_4), on the other hand, was a photosynthetic stimulant on only one occasion and produced inhibition in four other experiments. Additions of other compounds [$MgSO_4$, $Ca(NO_3)_2$, Fe citrate–citric acid, $MnSO_4$] also appeared to significantly affect photosynthetic rates at times, but none to the extent of silicon.

In attempting to relate his results to the phytoplankton ecology of Cayuga Lake, Hamilton (1969) followed the patterns of change in chlorophyll a, phytoplankton abundance as determined by cell counts, phosphorus, and silicon that occurred in the water column during the source of his experiments. He was unable to rationalize his experimental results in terms of the observed changes, especially in regard to silicon-requiring organisms. It is true that the concentrations of silicon in the lake as reported by him were exceedingly low (generally less that 30 μg/liter at the surface), but his analyses were done on frozen samples and concentrations were therefore probably underestimated.

The phytoplankton assays of Hamilton (1969) may be questioned on several grounds. First, the levels of Si and P added were unrealistically high. 164 μg at/liter (4592 μg/liter) for the former and 5 μg at/liter (155 μg/liter) for the latter, compared with realistic background levels of about 100 and 3 μg/liter Si and P, respectively. Second, the short time interval (8 hr) of the experiments may well have given spurious results (Barlow et al., 1973; Peterson et al., 1973, discussed below). Finally, the lack of rational correlations with phytoplankton and nutrient kinetics in the lake indicates that the algal assay data must be viewed with caution.

The second of the above considerations also provides grounds for questioning the *in situ* bioassays conducted by Mills and Oglesby (1971) on trace metals and vitamin B_{12}. These indicated an increase in net ^{14}C uptake when small amounts of zinc, copper, and cobalt were added to Cayuga Lake water containing natural populations of phytoplankton. However, lead, not known to be a metabolite for any alga, produced the same effect. No response was obtained from added cadmium or vitamin B_{12}.

Barlow et al. (1973) conducted an elegant series of chemostat algal assays on natural populations of phytoplankton in Cayuga Lake water collected in late summer of 1971. At this time soluble reactive phosphorus (SRP) was usually below 0.1 μg at/liter (3 μg/liter) and chlorophyll a was 2–3 μg/liter. Experiments were designed to test the effects of added phosphorus (enrichments of either 15 or 30 μg P/liter) and phosphorus plus trace elements. Phosphorus remained a limiting factor almost continuously in all of the chemostat experiments, and on one out of four occasions phosphorus plus trace elements produced some additional stimulation of photosynthesis.

Chemostat assays were paralleled by batch cultures in which response to the same nutrient additions was measured either by 4-hr ^{14}C uptake or pigment synthesis, as measured by fluorescence, over a longer period. The short-term ^{14}C uptake measurements produced no more than "a small and inconsistent stimulation of photosynthesis by any enrichment, while in the 3- to 5-day experiments phosphorus always produced a 2- to 3-times increase in pigment synthesis."

Barlow et al. (1973) explained the batch culture study results by detailing the course of a typical chemostat study. In this an initial (first 24- to 48-hr) decrease in pigments occurred, probably due to the loss of certain small flagellates. Subsequently, pigment in the control decreased more slowly while the P-enriched cultures showed a rapid increase, coming into equilibrium with the displacement rate in 5–8 days. Qualitative changes in the composition of the phytoplankton were minimal. Cyclotella grew more rapidly for the first 3 days, but at the end of 8 days, despite a 17-fold increase in numbers, the various species occurred in about the same relative abundances as in the original culture.

In the chemostat experiments added phosphorus was assimilated with great rapidity, but, despite this strong demand, levels were never reduced below about 3 μg/liter. Increase in particulate phosphorus accounted for the loss of P from solution. The maximum photosynthetic rate per unit of pigment occurred well before the population reached a maximum and then declined as the concentration of pigment increased.

Peterson et al. (1973, 1974) conducted a series of laboratory continuous culture experiments which, together with observations on the lake ecosystem, established the nutrient status of Cayuga Lake phytoplankton in the summers of 1971 and 1972 and in the spring of 1973. Their strategy was to maintain the mixed populations under controlled nutrient conditions in chemostats, which simulated the lake with low to moderate nutrient supplies, and in turbidostats which simulated the lake under nutrient-saturated conditions.

Their experiments were conducted with integrated trophogenic zone samples collected at mid-lake off Portland Point (Fig. 5). Large grazers were removed from the initial cultures and reservoirs were filled with membrane-filtered lake water for subsequent addition to cultures as either a nutrient-enriched (experimental) or -unenriched (control) supply. Preliminary experiments and in-lake measurements indicated that phosphorus was most likely the chemical element present at critically low levels, and this was the nutrient tested unless otherwise specified below. Initial experiments indicated that the most realistic mode of chemostat operation was at a displacement rate of 0.2/day (calculated rate of carbon turnover for the *in situ* lake populations) and phosphorus loadings of 15–60 μg/liter. The months of July and August were focused upon for most of the experiments since soluble reactive phosphorus was usually present in the lake at concentrations of less than 3 μg/liter, and hence was most likely to be limiting during this period.

The more extensive series of algal assay experiments, taken together with field observations by Peterson *et al.* (1973, 1974), further confirmed the conclusion of Barlow *et al.* (1973) that natural phytoplankton populations in Cayuga Lake were phosphorus limited in the summer of 1971, and they extended this to include a comparable period in 1972. An experiment conducted in the spring of 1973, when phosphorus was probably present in nonlimiting quantities, resulted in similar P uptake rates by chemostats and turbidostats, confirming the postulated situation in the lake.

Confirmatory evidence of the phosphorus limiting postulate was also provided by Peterson and his co-workers in the form of data relating to carbon–phosphorus ratios (C:P). These were typically high (120–200) during July and August compared to those existing in the turbidostats or to early spring lake ratios, both of which were about 60. As with the chemostat experiments reported by Barlow *et al.* (1973), species composition and evenness indexes paralleled those occurring in the lake, adding further credence to the validity of the continuous culture approach as a means of studying the nutrient physiology of lake phytoplankton populations.

In regard to other potentially limiting nutrients, Peterson *et al.* (1973, 1974) rejected nitrates as a possibility due to the high and relatively constant levels reported in the 1968 study (Wright, 1969i). Experiments with silicon posed some problems of interpretation, but they concluded that "It therefore seems unlikely that silicon was severely limiting. . . . "

The continuous culture experiments also yielded valuable data (Peterson *et al.*, 1973) on the growth rates of some of the more abundant phytoplankton species. These are summarized in Table 33.

TABLE 33

Growth Rates (Divisions per Day ± One Standard Error with *n* Given in Parentheses) of Some Cayuga Lake Phytoplankton as a Function of Nutrient Status[a]

Species	Control	15 μg P/liter (chemostat)	30 μg P/liter (chemostat)	77.5 μg P/liter (turbidostat)
Cyclotella glomerata	0.12 ± 0.17(6)	0.86 ± 0.16(11)	0.87 ± 0.19(21)	1.91 ± 0.22(9)
Cyclotella comta	—	0.11 ± 0.26(3)	—	0.83 ± 0.10(5)
Scenedesmus bijuga	—	0.58 ± 0.15(9)	—	1.18 ± 0.23(5)
Synedra sp.	—	1.02 ± 0.08(3)	—	1.91 ± 0.22(5)
Diatoma elongatum	—	0.63 ± 0.05(3)	—	1.07 ± 0.17(5)
Gomphosphaeria lacustris	0.08 ± 0.07(12)		0.42 ± 0.07(16)	

[a] Determinations are from continuous flow laboratory cultures.

Identification, Counts, and Distribution

Phytoplankton have been collected, identified, and counted on one or more occasions and at a number of locations in Cayuga Lake during 11 different years since the earlier samplings by Birge and Juday in 1910. However, levels to which taxonomic classifications were made, the degree of attention paid to the rarer forms, sampling gear used, methods of sample preservation and preparation, the temporal and spatial aspects of sampling, and other factors greatly complicate the task of comparing the results of these investigations. Some of the basic differences relating to sampling, sample handling, and data presentation in the published studies are given in Table 34.

As most times and places, and particularly when phytoplankton are abundant, the algal flora of Cayuga Lake is dominated by either diatoms or blue–greens. Findings of the various investigators relative to these two groups of phytoplankton are summarized below.

Cyanophytes.   Birge and Juday (1914) found a few *Anabaena* at the 0- to 5-m depth in their August 12, 1910 sampling. They noted a similar occurrence on July 30, 1918 and also found *Aphanocapsa* to be common from the surface (94 cells/ml) to a depth of 100 m (68 cells/ml).

Muenscher (1927) reported that his netplankton collections in 1927 contained *Anabaena* from late July to the end of August, and there were also traces of *Microcystis*. In the nannoplankton *Coelosphaerium* was the most common blue–green, and he reports this as being "rather abundant" in August. *Oscillatoria* appeared early in the sampling period, and *Gleocapsa*, *Aphanocapsa*, and *Merismopedia* were present but never abundant.

The collections of Burkholder (1931) overlapped those of Muenscher. The observations of the two researchers on the plankton composition during this period contain some significant differences. Burkholder reported that *Anabaena* was in the netplankton during August and September of 1927 and notes that *Coelosphaerium* was present during the sampling period. Between April, 1928 and March, 1929, he found *Anabaena*, *Aphanocapsa*, and *Gomphosphaeria* in the netplankton and *Gomphosphaeria*, *Oscillatoria*, and *Coelosphaerium* were present in the nannoplankton. The tables in which Burkholder reports cell counts were reproduced from handwritten drafts that are sometimes difficult to read, and annotations as to what species the counts refer are lacking in some cases.

Howard (1963) found that in 1957 blue–greens entered the phtyoplankton in significant number during early July, but levels remained relatively low through August. *Anabaena* was the dominant blue–green and from early to mid-October standing crops (averaged over the upper 20 m) attained levels

TABLE 34

Sampling and Sample-Handling Methodologies

Investigator(s)	Period sampled	Sampling frequency	Sampling gear
Birge and Juday (1914)	1910	Aug 12 only	Pump for upper 30 m, closing net for deeper
Birge and Juday (1921)	1918	Jul 30 only	Same
Muenscher (1927)	Jun–Sep 1927	2-week intervals	#20 mesh closing net and bottle
Burkholder (1931)	Jun 1927–Apr 1929	2-week intervals thru Sep 1927, then monthly	#20 mesh net and Kemmerer bottle
Howard (1963)	Apr 1957–Aug 1958	Generally at 1-week intervals	Van Dorn bottle
Barlow (1969)	Jul 1–Nov 26 1968	At 2-week intervals during stratified season, then monthly	Van Dorn bottle
Dahlberg (1973)	May 17–Nov 11, 1972	5 times at about monthly intervals	Kemmerer

for Studies on Cayuga Lake Phytoplankton

Kind of plankton sampled	Depths sampled	Sample preservation	Form of data	Station(s) location
Net	Entire water volume	Alcohol	Cell counts	Between Sheldrake and King Ferry
Net and nanno	Same	Same	Same and biomass	Same
Net and nanno	Same	70% alcohol	Graphic and biomass	3 and 39 km north of southern end
Net and nanno	To 50 m	Formalin	Cell counts (but not always defined as to species they represent) and photomicrographs	3 km north of southern end
All	To 20 m	Lugol's	Cell counts	Taughannock
All	Surface, base of epilimnion, mid-metalimnion, base of metalimnion, hypolimnion	10% formalin	Cell volume and cell counts (graphic) for surface only, evenness values, species listing	7 stations along north–south axis of lake
All	Surface, metalimnion, hypolimnion	Examined both alive and preserved (method not specified)	Cell counts, graphics, Shannon diversity indexes, evenness and richness values, species listing	9 stations along north–south axis of lake but detailed data presented for a central station only

of nearly 1000 cells/ml. Other "summer" forms were *Chroococcus, Coelosphaeria*, and *Gomphosphaeria* (all counted as undefined "units") and *Oscillatoria* (counted as 100 μm of filament). Blue–greens were the second most abundant group in the phytoplankton through the winter and into the spring of 1958 and exceeded the number of diatoms in April. Through August of 1958 blue–greens reached a maximum of 168/ml (*Oscillatoria*) on June 5, being present at relatively low levels for the rest of the summer. *Microcystis* (also counted as "units"), *Gomphosphaeria, Anabaena, Coelosphaerium*, and *Chroococcus* were the other cyanophytes occurring in 1958.

Barlow (1969) sampled seven stations on ten dates during 1968. Between July 1 and November 26 three blue–greens, *Coelosphaerium* sp., *Merismopedia* sp., and *Aphanizomenon flos-aquae*, comprised at least 10% by volume of at least one sample collected at the surface. *Coelosphaerium* exhibited a conspicuous bloom beginning with increases on August 1, reached a peak September 5 when it comprised 74–91% by volume of the phytoplankton, and then declined in importance during the rest of the sampling period. Barlow comments that in mid-July of 1969 a "conspicuous bloom" of *Anabaena* occurred.

In samples collected on six occasions from May to November, 1972, Dahlberg (1973) found blue–greens to be relatively unimportant members of the phytoplankton community on May 17 and June 14. On July 27 he found that a bloom (4635 cell/ml) of *Microcystis incerta* was in progress with *Aphanocapsa delicatissima* being a strong subdominant (1296 cells/ml). On August 30 *Coelosphaerium naegalianum* was numerically dominant (1171 cells/ml) with *Microcystis aeruginosa* and *Aphanocapsa delicatissima* being the two next most abundant forms. *Coelosphaerium kuetzingianum* was fairly common (69 cells/ml) on October 3. Blue–greens no longer constituted a signficant portion of the phytoplankton on November 11.

Diatoms.   On August 12, 1910 Birge and Juday (1914) reported that *Asterionella* was abundant. *Fragilaria* was the next most common diatom and then *Tabellaria*. Numbers averaged for the upper 15 m were, respectively, 4.2, 1.8, and 0.2 cells/ml. *Asterionella* and *Fragilaria* were again the dominant netplankton diatoms on July 30, 1918, but concentrations were an order of magnitude lower than during 1910. The 1918 nannoplankton contained over 5 cells/ml of *Stephanodiscus* throughout the water column, and there were about the same number of *Navicula* in the surface sample.

Muenscher (1927) found the great bulk of the net phytoplankton in 1927 consisted of diatoms belonging to the genera *Asterionella, Fragilaria*, and *Tabellaria. Asterionella* dominated in spring and early summer with a maximum at 0–5 m of about 4 cells/ml on April 16. *Fragilaria* was rare when *Asterionella* was abundant. Numbers of the former reached a maximum in August and early September. *Tabellaria* occurred at all times. In the nannoplankton a small *Cyclotella* was the most abundant diatom and *Synedra* and *Melosira* were generally present. *Stephanodiscus* occurred in a few samples at the more northerly station.

Burkholder (1931) also reported *Asterionella* as being common throughout most of the 1927 study period. Concentrations decreased with depth but cells were often present in substantial numbers as deep as 40–50 m. His figures agree with those of Muenscher relative to the spring maximum, and he found a secondary maximum (2 cells/ml) on November 10. *Fragilaria* was absent or only traces detected in the winter and spring. Cell counts showed a peak of abundance on September 13 (almost 5/ml), and substantial numbers were also present in the August 12 and October 27 samples. *Tabellaria* was ubiquitous but never very abundant. *Cyclotella, Stephanodiscus, Melosira,* and *Synedra* were among the diatoms found in the nannoplankton in the April 1927–March 1928 period. From April of 1928 to March 1929 only a few *Asterionella* were found prior to August 12, when concentrations in the surface sample reached over 5 cells/ml. Numbers then decreased to a minimum in December followed by an increase to 0.8 cells/ml in March. The seasonal pattern of change exhibited by *Fragilaria* closely paralleled that of *Asterionella* with a maximum of 41 cells/ml on August 12. *Tabellaria* was uncommon in July and August and reached its peak abundance November 29. *Cyclotella* was the dominant diatom in the nannoplankton. A maximum of 150 cells/ml was observed in the 10-m sample of May 19, and a surface concentration of 29 cells/ml occurred October 30. Large numbers were characteristically present to a depth of 50 m, and the maximum was usually below the surface. *Stephanodiscus* was fairly common in May and June. *Melosira* and *Synedra* were again among the diatoms encountered.

Howard's (1963) data for April 1957 through August 1958 indicate that at times many of the diatoms reported by earlier workers as dominating the netplankton were present and had increased greatly in abundance. During 1957 *Asterionella* attained a maximum of 176 cells/ml in late May and then abruptly declined. While generally present through the winter of 1957–1958, it was not a prominent member of the phytoplankton and was absent from March and April samples. *Asterionella* then began to increase, attaining a concentration of 110 cells/ml on June 5, 1958, after which numbers again declined to remain at a low level throughout the rest of the sampling period. *Fragilaria* was absent from samples taken through the twenty-first of June, but 6 days later Howard recorded a bloom of 2325 cells/ml and dominance was maintained through July 18. It was sporadically common until autumn when a concentration of 476 cells/ml was reached on November 8. *Fragilaria* was common among the winter and spring diatoms. In 1958 it attained a maximum level of 94 cells/ml on July 7 and then declined, not again becoming prominent during the summer. *Tabellaria* dominated (561 cells/ml) the phytoplankton on June 21, 1957 and then was not recorded again until October. It was subsequently present but generally not abundant until June 12, 1958, when it became the dominant diatom and remained so, although often not very abundant itself, until August 7. *Melosira* was abundant during April and May of 1957 (537 cells/ml on May 25). Although it occurred throughout the winter of 1957–1958 and into mid-June, it was not a prominent part of the community.

During 1968 Barlow (1969) found that four species of diatoms at some time accounted for as much as 10% by volume of a sample. These were a centric form tentatively identified as *Cyclotella glomorata, Stephanodiscus astrea, Asterionella formosa,* and *Fragilaria crotonensis. Asterionella* increased in numbers in October and November, and *Fragilaria* was abundant at some stations on July 1, decreased, and then concentrations rose in mid-September. It was one of the common forms in the phytoplankton through November. *Cyclotella* was a dominant phytoplankter on July 1.

Dahlberg (1973) found *Asterionella* to be fairly abundant (43 cells/ml) on June 14, present in October, and the dominant (106 cells/ml) member of the phytoplankton on November 11 of 1972. He lists three species of *Fragilaria* as being present in the Cayuga Lake flora, of which *F. crotonensis* was the most common. This alga was dominant (94 cells/ml) on June 14, declined somewhat in the July 27 and August 30 samples, again dominated (234 cells/ml) the phytoplankton on October 3, and was the second most abundant form on November 11. *Melosira* was not found by Dahlberg, but among other forms reported as common by earlier workers he reported the presence of four species of *Cyclotella,* three of *Stephanodiscus,* and six of *Synedra.* The cell counts reported by Dahlberg for dominant netplankton species are generally at the same level as those of Howard (1963), i.e., one or more orders of magnitude higher than those given by workers in the 1910–1929 period.

Barlow (1969), from his examination and historical comparison of phytoplankton data, concluded "There is strong evidence that there has been a change in the phytoplankton flora during the fifty years since the first measurements, a change that implies changes in fertility." To this statement he added some strong cautionary notes. Dahlberg (1973), in discussing Barlow's conclusion, expresses the view that "Other than the seemingly increased blue–green algal production in late summer months and the general increase in phytoplankton numbers, there appears to be little basis for this judgement." The qualifying phrase in this statement seems somewhat inappropriate since if both the relative abundance of blue–greens and the total abundance of phytoplankton have indeed increased, this would provide very strong evidence for a trend toward increasing eutrophication.

Examining the published evidence relative to cyanophytes, there indeed seems to be evidence suggesting an increase in both numbers and their relative importance in the phytoplankton during the three decades intervening between the studies of Muenscher and Burkholder and those of Howard and subsequent investigators. Quantitatively, *Anabaena* and possibly *Coelosphaerium* and *Microcystis* seem to be the forms that have exhibited an increased prominence. Changes in the less abundant genera are impossible to assess due to the many noncomparable features of the various studies, but it is noted that *Aphanizomenon flos-aquae* had not been reported prior to 1968.

Those diatoms described by earlier workers as dominating the netplankton seem to have increased rather dramatically in the concentrations associated with peak abundances during the same interval that produced changes in the blue–green flora. Qualitatively, there does not appear to have been any major shift in the relative abundance of the dominant netplankton diatoms, with the exception that *Tabellaria* was more prominent in 1957–1958 and, of all the years for which data exist, *Melosira* was the most abundant in 1957. Although not recorded by Birge and Juday (1921), *Cyclotella* appears to have dominated the diatoms of the nannoplankton during the years succeeding their study. Relative abundances between years are difficult to assess since, as demonstrated by Barlow (1969), this phytoplankter may at times undergo brief but rapid changes in concentration.

Of the other major algal taxa, the Chlorophyta appear to have played a rather minor role in the phytoplankton makeup through 1957–1958, although Muenscher (1927) comments that in 1921 *Botryococcus* "formed a very conspicuous bloom on the surface along the east shore of Cayuga Lake." In 1968, Barlow (1969) found that on a few occasions green algae dominated (volumetrically) the summer phytoplankton and that *Dictyosphaerium* was prominent in the autumn. Dahlberg's (1973) data show that from June 14 onward chlorophytes were numerically important in 1972, exceeding the number of diatoms in July and August and being intermediate between diatoms and blue–greens in October and November. Among the more important green algae in the phytoplankton of recent years are *Coelastrum microporum, Dictyosphaerium pulchellum, Scenedesmus bijuga,* and *S. quadricauda* and unidentified coccoid forms.

Flagellates and dinoflagellates were relatively abundant at the times of Birge and Juday's surveys and have continued to be so. Historical comparisons are particularly difficult to make for the smaller flagellates due to the adverse effects on them of many preservatives and the fact that few limnologists have bothered to master the taxonomy of this group. The data of Birge and Juday (1921) indicate that "monads" and an "unidentified, asymmetrical flagellate" were common, and *Cryptomonas* was present in the nannoplankton. Barlow reported forms such as *Cryptomonas erosa, Rhodomonas lacustris,* and *Trachelomonas* (*volvocina*) to be at times common among the phytoplankton of 1968. *Dinobryon* appears to have been consistently present throughout the period covered by published reports. Burkholder (1931) is the only one of the early investigators to identify to species, and he records *D. divergens* as being the species present in 1927–1929, while both Barlow (1969) and Dahlberg (1973) recorded *sertularia* as the only species of *Dinobryon* present in the 1968 and 1972 collections. *Ceratium* was an important member of the phytoplankton community through the 1920's, but although reported as being present by all investigators, subsequently decreased in relative importance.

In terms of the total number of cells, Birge and Juday (1921) record a concentration of about 2×10^5 cells/liter for their single sampling in 1918. Howard (1963) found ranges of about 2×10^5 to 27×10^5 cells/liter in 1957 and 0.2×10^5 to 4×10^5 cells/liter for 1958, although he seems to have ignored many of the smaller forms. Barlow (1969) reported concentrations from 5×10^5 to greater than 20×10^5 cells/liter in 1968, and Dahlberg (1973) indicates a range from a low of about 2×10^5 on May 17 to a high of more than 78×10^5 cells/liter on July 27, 1972.

The first investigator to provide information on the variations in phytoplankton along the long axis of the lake was Muenscher (1927). He concluded that the composition of collections from his "South" and "North" stations was very similar. Barlow (1969) found that at times there were very marked qualitative differences in the 1968 plankton. Describing the distribution of the surface plankton on July 1, he states that the northern end of the lake was dominated by *Aphanizomenon,* further south off Aurora *Peridinium cinctum* became the most abundant form, then proceeding southward *Cyclotella* increased in importance reaching a maximum off Glenwood, and at the most southerly station *Cryptomonas* and *Peridinium* replaced *Cyclotella*. Barlow (1969) also concluded that variations from station to station were greater for cell numbers, for which maxima were more pronounced near the southern end, than cell volume.

Most of the published reports contain data on the vertical distribution of phytoplankton in Cayuga Lake. In general, during the period of summer thermal stratification all of the netplankton have peaks of abundance between the surface and 10 m but generally are still numerous at depths of 15–20 m (the epilimnion and sometimes the upper metalimnion). Some forms, parlticularly the diatoms, are common to greater depths, and occasionally peak concentrations of a particular form (probably senescent cells) occur in the metalimnion and hypolimnion. Among the smaller phytoplankton Birge and Juday (1921) found that "monads," *Aphanocapsa,* and

Stephanodiscus showed no or only small decreases down to a depth of 100 m. Muenscher (1927) describes the great bulk of the phytoplankton as occurring in the upper 10–15 m. Burkholder (1931) comments that diatoms appeared capable of "developing" at considerable depths but indicates that other forms, such as *Ceratium,* were largley confined to the upper strata. In May, prior to thermal stratification, Dahlberg (1973) found a heterogenous distribution within the water column of both kinds and numbers of phyto-plankton. During the summer, cell concentrations increased in the surface samples and decreased in those from the metalimnion and hypolimnion. The flora of the latter was dominated by chrysophytes, with green algae being the next most abundant through the August 30 sampling. Blue–greens were the dominant metalimnetic algae on July 27. On October 3 there were few chlorophytes in the deeper samples, which contained approximately equal numbers of chrysophytes and blue–greens.

Godfrey (1977) carried out a detailed study of the Cayuga Lake phyto-plankton community during 1972–1973. It might be supposed that taken together with the comprehensive studies of Barlow (1969) and Dahlberg (1973), a general pattern of present species distribution could be arrived at. This is not the case, a comparison of these surveys seeming to raise more questions than it answers. For example, of the phytoplankton identified to species level, the three investigators report only three cyanophytes, fifteen chlorophytes, one euglenophyte, three pyrrophytes, and nine chrysophytes in common. Some of those reported as common, abundant, or even dominant by one worker are not even recorded as present by one or both of the others. It is impossible to sort out how much of the difference is due to sampling regima and methodology, how much to differences in taxonomic approaches and expertise, and to what extent methods used to preserve and examine sam-ples affected their results relative to real variations in community structure. Considering the extent of sampling and the methods used in preserving and processing samples, the investigation of Godfrey should give the most com-plete picture of the phytoplankton community. Of the 217 forms of algae which he found in the limnetic zone of Cayuga Lake, those listed in Table 35 constituted species of relatively great importance at some time during the course of his study.

Zooplankton

The zooplankton population of Cayuga Lake has yet to be the subject of a truly definitive study. However, a considerable amount of descriptive data does exist for portions of a number of years extending back to 1910. These studies are summarized in Table 36. Several authors have attempted to draw conclusions from historical comparisons.

TABLE 35

Prominant Phytoplankton Found during 1972–1973 by Godfrey (1977)

Chlorophyta
Ankistrodesmus falcatus
Carteria klebsii
Chlamydomonas spp.
Chlorococum sp.
Closteriopsis longissima
Coelastrum microporum
Coelastrum sphaericum
Cosmarium botrytis
Cosmarium nitidulum
Cosmarium reniforme
Dimorphococcus lunatus
Golenkinia radiata
Kirchneriella lunaris
Lagerheimia ciliata
Lagerheimia quadriseta
Lagerheimia subsala
Nephrocytium limneticum
Oocystis pusilla
Pediastrum boryanum
Pediastrum duplex
Pediastrum simplex
Scenedesmus armatus var. bicaudatus
Scenedesmus bijuga
Scenedesmus dimorphus
Scenedesmus opoliensis
Scenedesmus quadricauda
Selenastrum minutum
Sphaerocystis schroeteri
Staurastrum gracile
Staurastrum natator var. crassum
Staurastrum paradoxum
Staurastrum polymorphum
Trachelomonas sp.
Tetraedron hastatum
Tetraedron minimum
Chrysophyta–Bacillariophyceae
Asterionella formosa
Cyclotella sp. 5 μm
Cyclotella sp. 10 μm
Cyclotella sp. 15 μm
Cyclotella sp. 25 μm
Diatoma tenue var. elongatum
Diatoma vulgare
Fragilaria brevistriata
Fragilaria crotonensis
Nitzschia vermicularis
Stephanodiscus astraea
Stephanodiscus hantzschii
Synedra amphicephala
Synedra delicatissima
Synedra radians
Synedra rumpens
Synedra ulna
Tabellaria fenestrata
Chrysophyta–Chrysophyceae
Biocoeca socialis
Chromulina minuta
Chromulina ovalis
Cladomonas fruticulosa
Dinobryon bavaricum
Dinobryon sertularia
Dinobryon sociale
Ochromonas sp.
Cryptophyta
Cryptomonas erosa
Cryptomonas nasuta
Cryptomonas pusilla
Cyanophyta
Anabaena flos-aquae
Aphanizomenon flos-aquae
Chroococcus sp.
Coelosphaerium pallidum
Gomphosphaeria lacustris
Lyngbya contorta
Lyngbya limnetica
Merismopedia tenuissima
Microcystis aeruginosa
Pyrrophyta
Ceratium hirundinella
Glenodinium armatum
Glenodinium palustre
Glenodinium pulvisculus
Glenodinium quadridens
Glenodinium limneticum
Peridinium cinctum

TABLE 36

Summary of Cayuga Lake Zooplankton Studies

Reference	Sampling date(s)	Sampling gear and methodology	Location	Summary of observations
Birge and Juday (1914)	Aug 11, 1910	Pump and hose to 30 m, closing net for deeper samples	Sheldrake, King Ferry	Identified to genus only. Vertical distribution described. *Bosmina* sp. and *Polyarthra* sp. were dominant, averaging over 0–15 m deep 41.6 and 85.9 organisms per liter, respectively
Birge and Juday (1921)	Jul 30, 1918	Closing net	Sheldrake, King Ferry	*Bosmina* sp. and *Synchaeta* sp. were dominant averaging 12.8 and 36.0 per liter, respectively, for the upper 15 m
Muenscher (1927)	Jun 15–Sep 15, 1927 at 2-week intervals	#20 mesh closing net drawn vertically thru 5–10 m of water	2 miles from south end and 15 miles from north end	Identified to genus only. Seasonal and vertical patterns described. *Bosmina* sp. the most common crustacean and *Pleosoma* sp. the most abundant rotifer
Bradshaw (1964)	Jun 1950–Sep 1951	Juday trap	Taughannock	Cladocerans and copepod concentrations by depth compared for 1910, 1918, 1951, and 1961. *Bosmina longirostris* the dominant cladoceran and *Cyclops* sp. the most abundant copepod. No data on rotifers

(Continued)

TABLE 36 *(Continued)*

Reference	Sampling date(s)	Sampling gear and methodology	Location	Summary of observations
Bradshaw (1964)	Aug 1961	Juday trap	King Ferry	Dominant cladoceran was *Ceriodaphnia* sp.
Hall and Waterman (1967)	Spring and early autumn 1965	#20 mesh plankton net hauled vertically	Not given	Lists 16 species. Compare zooplankton present in 1965 with those described by Birge and Juday and conclude "no major changes in the intervening half century
Youngs (1969)	Summer 1968	#20 mesh Wisconsin net	Not given	Lists 16 species of Crustacea. *Cyclops bicuspidatus thomasi* the dominant crustacean, *Bosmina longirostris* the most abundant cladoceran, and *Diaptomus minutus* the most common calanoid
Dahlberg (1973)	May 10, Jul 25, and Oct 3, 1972	Kemmerer bottles and submersible pumps, Isaacs–Kidd used for additional surface samples	16 stations with 7 located from south to north end and 9 concentrated near Milliken power plant	Lists 57 zooplankton species (21 Crustacea, 27 rotifers, 9 protozoans). *Bosmina longirostris* most common cladoceran, *Cyclops bicuspidatus* most common copepod, *Keratella cochlearis* and *Polyarthra major* dominant among rotifers, and *Codonella cratera* dominant protozoan

Hennick (1973)	ca. 1-week intervals from Jul 5 to Oct 26, 1972	#10 mesh half-meter net hauled vertically, replicate samples of epilimnion and from 150 ft to surface, samples taken at night	Myers Point	Concentration greatest in July. Decrease in numbers from south to north end Crustacea sorted to major groups subdivided by size categories. Concentrations calculated on a weight basis. Maximum epilimnetic concentration of 3078 mg/m^3 obtained on Jul 21 and secondary peak of 1283 mg/m^3 observed on Sep 7
Behrman (1975)	ca. 1-week intervals from Jun 12 to Nov 10, 1972	#10 mesh half-meter net hauled vertically from 30 m above lake bottom to surface	Aurora, Sheldrake, Milliken, Myers Point, and Glenwood	Seven zooplankton species associations defined for this period. All except the nongravid female cyclopoids were concentrated above the thermocline during the day. Cyclopoid and cladoceran abundance appeared to be correlated with temperature and daylength. The importance of alewives and cyclopoid copepods as zooplankton predators was inferred

Bradshaw (1964) collected extensively off Taughannock Point in 1950–1951 and off King Ferry on a single occasion in 1961. In comparing his zooplankton data with that of Birge and Juday (1914, 1921) from 1910 and 1918, he concluded "There is no evidence of qualitative change." Hall and Waterman (1967) also reviewed the earlier data concerned with species composition and, in comparing this with their 1965 collections, agreed with Bradshaw. They point out that Birge and Juday's failure to identify the species of *Bosmina* precluded an analysis of a possible shift from *B. coregoni* to *B. longirostris* as a potential indicator of eutrophication. Later collections in 1968 (Youngs, 1969), 1972 (Dahlberg, 1973), and 1973 (Chamberlain, 1975) tend to support the postulate of no major shifts in zooplankton composition.

The most comprehensive descriptions of species composition are provided by Dahlberg (1973) and Chamberlain (1975), especially in terms of the smaller zooplankters. The former was the first to identify noncrustacean forms to the species level. His data and that of Youngs' (1969) indicate that in both 1968 and 1972 the most common cladoceran was *Bosmina longirostris*. Bradshaw states that in his 1950–1951 collections this species accounted for 72% of the cladocerans present, although *Ceriodaphnia quandrangula* outnumbered it in his single set of samples from 1961. In both 1968 and 1972 *Cyclops bicuspidatus* is identified as the most abundant copepod, and Bradshaw states that "Cayuga was, and remains, a *Cyclops* lake." Dahlberg reported that in 1973 the most common rotifer and protozoan were, respectively, *Keratella cochlearis* and *Codonella cratera.* The species of zooplankton found by Dahlberg (1973) and Chamberlain (1975) are given in Table 37.

The vertical distributions of zooplankton in Cayuga Lake have been delineated by the sampling methods used in several of the investigations. Birge and Juday (1914, 1921) found that both copepods and cladocerans attained maximum abundance in the epilimnion, but there were still moderate numbers of the latter present to depths of 50–75 m. Nauplii were most abundant in the metalimnion during 1918, Bradshaw's (1964) 1950–1951 and 1961 data show a cladoceran maximum in the epi- and metalimnia, but they were still common in samples to 100 m. During 1950–1951, copepods were at maximum concentrations in the epilimnion but were moderately abundant to 100 m, while in 1961 they exhibited no apparent vertical zonation, the maximum concentration being at 70 m and a secondary maximum occurring at 20 m. Dahlberg (1973) sampled at the surface, at one selected mid-depth plane, and another at 100 m. Behrman (1975) studied the zooplankton distribution at five stations during the summer of 1972. She concluded that rotifers, cladocerans, copepodids, and nauplii

TABLE 37

Zooplankton Found in Cayuga Lake[a]

Copepoda (calanoid)	Malocostraca
* *Diaptomus minutas* (C,D)	*Mysis relicta* (C)
Diaptomus sicilis (C,D)	Rotifera
Diaptomus oregonensis (D)	* *Asplanchna priodonta* (C,D)
Diaptomus ashlandi (D)	** *Ascomorpha sultans* (C,D)
Diaptomus siciloides (D)	*Ascomorphella volvociola* (D)
Senecella calanoides (C,D)	*Brachionus angularis* (D)
Copepoda (cyclopoid)	*Brachnionus caudatus* (D)
** *Cyclops bicuspidatus* (C,D)	*Brachionus calyciflorus* (C,D)
* *Cyclops scutifer* (C,D)	*Chromogaster ovalis* (C,D)
Cyclops vernalis (C,D)	*Colurella* sp. (D)
Mesocyclops edax (C)	** *Conochilus unicornis* (C)
* *Tropocyclops prasinus* (C,D)	*Filinia longiseta* (C,D)
Cladocera	*Gastropus hyptopus* (C,D)
Alonella sp. (D)	*Gastropus minor* (C)
** *Bosmina longirostris* (C,D)	*Hexarthra mira* (C,D)
* *Ceriodaphnia laticaudata* (C,D)	* *Kellicottia longispina* (C,D)
* *Ceriodaphnia quadrangula* (C,D)	** *Keratella cochlearis* (C,D)
Chydorus sphaericus (C,D)	*Keratella quadrata* (C)
Chydorus gibbus (D)	*Keratella valga* (D)
Daphnia dubia (D)	*Leptadella patella* (D)
Daphnia galeata mendotae (D)	*Monostyla pygmea* (D)
Daphnia retrocurva	*Monostyla quadridentata* (C)
Leptodora kindtii (C,D)	* *Notholca acuminata* (C,D)
Polyphemus pediculus (C)	*Platyias quadricornis* (D)
Protozoa	*Ploesoma hudsoni* (D)
** *Codonella cratera* (C,D)	*Ploesoma lenticulare* (C)
Difflugia oblonga (D)	*Ploesoma tricanthum* (C,D)
Difflugia lebes (C)	Ploesoma truncatum (C,D)
Difflugia tuberculata (C,D)	*Polyarthra dissimulans* (D)
Difflugia acumineta (D)	* *Polyarthra euryptera* (C,D)
Glaucoma scintillans (C,D)	** *Polyarthra major* (D)
Sphenoderia lenta (D)	*Polyarthra vulgaris* (C,D)
Thecacineta cothurnoides (C,D)	*Synchaeta stylata* (C,D)
Vorticella canvanella (D)	*Trichocerca multicrinis* (C,D)
Vorticella microstoma (D)	

[a] Those species reported by Dahlberg (1973) are indicated by (D) and those by Chamberlain (1975) with (C). The most important forms are designated with a double asterisk and less prominent, but still abundant species by a single asterisk.

concentrated in the epilimnion during daylight hours, while copepods were more evenly distributed throughout the water column. However, within the latter group *Cyclops bicuspidatus* males and gravid females and *Tropocyclops prasinus* gravid females remained above the thermocline.

In addressing the question of horizontal zonation along the long axis of the lake, Dahlberg (1973) concluded that "There was a distinct trend for a marked reduction during the summer in numbers of protozoans, rotifers, and cladocerans proceeding in successive stations from the southern end to the northern end of the lake. This trend was not found with copepods if both surface and metalimnetic samples are considered."

Seasonal changes in abundance of zooplankton in Cayuga Lake were first described by Muenscher (1972), whose data for the summer of 1927 indicated a maximum of crustaceans in mid-September at his station located 3 km from the south end, and no pronounced seasonal change at his other sampling point 24 km from the north end of the lake. Rotifera showed little seasonal variability in epilimnetic concentration at the southern station, but a pronounced maximum in early July occurred at the more northerly sampling location. In discussing his collections for 1950–1951, Bradshaw (1964) simply states that "there was rarely any marked difference in numbers at the various depths."

For the three occasions on which he sampled, Dahlberg (1973) found that total numbers of organisms were greatest (730–734, 796 organisms/liter) in late July of 1972, next most abundant (86–543 organisms/liter) in early October, and least abundant (73–175 organisms/liter) on May 10. He states that in May the dominant species were the protozoan *Codonella cratera*, the rotifers *Keratella cochlearis* and *Notholca acuminata*, and the copepod *Cyclops bicuspidatus*. The first two of these and *C. bicuspidatus* were joined by *Polyarthra euryptera, P. major,* and *Bosmina longirostris* as the dominant zooplankters found in July, and nauplii were abundant. The largest number of species (33) was found in the October collections. Major forms were *K. cochlearis, C. cratera,* and *B. longirostris,* with *C. bicuspidatus* and *Diaptomus minutus* being common.

In describing the qualitative changes which took place in the zooplankton population from mid-June through mid-November 1972, Behrman (1975) identified, in most cases only to the level of genus, seven seasonal associations. In general, rotifers reached their greatest importance in early summer, cladocerans in mid-summer, and copepods during the autumn.

Hennick (1973) sampled the crustacean zooplankton of Cayuga Lake at weekly intervals off Myers Point from July through October 26, 1972. Zooplankton were categorized according to major taxa and size. Concentrations were then converted to biomass estimates using the equations of Edmondson (1971). His estimates of epilimnetic biomass for the more important forms and for the total are shown in Fig. 12. All of the forms not shown together accounted for less than 15% of the total biomass until August 24, when an increase in *Ceriodaphnia* sp. began. This peaked in mid-September (over 18% of the total biomass) and declined to relatively low levels by the last sampling. Averaging 1072 mg/m³ over the period, concentrations ranged from a high of 3078 mg/m³ (due largely to copepodites) on July 5 to a low of 508 mg/m³ on October 19. Adult and immature copepods usually made up the bulk of the biomass, with *Bosmina* sp. accounting for most of the rest. Hennick's study also showed that a rapid response in alewife feeding rate followed increases in zooplankton biomass. Through multiple regression analysis, he determined that about 92% of the variance in alewife growth rate could be accounted for in terms of zooplankton standing crop.

The hypolimnetic crustacean zooplankton of Cayuga Lake contains *Senecella calanoides* and small numbers of other copepods and of cladocerans, but is almost certainly dominated in terms of biomass by the opossum shrimp, *Mysis relicta*. Noted in the surveys of Birge and Juday (1914, 1921), *Mysis* has not been found by all investigators. In some cases this appears to be due to the use of inappropriate sampling gear, e.g., the sampling bottles employed by Dahlberg (1973). Brownell (1970) conducted a study of *Mysis* ecology during 1969, and caught them more consistently

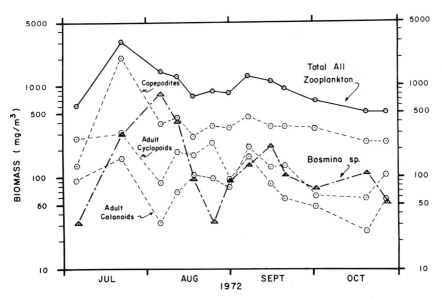

Fig. 12.   Biomass of the major crustacean zooplankton and the total of all in the epilimnion of Cayuga Lake during July–October 1972.

and in greater numbers than any other hypolimnetic crustaceans. Catches varied from a very few to as many as 2100 individuals in a 5-min meter net tow, indicating probable patchy distribution. From 1970 to 1973 vertical net tows by the author and his graduate students have almost always contained at least a few, and sometimes many, *Mysis*. The opossum shrimp is cited by Youngs (1969) as being "an extremely important food item for many fishes of Cayuga Lake, particularly young lake trout, smelt, and cisco." Daytime fathometer tracings frequently indicate a deep scattering layer off the bottom of Cayuga Lake. Whether or not this is caused by large shoals of *Mysis* remains to be verified.

Benthos

Birge and Juday (1921) collected two samples of the Cayuga Lake benthos on July 28, 1918 at depths of 34 and 113 m. Both numbers and biomass of organisms were greater in the deep profundal collection. They stated that *Chironomus* sp., *Pontoporeia,* and *Oligochaeta* "constituted by far the greater portion of the material obtained." Organism densities and standing crop estimates in terms of dry weights and organic matter were calculated (see below).

Henson (1954) collected (five replicates) of the deep profundal benthos off Taughannock on 14 occasions in the summer and autumn of 1952 and five times during the summer of 1953. He reports the following average relative abundances for the dominant forms: *Pontoporeia affinis* (32% by dry weight and 61% by number), *oligochaetes* (32% by dry weight and 6% by number), *Metriocnemus** sp. (8% by dry weight and 5% by number), and *Ostracoda* (0.7% by dry weight and 2% by number).

In 1952 the standing crop of the benthic fauna associated with the deep profundal zone averaged nearly 6000 individuals/m^2, 878 mg/m^2 dry weight, and 688 mg/m^2 organic matter (Henson, 1954). During the same year, biomass increased at the rate of 10 mg/m^2/day through November 1. The net gain in the summer standing crop of 1952 carried over into 1953. However, based on more limited sampling, the pattern of change observed in the earlier year did not appear to be repeated.

Henson (1954) compared the standing crops of profundal benthos of Cayuga with those of a number of other lakes for which data existed in the literature. While recognizing the tenuous nature of many of these estimates, it is probably worth noting that, in terms of dry weight of benthos per square meter of bottom, Cayuga in 1952 had about one-eighth that of

---

* Dahlberg (1973) believes that the midge larvae which Henson identified as *Metriocnemus* were actually *Heterotrissocladius* sp.

Mendota in 1921, one-fifth that of Linsley Pond in 1941, approximately the same as Lake George in 1922, and approximately three times that of Lake Athabaska in 1947. However, based on numbers alone, Henson notes that most oligotrophic lakes have densities of less than 2000 organisms/m^2 compared with the 1952 average for Cayuga of nearly three times this.

Green (1965) conducted a study on the ecology of *Pontoporeia affinis* in Cayuga Lake. He defined growth and mortality rates and determined that the population was nonrandomly distributed. In Cayuga Lake these amphipods appear to have the unique characteristic of producing two annual broods, a major one in late March and early April which concurs with the time when brood production has been observed in other lakes, and a secondary one in September. Green verified that *P. affinis* is an alternate host of the acanthocephalan parasite *Echinorychus salmonis*. This agrees with the findings of Brownell (1970) that young, uninfected lake trout fed *Pontoporeia* from Cayuga Lake subsequently harbored this parasite.

Dahlberg (1973) has provided the most extensive listing of Cayuga Lake benthos, with collections made on two occasions during 1973 in the inner (10–25 ft) and outer (50–100 ft) littoral as well as the profundal (200–400 ft) zone. In all, he notes the presence of 48 different benthic organisms identified to the genus or species level (including 23 Chironomidae) plus 11 others cited only as belonging to broader taxonomic groupings (Table 38). The most abundant littoral forms were *Hexagenia, Gammarus, Pisidium,* Nematoda, and Chironomidae.

Littoral samples, taken by Dahlberg in May and July, showed a pronounced increase in *Gammarus* and a decrease in *Pisidium* as the summer progressed. At the inner littoral stations there were markedly fewer *Hexagenia* in July than in May. Chironomids were abundant in July but were not counted in May. Among the gastropods there was also a shift from *Gyraulus, Viviparus,* and *Amnicola* in May to *Physa* in July.

Dahlberg is also the only investigator to sample the benthos at a number of stations along the length of the lake. Considerable variations from station to station were found, with an indication of greater abundance at the most northerly and most southerly locations. His collections indicated that *P. affinis* was dominant in the profundal benthos. Numbers increased from an average for all stations of 1058/m^2 in May to 3145/m^2 in July.

A historical comparison of the species composition and abundances of the dominant invertebrates in the profundal zone (Table 39) suggests that no major changes in overall biomass have taken place between 1918 and 1972. Qualitatively, *Metriocnemus* (?*Heterotrissocladius*) may have decreased between 1952–1953 and 1972, and *P. affinis* may have increased from 1918 to 1952–1953. The lack of replicate samples for 1918 and 1972, different sampling methodologies and locations, and the nonrandom distribution of

TABLE 38

Benthic Invertebrates Found by Dahlberg (1973) in the Inner and Outer Littoral and Profundal Zones of Cayuga Lake During Summer of 1972[a]

Organism	Inner littoral	Outer littoral	Profundal
Turbellaria			
Dugesia			x
Hirudines			
Helobdella	x	x	x
Pisicola		x	
Cladocera			
Leptodora kindtii			x
Mysidea			
Mysis relicta	*	*	*
Isopoda			
Asellus	*	*	*
Amphipoda			
Pontoporeia affinis	x	*	*
Gammarus linnaeus	*	*	*
Gammarus fasciatus	*	x	
Plecoptera			
Unidentifiable immature	x	x	
Ephemeroptera			
Hexagenia	*	*	
Caenis	x	x	
Trichoptera			
Hydroptila	x		
Oecetis	*	x	x
Leptoceridae immature	x		
Coleoptera			
Stenelmis	x	x	
Heteroceridae	x		
Hemiptera			
Sigara	x		
Corixidae immature		x	x
Lepidoptera			
Pyralidae	x		
Diptera			
Hexatoma	x	x	
Chironomidae			
Ablabesmyia c.f. mallochi	x	x	
Ablabesmyia (3 unidentified species)	x	x	

TABLE 38 (*Continued*)

Organism	Inner littoral	Outer littoral	Profundal
Procladius	*	*	
Chironomus staegeri?	x	*	
Chironomus		*	
Cryptochironomus amachaerus		x	
Cryptochironomus nais		x	
Cryptochironomus		x	x
Dicrotendipes	x	x	
Paralauterborniella nigrohalterale	x	x	
Paratendipes albimanus	x	*	x
Polypedilum c. f. *fallax*	x	x	
Pseudochironomus fulviventris	x	x	
Stitochironomus devinctus		x	
Tribelos jucundus	*	*	
Microspectra deflecta	*	*	x
Prodiamesa bathyphila		x	
Cricotopus	x		
Heterotrissocladius		*	x
Psectrocladius	x		
Trichocladius	*	x	
Megaloptera			
Sialis	x		
Gastropoda			
Gyralus	*		
Valvata	x	x	
Helisoma	x	x	
Viviparus	*	x	
Amnicola	x	*	*
Physa	x	*	*
Viviparidae	x		
Pelecypoda			
Pisidium	*	*	*
Sphaerium		x	x
Nematoda	*	*	x
Oligochaeta	x	x	x
Acari	*	x	
Ostracoda	x		

An (x) indicates presence and an asterisk (*) that the organism was common or abundant in a particular zone at one or more of the nine sampling stations.

TABLE 39

Comparison of the Profundal Benthos Collected in 1918, 1952, 1953, and 1972

	Organisms/ m^2	Dry weight (mg/m^2)	Organic matter (mg/m^2)
1918 (Birge and Juday, 1921)			
July 28, King Ferry, 113 m			
Chironomus	3863	1058.5	784.4
Pontoporeia	710	340.8	255.6
Oligochaeta	1288	474.0	416.2
	5861	1873.3	1456.2
1952–1953 (Henson, 1954)			
July 23, 1952, Taughannock, 98–105 m			
Pontoporeia	1459	407.2	355.5
Oligochaeta	2525	82.1	67.2
Pisidium	460	168.2	66.4
Metriocnemus	325	65.1	48.8
(*?Heterotrissocladius*)			
Ostracada	98	6.4	3.1
	4867	729.0	541.0
August 6, 1953, Taughannock, 98–105 m			
Pontoporeia	949	421.9	368.6
Oligochaeta	7593	520.5	421.7
Pisidium	221	64.3	23.7
Metriocnemus	161	57.3	47.6
(*?Heterotrissocladius*)			
Ostracada	304	12.8	6.6
	9228	1076.8	868.2
1972 (Dahlberg, 1973)			
July 25, Station 14 (= King Ferry), sampling depth not given by station			
Pontoporeia	2016	562.5[a]	491.9[a]
Chironomidae	66	?	?
(includes *Heterotrissocladius*)			
Oligochaeta	Present but not counted	?	?
	2082(?)	562.5(?)	491.9(?)
July 25, Station 12 (= Taughannock), sampling depth not given by station			
Pontoporeia	4506	1257.2[a]	1,099.5[a]
Chironomidae	20	?	?
(includes *Heterotrissocladius*)			
Oligochaeta	Present but not counted	?	?
	4526(?)	1257.2(?)	1,099.5(?)

[a] Calculated using Henson's data of July 23, 1952 on dry and organic weights to obtain conversion factors.

forms such as *Pontoporeia* prevent any very firm conclusions from being drawn.

Fish

The fish and fisheries of Cayuga Lake have been reviewed by Youngs and Oglesby (1972). They describe the present distribution of species in terms of a distinctive split between typically limnetic and littoral populations of post-juvenile fish. Prominent among the former are the lake trout (*Salvelinus namaycush*), the smallmouth bass (*Micropterus dolomieui*), smelt (*Osmerus mordax*), alewives (*Alosa pseudoharengus*), the rainbow trout (*Salmo gairdneri*), the yellow perch (*Perca flavescens*), and the sea lamprey (*Petromyzon marinus*). Cisco (*Coregonus artedii*) and the salmon (*Salmo salar*) are present but not abundant.

Youngs and Oglesby (1972) summarized available data on depth distribution and food habits of the more important game and forage species (Table 40). The central role of the alewife as a source of food for piscivores is a striking feature of the food web. The growth of yearling alewives during the warmer months is, in turn, highly dependent upon zooplankton concentration (see above).

The dynamics of the lake trout population have been studied by Youngs (1972). He approximates the minimum annual harvest of lake trout by sport fishermen as being 5675 kg during recent years and tentatively estimates that annual total mortality varies between 25 and 60% from year to year. The lake trout is a particularly valuable subject for study since there is no natural spawning in the lake and all stocked fingerlings (28,000/year) are

TABLE 40

Summary Depth Distribution and Principal Food of Important Limnetic Fish Species

Fish	Depths (m) at which most commonly found	Maximum depth (m) at which still found	Principal food of adults
Lake trout	13–30	60	Alewife
Smallmouth bass	2–9	13	Alewife, other species, crayfish
Rainbow trout	?	?	Alewife?
Smelt	20–45	—	*Mysis relicta, Pontoporeia affinis*
Alewife	10–20[a]	25	Zooplankton
Cisco	25–43	—	*M. relicta*
Yellow perch	12	23	Alewife

[a] From Hennick (1973).

marked for future study. The adult lake trout gather from throughout the lake during the fall as a spawning population off Taughannock Point (their former spawning ground). Here many are netted by personnel from New York's Department of Environmental Conservation in order to obtain fertilized eggs for hatcheries. This provides an additional opportunity to tag adults and to obtain statistics on growth, mortality, lamprey attack, and other parameters. The biology of the lake trout in Cayuga has been under study for many years by fishery biologists at Cornell (Webster *et al.*, 1959).

Important gamefish in the population associated with the large littoral area at the north end of Cayuga Lake are northern pike (*Esox lucius*), chain pickerel (*Esox niger*), largemouth bass (*Micropterus salmoides*), yellow perch, and bullheads (*Ictalurus nebulosus*). A fishery existed for most of these species at the south end of the lake but is now greatly decreased, probably as a result of the extensive filling of marshes over the last 45 years.

Species of fish that are common in Cayuga Lake but about which little is known are the white sucker (*Catostomus commersoni*), the rock bass (*Ambloplites rupestris*), and the carp (*Cyprinus carpio*). The latter was probably introduced in the late nineteenth century and the rainbow trout at about the same time. Smelt were introduced in the late 1920's. When and how two important species, the sea lamprey and the alewife, were established in Cayuga Lake is not known. They could have been residents since the last glacial retreat or may have been introduced with the connection of the Erie Canal to Lake Ontario in 1828 (Whitford, 1906).

The earliest records referring to fish occurring in Cayuga Lake are provided by Raffieux (1671–1672), who writes about the abundance of salmon and eels (now very uncommon). In 1879 a writer (Anonymous, 1879) described how "Cayuga Lake abounds with fish; salmon-trout, whitefish, bass, pike, and pickerel being the chief varieties." Youngs and Oglesby (1972) concluded that changes in the fish population have probably been affected in only a minor way by exploitation. Major changes in gamefish are associated with the stocking of lake trout and with the introduction, in the 1950's, of self-perpetuating populations of rainbow trout. Physical alterations in the form of landfilling (south end), the silting in of reefs suitable for lake trout spawning, and the connection of Cayuga Lake with Lake Ontario together with downstream pollution of rivers through which fish could formerly migrate to and from the sea may also have been substantial factors in modifying the structure of the fish population, the relative abundance of the various species, and the total productivity of fish.

Bacteria

As is the case with most lakes, the bacterial flora of Cayuga is virtually undefined. Dahlberg (1973) has provided a small amount of data for 1972.

TABLE 41

Bacteria (Colonies/ml) as Determined from Samples Taken in May of 1972 at Seven Mid-Lake Stations Located along the Center Line of the Lake

Depth (m)	Plate count $\times 10^4$	Coliform	Fecal coliform	Fecal streptococci
0	11.4	219.9	72.3	6.6
30.5	0.7	1.4	1.1	7.6
91.4	0.5	2.6	3.3	8.0

Total plate count and concentrations of those forms used as indicators by public health authorities are given in Table 41 as defined by a single sampling. Means for indicator organisms at the surface were heavily influenced by the large numbers found at the most southerly station. This location would be sensitive not only to possible additions from sewage treatment plant discharges but is also the closest station to an area containing a small zoo and large natural populations of birds and wildlife. The plate count average was similarly biased by a single high concentration, in this instance occurring off the Milliken power plant. Bacterial concentrations for samples taken in May, July, and November at a single location (Milliken) are given in Table 42.

Submerged Macrophytes

Along the shore of Cayuga Lake the littoral zone is generally rocky, and water depth frequently increases precipitously so that the environment is unsuitable for the growth of submerged macrophytes. However, at either end and at a few locations along the east and west shores, especially on the southern sides of points, areas of shallow depths and sediment-covered bottoms provide a habitat that often supports luxuriant beds of these plants. Growth is sparse at depths exceeding 3.5–4 m. Beds are extensive north of Union Springs and in a much smaller area at the south end.

TABLE 42

Mean Concentrations (Cells/100 ml) of Bacteria in May, July, and November at a Series of Stations off Milliken

	Plate count $\times 10^4$	Coliform	Fecal coliform	Fecal streptococci
May	43,125	4,298	1,640	1,157
Jul	74,533	8,338	0	13.3
Nov	2,589	11	0.3	9.5

Early workers (Dudley, 1886; Weigand and Eames, 1925; Muenscher, 1927) have provided data on the occurrence and location of submerged macrophytes in Cayuga Lake and sometimes offered comments on relative abundances. In 1929, 1942 1943, 1970, and 1972–1973 the distribution of rooted aquatics was mapped for an area of over 150 hectares at the head end of the lake, and quantitative measurements of standing crop at selected stations were made in all but the 1929 study.

Data from all studies through 1970 at the south end have been compared by Vogel (1973) to assess possible changes in the flora. He concluded that since the early 1940's marked decreases have occurred in the abundance and/or the areal distribution of *Anacharis canadensis, Potamogeton Richardsonii, Najas flexilis, Nitella* sp., *Vallisneria americana, Potamogeton pectinatus, Zannichellia palustris,* and the alga *Chara* sp. Increases occurred in *Myriophyllum exalbescens, Heteranthera dubia, Potamogeton crispus,* and *Ceratophyllum demersum* and were particularly large for the first two.

Leister (1929) described 21 species as being present at the head end of the lake in 1929, Hewitt (1944) found 18 in 1942–1943, and Vogel (1973) only 10 in 1970. Qualitative changes are particularly striking for members of the genus *Potamogeton,* with decreases from 13 to 11 to 4 species, respectively, being recorded in the three studies.

Vogel duplicated the standing crop sampling stations of Hewitt and concluded that plant density had increased. The highest values in 1970 were for stations off the west shore, where *Heteranthera dubia* and *Myriophyllum exalbescens* form dense beds. The former produced the highest standing crop (700 gm dry weight/m^2). The mean value for this species as determined from 66 samples taken on the west side was 133 gm/m^2. *Myriophyllum exalbescens* at the same stations reaches a maximum of 400 gm/m^2 and averaged 71 gm/m^2. Except at the extreme southwest and southeast corners of the lake, where *V. americana* and *P. pectinates* were, respectively, the most abundant species, *H. dubia* and *M. exalbescens* accounted for 80% of the sampled standing crop at the head end in 1970.

Tropical Storm Agnes deposited some 17 cm of rainfall in the Cayuga Lake basin on June 21 and 22, 1972. Low transparencies due to allochthonous suspended materials persisted well into July, and substantial quantities of new sediment were deposited in the shallows at the south end. A survey of submerged aquatic plants (Oglesby *et al.*, 1976) indicated that growth was severely affected. Of the 11 species described by Vogel, only five were found in late August of 1972. At the standing crop sampling stations biomass had been reduced by 44–100%, with the area along the west shore being the most severely affected. The storm impact varied somewhat according to species. *Heteranthera dubia* was the most adversely affected plant and *Myriophyllum exalbescens* the least. By the summer of 1973 the extent and density of rooted plant growth at the head end of Cayuga Lake was only slightly lower than that found in 1970. *Myriophyllum exalbescens* had increased even further in relative dominance.

Casual observation indicates that beds of macrophytes are both dense and widespread over the large, shallow northern end of the lake. *Myriophyllum* is prominent but stands of plants dominated by other species are also present. The submerged plant growths gradually merge with and then are replaced by emergent vegetation as the lake grades into very shallow depths adjacent to the Montezuma Marsh. A study of the macrophyte flora at the north end of the lake by researchers from Cornell University and Eisenhower College was begun in the summer of 1973 and has included aerial mapping with field study verification.

Cladophora sp. is found along the sides of the lake attached to stones. Shoreline residents believe that growths of this alga have increased in extent and density in recent years, but there are no data to either substantiate or refute this contention.

NUTRIENT BUDGETS

Phosphorus

Likens (1972, 1974a,b) has reported nutrient inputs and outputs for Cayuga Lake as determined from a 1-yr study of tributary flow and precipitation (Fig. 13). Tributary input was derived from samples taken in 25 streams, whose combined watershed areas equaled 77.6% of the lake basin drainage area, with that from unmonitored streams being estimated. Measurements did not include bedload.

Precipitation was collected at sites chosen to define the basin-wide pattern of input. Likens (1972, 1974a) determined that the annual contribution of phosphorus by precipitation directly on the lake's surface was only 2–3% of the total from all sources. Nine percent of the total water income to the lake was from precipitation.

According to Likens' calculations, some 73.9% of the molybdate reactive phosphorus (MRP) and 64.1% of the total phosphorus added annually are retained in the lake. Adjusting his input figures for additional sources of phosphorus (see below), the percentage retained becomes 76.6 and 66.4% for MRP and total P, respectively. Oglesby *et al.* (1973) estimated that another of the Finger Lakes, Owasco, retained 85% of the soluble reactive phosphorus added annually.

Likens (1974a) did not include the inputs of sewage directly to the lake from several small communities (Cayuga Heights, Aurora, and Cayuga) nor

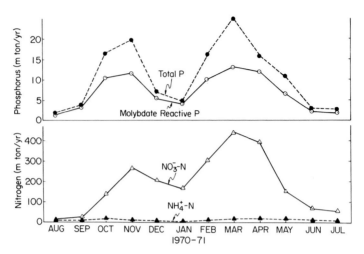

Fig. 13. Inputs of phosphorus (molybdate reactive and total) and nitrogen (nitrate and ammonium) to Cayuga Lake during the 1970–1971 hydrologic year from tributaries and atmospheric fallout.

from lakeside homes and cottages in his phosphorus balance. Measurements of phosphorus discharges from sewage treatment plants are not of sufficient number to provide a basis for direct calculation of inputs from these sources. If it is assumed that the yearly per capita discharge of phosphorus in household wastes is 1.5 kg (Van Wazer, 1961; Engelbrecht and Morgan, 1959) and that 20% of this is removed by primary and secondary treatment, the loading produced by that fraction of the population served by treatment plants (Table 9) is 56.3 metric tons P/year. However, this must be decreased by about 50% for the period after June, 1973, when New York State banned the use of household laundry detergents containing phosphate builders. The input of phosphorus represented by Aurora's discharge of P to the lake should be further reduced to about 5–10% of that entering the treatment plant since tertiary treatment was instituted between 1970–1971 and 1973. The unsewered population living in the Cayuga Lake basin numbers about 43,000. Oglesby *et al.* (1973), in a study of the phosphorus contributed to Owasco Inlet by several small, unsewered villages, have provided evidence that for the Owasco Lake basin 50% of the P from unsewered households entered the surface waters. The estimate of 50% is undoubtedly too high for dwellings located in the midst of a typically rural landscape. However, concentration of housing adjacent to watercourses and along the lake shore is a pronounced demographic trait of the Cayuga basin. For lack of more definitive information, the simplifying assumption is therefore made that all of the rural population falls into this category.

Extrapolating this percentage to the Cayuga Lake population not served by sewage treatment facilities would give a calculated yearly loading of 32.3 metric tons P prior to June, 1973 and 15.7 metric tons thereafter from this source.

Molybdate-reactive (unfiltered samples) and total phosphorus loadings, with sewage inputs calculated and the fraction from land runoff determined by difference (total loading from Likens adjusted for effluents discharged directly to the lake and P measured in precipitation), are given in Table 43. As of June, 1973, the P input from waste discharges should be decreased to some 43.8 metric tons/year.

The partitioning of total phosphorus between sources as given in Table 43 appears to be reasonable. Land runoff is separately calculated, adapting estimates of yield per unit area for various land use categories as given by Dillon and Kirchner (1975) in Table 44. The total of 57.4 metric tons/year compares well with the 47.4 metric tons/year estimated by the difference method. Unfortunately, the estimate of molybdate-reactive phosphorus from land runoff cannot be subjected to such a comparative analysis due to a lack of definitive data in the literature. The loading fraction represented by land runoff is almost certainly too low. The question of whether the principal source of error lies in the total (from Likens but modified for direct input of sewage discharges) or in the calculation of P in waste discharge cannot be resolved with presently available information.

A third approach to the determination of phosphorus loading (Schaffner and Oglesby, in press; Oglesby and Schaffner, 1977) involves the calculation of all inputs except precipitation, for which the values of Likens (1972) are

TABLE 43

Phosphorus Inputs and Loadings (Excluding the Seneca River) for 1970–1971 with Waste Discharges Calculated and Land Runoff Determined by Difference

Source	Total P (kg × 10^{-3}/yr)	Molybdate reactive P (kg × 10^{-3}/yr)
Waste discharges	88.6	88.6
Land runoff	47.4	2.0
Precipitation	3.4	3.4
Ground water	?	?
Other	?	?
	139.4	94.0
Volumetric loading (μg/liter/yr)	14.9	10.0
Areal loading (gm/m²/yr)	0.81	0.54

TABLE 44

Computational Data for Nonpoint Source Total P Runoff in the Cayuga Lake Basin

Land use	Area (km²)	Runoff factor (gm/km²/yr)	Runoff of total P (kg × 10⁻³/yr)
Active agriculture	904	46	41.6
Forest	582	11.7	6.8
Inactive agriculture (91%) and other	386	23.3	9.0
			57.4

used. This was made possible through the development of land runoff coefficients based on an extensive study of phosphorus movement in Fall Creek and its tributaries (Bouldin et al., 1977). Loadings are expressed in terms of total P and of a composite form (soluble reactive plus soluble unreactive plus the P associated with particulate matter that desorbs in aqueous solution) which they term "biologically available phosphorus." Areal loading of the former was calculated to be 0.86 kg × 10⁻³/m²/year and for the latter was 0.67 kg × 10⁻³/m²/year.

The estimated water renewal time for the 1970–1971 12-month period during which Likens monitored input and output is 8.6 years. This was calculated by dividing the discharge, minus the flow of the Seneca River as it enters the lake, by the volume of the lake.

Nitrogen

Likens' (1972, 1974a,b) determinations of inorganic nitrogen input were given in Fig. 13. Inputs and outputs of organic nitrogen were not monitored. For the inorganic forms he estimated that 36.1% of the ammonium and 86.3% of the nitrate added annually were retained (or lost to the atmosphere). Precipitation was the source of 6.1% of the nitrate and 34.9% of the ammonium.

If nitrogen (organic, ammonium, and nitrite) in household wastes is taken to be 4.44 kg/cap/year (Olsson et al., 1968), calculations similar to those for phosphorus (see above) can be carried out to estimate the nitrogen associated with waste discharges. The quantity of N removed by either sewage treatment plants or household disposal systems is difficult to estimate. An assumed decrease of 50% for both is probably reasonable. Thus, for a total basin population of 90,221 in 1970, nitrogen would equal 200.3 metric ton/year (90,221 cap × 4.44 kg/cap × 0.5 × 10³ kg/metric ton). Data on N input are summarized in Table 45. Since Likens did not measure organic N, the total calculated input errs on the low side.

TABLE 45

Nitrogen Inputs and Loadings (Excluding the Seneca River) for 1970–1971 with Waste Discharges Calculated and Land Runoff Determined by Difference

Source	Soluble N (kg × 10^{-3}/yr)
Waste discharges	200.3
Land runoff	1694.1
Precipitation	565.6
Ground water	?
Other	?
	2460.0
Volumetric loading (μg/liter/yr)	262
Areal loading (g/m²/yr)	14.2

Other Nutrients

Likens (1972, 1974a,b) also monitored the input of other macronutrients, and these data are summarized in Table 46.

DISCUSSION

Limnology

As a moderately large and deep cold water lake affected by a variety of human influences, Cayuga is representative of many important bodies of water formed in north temperate latitudes by the action of glaciers. Its

TABLE 46

Sulfur, Silicon, Calcium, Magnesium, and Bicarbonate Inputs and Loadings from Precipitation and Tributary Inflow (Excluding the Seneca River) during 1970–1971

Nutrient	Input (kg × 10^{-3}/yr)	Volumetric loading (μg/liter/yr)	Areal loading (gm/m²/yr)
SO_4^{2-}–S	13,253	1,410	76.6
SiO_2–Si	2,147	229	12.4
Ca^{2+}	63,802	6,800	369
Mg^{2+}	15,263	1,630	88.2
HCO_3^-–C	31,523	3,360	182

limited shoreline development, restricted littoral zone (except at the tail end), and relatively uncomplicated hydrology are factors that enhance the value of Cayuga as a subject for limnological investigation. The existence of a substantial body of knowledge on both the lake and its drainage basin accumulated over the last century places Cayuga among the limnologically best-defined lakes in the world.

The status of descriptive information on Cayuga relative to that available for other lakes is evaluated (somewhat subjectively) in Table 47. For the present emphasis on examining primary production in an ecological context, adequate data are available on the more static properties, e.g., geology and morphometry, and on other parameters, such as the concentra-

TABLE 47

Present State of Knowledge Concerning the Descriptive Limnology of Cayuga Lake

	Good	Moderate	Poor	Comments
Basin morphometry	X			
Geology and soils	X			
Climate	X			Over 100 years characterizing basin but little for condition at immediate surface of lake
Vegetation	X			
Population	X			
Land use	X			Uniquely quantitative description
Water use		X		
Waste discharge	X			
Hydrology				
Lake volume and area	X			Based on data over 100 years old
Sediments			X	
Stream flow and retention time		X		Limited gauging of inflow and questionable evaporation rates
Precipitation	X			
Currents			X	
Temperature	X			
Specific conductance		X		
Light		X		
Color			X	
pH	X			

TABLE 47 (*Continued*)

	Good	Moderate	Poor	Comments
Dissolved oxygen	X			
Phosphorus	X			
Nitrogen		X		For post-1967 period only; good for No_3^- since 1964 but very limited for other forms
Alkalinity	X			
Major ions	X			
Trace metals		X		
Phytoplankton				
Pigments	X			For post-1967 period only
Primary production		X		
Algal assays	X			Unusually good continuous culture data but still not sufficient to permit good historical comparisons
Population description	X			
Zooplankton		X		
Benthos		X		
Fish	X			Especially for salmonids
Bacteria			X	
Submerged macrophytes		X		Good for southern portion but lacking for northern end

tions of major ions, for which important links to phytoplankton production have not been established.

If the empirical approach of Vollenweider (1968) and Dillon (1974) is taken toward defining trophic state as a function of nutrient loading, Cayuga maintains its status as a relatively well-defined system thanks to the nutrient input–output balances of Likens (1974a), the strong evidence that has been developed indicating phosphorus as the primary factor controlling phytoplankton production, and the description of standing crop, transparency, and other indicators of trophic condition at a number of points in time. The latter is stated with some reservations since differences in sampling locations, methods used to collect and analyze samples, and equipment employed to make *in situ* measurements occur between the various investigations, making historical comparison of many parameters difficult.

Another empirical approach toward defining a relationship between nutrients and trophic state was that proposed by Sawyer (1947) and later expanded on by Sakamoto (1967) and Dillon and Rigler (1974). Sawyer sought, with some success, to relate the winter concentration of nutrients to summer standing crops of algae for a group of Wisconsin lakes. Despite the clarity with which he cited the caveats associated with this work, it has often been both misquoted and misinterpreted, the most common fallacy being the extrapolation of his "critical" winter concentrations as if they were applicable on a year-round basis. It was not until Edmondson (1972) tested winter total phosphorus concentrations in Lake Washington against summer chlorophyll *a* levels that this correlation could be accepted as a valid working relationship in aquatic ecology. However, it must be realized that Edmondson's highly correlated relationship was based on points spanning a period of 10 years, during which phosphorus levels underwent extensive changes. Oglesby *et al.* (1975) have shown that over a 2-year period winter total P was an excellent index of mean summer standing crops of phytoplankton for a group of deep New York lakes, including Cayuga. When the linear regression expressing this relation was extended to cover the range of winter phosphorus and summer chlorophyll values exhibited by Lake Washington, Edmondson's data fitted the line well. This regression should prove especially useful in tracing the changes that should occur as a result of New York's 1973 ban on phosphates in household laundry detergents.

So far two "macromodels" regulating primary production to nutrients have been discussed. When either an holistic model, one designed to describe all components and functions, or micromodels, for example the relation of daily phosphorus supply to carbon fixation rate, are considered for Cayuga Lake, the review of data available relative to needs becomes extremely complex and is certainly beyond the scope of this paper. However, there are a number of specific areas where it is recognized that further information would be especially valuable in delineating limnological processes within the lake. These are listed below together with brief notations as to their relevance.

1. Nutrient, pigment, and zooplankton exoskeleton content of sediments deposited over the past 200 years: Land use and population histories in the basin are well defined for this period, and such information should provide worthwhile insights into human impacts on the lake. Such a study is now being undertaken by James McKenna, a Cornell graduate student.

2. Quantitative relationships in the food web: The effect of zooplankton grazing on phytoplankton populations must be an integral part of any attempt to define cause and effect relationships associated with

plankton population dynamics. The relation of phytoplankton to fish production should be determined before management schemes to control nutrient input are finalized.

3. Abiotic factors that may remove P from solution in the lake: Evidence was cited that some of the soluble P introduced with clay during the periods of heavy runoff may be removed to the bottom following physical adsorption and settling of the clay particles. At the high pH's and high calcium levels characteristic of Cayuga Lake, chemical precipitation may be a significant factor, particularly during the summer when allochthonous P input is minimal and demand is greatest.

4. Currents and turbulent diffusion: The latter may be a mechanism that provides significant "internal" supplies of phosphorus from the deeper waters via upwelling during the summer-stratified period. A knowledge of the current pattern would permit the calculation of more meaningful hydraulic retention times for the nutrients and plankton existing in specific physical compartments of the lake.

Delineation of Trophic State

At the present time, Cayuga Lake appears to fall in the mesotrophic category, yet for given parameters and at specific times, it could be termed either oligotrophic or eutrophic. This is partly due to the fact that trophic state terminology represents a continuum with no well-defined boundaries, but perhaps even more to the broad array of environments represented in the lake.

The composition of the phytoplankton is especially illustrative of the intermediate trophic status of Cayuga. In comparing major associations with those given by Naumann, Pearsall, Järnfelt, and Rawson (from a review by Hutchinson, 1967) as characteristic of various enrichment states, Cayuga Lake is seen to contain the dominant groupings cited for both oligotrophic and eutrophic categories. While Myxophycean types and diatoms such as *Melosira,* cited as typical eutrophic indicators, are abundant at times, so are *Cyclotella, Tabellaria,* chrysomonads, *Sphaerocystis,* and others given as indicators of oligotrophy.

The occurrence of blue–green blooms is often pointed to as a general characteristic of the phytoplankton indicating eutrophy. Here also the situation is somewhat ambiguous for Cayuga Lake. Growths of blue–greens, particularly of *Anabaena,* sufficient to produce visual nuisances certainly occur at times during the summer. However, these are generally short-lived, and cell densities are lower than those encountered in very highly eutrophied lakes.

Summer pigment values for Cayuga Lake provide further evidence for mesotrophy. Mean levels of chlorophyll *a* plus phaeophytin in the upper 10 m of the water column were, for July through September 1968, 6.1 μg/liter; for June through September 1972, 9.2; and for June through August 1973, 7.5 μg/liter. Actual measurements of annual productivity rates are available for 1957 and 1958 (Howard, 1963). In the former year this was quite high (341 mg C/m²) and in the latter very low (66 gm C/m²). Based on these measurements alone, Cayuga was in a state of eutrophy in 1957 and oligotrophy in 1958. From the data of Peterson (1971) for 1968–1970, daily productivity appears to average about 160 mg C/m² over a 1-year period. This value is typical of lakes regarded as being mesotrophic.

The littoral and profundal benthos and the hypolimnetic zooplankton are those expected for a cold, well-oxygenated environment. The standing crop of profundal benthos is between 500 and 1000 mg/m² organic matter, about the level found elsewhere in mesotrophic lakes.

Evidence has been presented that a trend to more eutrophic conditions has probably occurred between the 1910–1930 period, when the first series of investigations on Cayuga were carried out, and 1950–1973, when limnologists resumed their studies of this lake. Summer phytoplankton standing crops appear to have increased with blue–greens constituting a more prominent part of the community. Comparisons of pH and epilimnetic dissolved oxygen extremes tend to support the view that the lake has become more productive.

However, hypolimnetic dissolved oxygen shows no parallel decrease, nor is there evidence that the zooplankton or benthic community structures have undergone any significant alterations. Unfortunately, there is not sufficient data to define the changes in faunal biomass that must have accompanied increased phytoplankton production. The constancy in profundal oxygen tensions over the 1910–1973 period indicates that much of the summer increase in phytoplankton must have been consumed within the epilimnion.

Given the demonstrated importance of phosphorus in controlling Cayuga Lake phytoplankton production, it is tempting to postulate that the advent of synthetic detergents with phosphate builders during the 1940's was a prime eutrophying influence. Neither population increase nor accelerated agricultural activity, the two other most obvious potentials for increased phosphorus input, seem to provide viable alternatives as major causative agents. A current nutrient budget also emphasizes the importance of detergent phosphates as sources for the lake. The effect of New York State's ban on phosphorus in household laundry products will provide an interesting test of this hypothesis during the next few years.

Trophic State vs. Nutrient Budgets

The values shown in Table 48 are calculated for Cayuga Lake using the definitions of Dillon (1974). Estimates of nutrient loadings and water residence time apply to August 1970 through July 1971 only.

Referring to the graphic presentations of trophic state as a function of total loading vs. mean depth (Vollenweider, 1968) and versus the ratio of mean depth to water residence time (Vollenweider, as given by Dillon, 1974), Cayuga is seen to fall into the eutrophic category (about the same as Lake Mälaren), with total P loadings above the so-called "dangerous" level. Precipitation during 1970–1971 was relatively high. If the driest year on record is considered, hydraulic retention time would increase almost threefold (see above) and $\bar{z}/T_w$ by a factor of four. This would effectively move Cayuga even further into the eutrophic category.

Based on biological and chemical characteristics, Cayuga has been judged to be in a mesotrophic state (see above). Thus, it would seem that it does not fit the relationships defined in the foregoing "models." The reason for this can only be speculated on at present. One of the more plausible explanations is the probable low rate of nutrient recycling to the euphotic zone during the stratified season relative to that which occurs in many lakes. Cayuga's aerobic hypolimnion, small degree of shoreline development, and its basin morphology are factors that would keep phosphorus cycling to a minimum. Adsorption of phosphorus on clay particles followed by settling from the water column and chemical precipitation as a function of high calcium levels and high pH may also be factors that play a significant role in diminishing the effects of phosphorus inputs to Cayuga Lake.

Dillon (1974) presents a graph (after Vollenweider et al., 1974) in which yearly primary production for the various Great Lakes is plotted against phosphorus loading. If the two annual productivity values of Howard (1963) are averaged, and the 1970–1971 P loading data used, Cayuga fits their

TABLE 48

Parameters Relating to Annual Loading of Total Phosphorus

Parameter	Value
Mean water depth ($\bar{z}$)	54.5 m
Water residence time (T_w)	8.6 yr
Ratio of mean depth to water residence time ($\bar{z}/T_w$)	6.34 m/yr
Loading of total P (L)	0.81 gm/m²/yr
Flushing rate ($1/T_w = \rho$)	0.12 yr^{-1}
Retention time of total P excluding Seneca River (R)	0.56 yr^{-1}

curve quite well, falling between Lake Ontario and the central basin of Lake Erie.

SUMMARY

Cayuga Lake lies in an elongated, steep-sided basin surrounded by forested rolling hills at its southern end with flatter topography and a greater amount of agriculture to the north. Some 90,000 people inhabit the basin, with about half of these living in Ithaca and its immediate vicinity. The lake is a major recreational resource for the region, and its waters are also used for both drinking and waste disposal. Industrial development is low and does not constitute a substantial pollution hazard.

The bedrock of the basin consists mostly of acid shale and sandstone to the south, but northward is intersected by two major limestone formations. The soils of the northern two-thirds are dominated by moderately coarse-textured types with calcareous substrata. Those of the major tributaries and highlands surrounding the southern part of the basin are composed of a diverse and complex assemblage and, in general, are less well drained and more acid.

The climate is of the humid continental type with warm summers and long, cold winters. The area lies on the main west–east track of cyclonic storms, and hence its weather is highly variable and is characterized by considerable cloudiness. Annual precipitation ranges from 71 to 117 cm, with half of this normally falling from May through September.

The most intense period of surface runoff into Cayuga Lake usually occurs in late March or early April, when spring rains combine with melting snow and water released from thawing soils. Drainage from only 40% of the land in the basin has been gauged over a long period of time. Outflow has been measured since 1932. It is estimated that water retention time may vary from about 8 to 24 years depending on climatological variables.

Cayuga Lake has a single basin and 5% of its volume is at depths greater than 90 m. Mean depth is 54.5 m and the littoral zone is limited in extent except for an extensive area at the northern (tail) end. The lake is long (61.4 km) and narrow (mean width of 2.8 km) and shoreline development is low (3.35). The length of the shoreline is 153.8 km, the surface of the water area is 172.1 km^2, and the volume amounts to 9379.4×10^{-6} m^3.

The temperature regime of the lake is characterized by minimum homothermy at $1.5°–3.3°C$ sometime between late February and early April. Between mid-May and early June the bottom waters become effectively isolated from the surface, although the classic three-layered system is not present until mid-June or early July. Maximum summer bottom temperatures range from 4.1° to 5.5°C during various years, and highest surface temperatures vary from about 20° to 27°C. Maximum bottom temperature (6.6°–9.6°C) for the year is associated with fall homothermy, which occurs between early November and December.

Average monthly (1968–1970) light extinction coefficients vary from a low of about 0.25 in the winter to a pronounced July maximum of 0.85. Secchi disc transparency is typically 2–3 m during the summer and 5–6 m from late October through the spring.

Cayuga is a moderately hard water lake with a well-developed calcium carbonate–bicarbonate buffering system and a higher than normal NaCl level. Specific conductance is generally between 500 and 600 μmhos/cm^2 and is lower in the summer than in the winter. The level of trace metals is uniformly very low. Hydrogen ion concentrations reach a minimum (pH 7.7–8.0) throughout the water column in the winter. By the onset of thermal stratification pH has increased to about 8. During the summer, pH decreases rather quickly in the hypolimnion to a low of about 7.7, and at the same time, epilimnetic values are maximum (as high as 9.0), averaging 8.2–8.4.

Dissolved oxygen follows a pattern to be expected for a cold, deep, moderately productive lake. During the summer, daytime supersaturation is fairly common in the epilimnion, and hypolimnetic levels decrease seasonally, reaching a minimum of about 6 mg/liter in the deepest portion of the lake just before fall overturn. The water column is only 80–90% saturated at the time of autumnal mixing. Dissolved oxygen increases gradually during the winter and reaches 90–95% of saturation by the time thermal stratification is reestablished in the summer. Hypolimnetic minima do not appear to have changed since the early part of this century.

Nitrate nitrogen concentrations vary from year to year, seasonally, and with depth, but appear during most years to be substantially in excess of the minimum levels needed for unrestricted phytoplankton growth. Maximum input is via tributary inflow during the spring, and concentrations are still high in mid-May. Following stratification, nitrate decreases erratically in the epilimnion, and a slight vertical cline of increasing concentrations becomes apparent. Organic nitrogen increases in the epilimnion as nitrate levels decrease.

Seasonal variations in the form and concentration of phosphorus, an elegant series of continuous culture bioassays, and alkaline phosphatase release (Griffin 1974) all indicate that phosphorus is a critical element in controlling the level of summer phytoplankton production. Summer total phosphorus ranges between about 15 and 20 μg/liter throughout the water column, with an epilimnetic shift from the soluble-reactive to the soluble-unreactive and particulate forms taking place as the stratified season progresses. Soluble-reactive phosphorus is generally below 5 μg/liter, and sometimes less than 1 μg/liter, in the epilimnion.

A materials balance for total phosphorus indicates a total input of about 139 metric tons/year, which results in a volumetric loading of 14.9 μg/liter and an areal loading of 0.81 gm/m². Over 60% of this input is thought to originate as domestic waste discharges. Only about 67% of the total phosphorus entering the lake is in the molybdate-reactive form. As of June 1, 1973, New York State banned the sale and use of household laundry detergents containing phosphates. It is estimated that this will reduce the annual total phosphorus input by about 30%, with consequent new loadings of 10.4 μg/liter and 0.57 gm/m²/year.

The phytoplankton of Cayuga Lake is comprised of a mixture of associations, some of which have been described in the literature as being indicative of oligotrophy and others as typifying eutrophic conditions. Myxophycean "blooms" occur at times during the summer but do not persist for long. Seasonal patterns of succession and peaks of abundance as indicated by cell counts and pigment concentration are highly variable from year to year. A general pattern of maximum standing crop from late June into early October exhibits large week to week fluctuations, with surface chlorophyll a ranging from a low of near 1 μg/liter to over 20 μg/liter on one occasion in 1972. Average summer chlorophyll a plus phaeophytin in the epilimnion has ranged between 6 and 10 μg/liter during recent years. Both cell counts and the species composition of the phytoplankton indicate a probable trend to more eutrophic conditionns when data from 1910–1930 are compared with those for 1950–1973.

The zooplankton and benthic fauna are typical of deep, moderately productive, north temperate latitude lakes, and there is no evidence of qualitative changes having taken place in either over the last 60 years. *Mysis relicta* appears to be quite abundant and is an important food resource for some species of fish. The fish population is managed to maximize salmonid production. Significant sport fisheries also exist for smelt, smallmouth bass, and other species. The principal food chain associated with the limnetic zone is

Phytoplankton→zooplankton→alewives→salmonids

Dense growths of rooted macrophytes occur in a limited area of shallow water at the southern end of the lake and over a much larger area at the northern end. Historical data indicate a possible increase in plant density and a shift to species that constitute more of a nuisance.

When phosphorus loading, mean depth, and water renewal time are used to assess the trophic state of Cayuga relative to other lakes, a eutrophic rather than mesotrophic state is indicated. The reason for this variation from existing "models" is not known.

ACKNOWLEDGMENTS

In addition to support from New York's Department of Environmental Conservation, funds used for the preparation of this manuscript were provided by the United States Environmental Protection Agency as part of the North American Lake Project, which in turn is part of the Comparative Projects for Monitoring Inland Water Programme of the Organization for Economic Cooperation and Development (Paris).

For much of the current information on Cayuga Lake attributed to the author and his colleagues, the sponsorship of the Office of Water Resources Research (Matching Grant B-038-NY) and the Rockefeller Foundation is gratefully acknowledged.

REFERENCES

Allee, D. J. (1969). Uses and values of Cayuga Lake. *In* "Ecology of Cayuga Lake and the Proposed Bell Station (Nuclear Powered)" (R. T. Oglesby and D. J. Allee, eds.), Publ. No. 27, Chapter XXI. Cornell Univ. Water Resour. and Mar. Sci. Cent., Ithaca, New York.

Anonymous. (1879). "History of Tioga, Chemung, Tompkins and Schuyler Counties, New York." Everts & Ensign, Philadelphia, Pennsylvania.

Anonymous. (1972). "Monthly Meteorological Summary," Vol. 8, No. 13, pp. 1–6. Div. Atmos. Sci., Dep. Agron., Cornell University, Ithaca, New York.

Barlow, J. P. (1969). The phytoplankton. *In* "Ecology of Cayuga Lake and the Proposed Bell Station (Nuclear Powered)" (R. T. Oglesby and D. J. Allee, eds.), Publ. No. 27, Chapter XVI. Cornell Univ. Water Resour. and Mar. Sci. Cent., Ithaca, New York.

Barlow, J. P., Schaffner, W. R., deNoyelles, F., Jr., and Peterson, B. J., (1973). Continuous flow nutrient bioassays with natural phytoplankton populations *In* "Bioassay Techniques and Environmental Chemistry" (G. E. Glass, ed.), p. 299. Ann Arbor Sci. Publ., Ann Arbor, Michigan.

Behrman, V. L. (1975). The seasonal succession of the major zooplankton species in Cayuga Lake with reference to some possible causal factors. M.S. Thesis, Cornell University, Ithaca, New York.

Benoit, J. R. (1957). Preliminary observations on cobalt and vitamin B_{12} in freshwater. *Limnol. Oceanogr.* **2**, 233–240.

Berg, C. O. (1966). Middle Atlantic States. *In* "Limnology in North America" (D. G. Frey, ed.), p. 191. Univ. of Wisconsin Press, Madison.

Birge, E. A., and Juday, C. (1914). A limnological study on the Finger Lakes of New York. Doc. No. 791. *Bull. Fish.* **32**, 525–609.

Birge, E. A., and Juday, C. (1921). Further limnological observations on the Finger Lakes of New York. Doc. No. 905. *Bull. Fish.* **37**, 210–252.

Bouldin, D. R., Johnson, A. H. and Lauer, D. A. (1977). The influence of human activity on the export of phosphorus and nitrate from Fall Creek. *In* "Nitrogen and Phosphorus—Food Production, Wastes and the Environment" (K. S. Porter, ed.), Chapter 3. Ann Arbor Sci. Publ., Ann Arbor, Michigan.

Bradshaw, A. S. (1964). The crustacean zooplankton picture: Lake Erie 1939-49-59; Cayuga 1910-51-61. *Verh. Int. Ver. Theor. Angew. Limnol.* **15**, 700–708.

Brownell, W. N. (1970). Studies on the ecology of *Mysis relicta* in Cayuga Lake. M.S. Thesis, Cornell University, Ithaca, New York.

Burkholder, P. R. (1931). Studies in the phytoplankton of the Cayuga basin, New York. *Bull. Buffalo Soc. Nat. Sci.* **15**, 21–181.

Chamberlain, H. D. (1975). A comparative study of the zooplankton communities of Skaneateles, Owasco, Hemlock and Conesus Lakes. Ph.D. Thesis, Cornell University, Ithaca, New York.

Child, D., Oglesby, R. T., and Raymond, L. S., Jr. (1971). "Land Use Data for the Finger Lakes Region of New York State," Publ. No. 33. Cornell Univ. Water Resour. and Mar. Sci. Cent., Ithaca, New York.

Cline, M. G., and Arnold, R. W. (1970). Working draft soil association maps for New York. Unpublished.

Dahlberg, M. (1973). "An Ecological Study of Cayuga Lake, New York," Vol. 4. Report to New York State Electric and Gas Corporation. NUS Corp., Pittsburgh, Pennsylvania.

Daniell and Long Engineers. (1969). "Comprehensive Sewerage Study of Seneca County." Boston, Massachusetts.

Dethier, B. E. (1966). Precipitation in New York State. N. Y., *Agric. Exp. Stn., Ithaca, Bull.* **1009**, 1–78.

Dethier, B. E., and Pack, A. B. (1963a). "Climatological Summary," RURBAN Climate Ser. No. 1, Ithaca, New York. N.Y. State College of Agriculture, Ithaca, New York.

Dethier, B. E., and Pack, A. B. (1963b). "Climatological Summary," RURBAN Climate Ser. No. 3, Geneva, New York. N.Y. State College of Agriculture, Ithaca, New York.

Dillon, P. J. (1974). Progress report on the application of the phosphorus loading concept to eutrophication research. A report prepared on behalf of R. A. Vollenweider for NRC Associate Committee on Scientific Criteria for Environmental Quality Subcommittee on Water, Canada Centre for Inland Waters, Burlington, Ontario.

Dillon, P. J., and Kirchner, W. B. (1975). The effects of geology and land use on the export of phosphorus from watersheds. *Water Res.* **9**, 135–148.

Dillon, P. J., and Rigler, F. H. (1974). The phosphorus-chlorophyll relation in lakes. *Limnol. Oceanogr.* **19**, 767–773.

Dudley, W. R. (1886). "The Cayuga Flora," Vol. II. Bull. Cornell University (Science).

Edmondson, W. T. (1971). A manual on methods for the assessment of secondary productivity in fresh waters. *IBP Handb.* **17**, 1–358.

Edmondson, W. T. (1972). Nutrients and phytoplankton in Lake Washington. *Am. Soc. Limnol. Oceanogr., Spec. Symp.* **1**, 172–188.

Engelbrecht, R. S., and Morgan, J. J. (1959). Studies on the occurrence and degradation of condensed phosphates in surface waters. *Sewage Ind. Wastes* **31**, 458–478.

Ferguson, R. H., and Mayer, C. E. (1970). The timber resources of New York. *U.S., For. Serv., Res. Pap.* NE 20, 1–193.

Gates, C. D. (1969). Biochemical oxygen demand. *In* "Ecology of Cayuga Lake and the Proposed Bell Station (Nuclear Powered)" (R. T. Oglesby and D. J. Allee, eds.), Publ. No. 27, Chapter XX. Cornell Univ. Water Resour. and Mar. Sci. Cent., Ithaca, New York.

Godfrey, P. J. (1977). Ph.D. Thesis. Cornell University, Ithaca, New York (in preparation).

Green, R. H. (1965). The population ecology of the glacial relict amphipod *Pontoporeia affinis* Lindstrom in Cayuga Lake, New York, Ph.D. Thesis, Cornell University, Ithaca, New York.

Greeson, P. E., and Williams, G. E. (1970). Characteristics of New York lakes. Part 113— gazateer of lakes, ponds, and reservoirs by drainage basins. *U.S., Geol. Surv. N.Y. State Dep. Environ. Conserv. Bull.* **68B**, 1–122.

Griffin, K. C. (1974). Alkaline phosphatase as an ecological parameter in Cayuga Lake. M.S. Thesis, Cornell University, Ithaca, New York.

Hall, D. J., and Waterman, G. C. (1967). Zooplankton of the Finger Lakes. *Limnol. Oceanogr.* **12**, 542–544.

Hamilton, D. H., Jr. (1969). Nutrient limitation of summer phytoplankton growth in Cayuga Lake. *Limnol. Oceanogr.* **14**, 579–590.

Hennick, D. G. (1973). Alewife growth rate and foraging effort in Cayuga Lake as related to zooplankton standing crop. M.S. Thesis, Cornell University, Ithaca, New York.

Henson, E. B. (1954). The profundal bottom fauna of Cayuga Lake. Ph.D. Thesis, Cornell University, Ithaca, New York.

Henson, E. B., Bradshaw, A. S., and Chandler, D. C. (1961). "The Physical Limnology of Cayuga Lake, New York," Mem. No. 378. N.Y. State Coll. Agric., Ithaca, New York.

Hess, A. D. (1940). "A Preliminary Study of the Annual Temperature Cycle in Cayuga Lake," Mimeo.

Hewitt, O. H. (1944). Waterfowl food plants and the duck population at the head of Cayuga Lake. Ph.D. Thesis, Cornell University, Ithaca, New York.

Howard, H. H. (1963). Primary production, phytoplankton, and temperature studies of Cayuga Lake, New York. Ph.D. Thesis, Cornell University, Ithaca, New York.

Hutchinson, G. E. (1967). "A Treatise on Limnology," Vol. II. Wiley, New York.

Johnson, H. (1936). The New York State flood of July 1935. *U.S., Geol. Surv., Water-Supply Pap.* **773-E,** 233-268.

Leister, C. W. (1929). Food and feeding habits of the ducks of the Cayuga Lake region. Ph.D. Thesis, Cornell University, Ithaca, New York.

Likens, G. E. (1972). "The Chemistry of Precipitation in the Central Finger Lakes Region," Tech. Rep. No. 50. Cornell Univ. Water Resour. and Mar. Sci. Cent., Ithaca, New York.

Likens, G. E. (1974a). "Water and Nutrient Budgets for Cayuga Lake, New York," Tech. Rep. No. 82. Cornell Univ. Water Resour. and Mar. Sci. Cent., Ithaca, New York.

Likens, G. E. (1974b). "The Runoff of Water and Nutrients from Watersheds to Cayuga Lake, New York," Tech. Rep. No. 81. Cornell Univ. Water Resour. and Mar. Sci. Cent., Ithaca, New York.

Liu, C., and Tedrow, A. C. (1973). Multilake river-systems operation rules. *J. Hydraul. Div., ASCE* **99** (HY9), 1369-1381.

Ludlam, S. T. (1967). Sedimentation in Cayuga Lake, New York. *Limnol. Oceanogr.* **12,** 618-632.

McKenna, J. Z. (1977). M.S. Thesis, Cornell University, Ithaca, New York (in preparation).

Mills, E. L., and Oglesby, R. T. (1971). Five trace elements and vitamin B_{12} in Cayuga Lake, New York. *Proc. Conf. Great Lakes Res.* **14,** 256-267.

Muenscher, W. C. (1927). Plankton studies of Cayuga, Seneca and Oneida Lakes. *In* "A Biological Survey of the Oswego River System." Suppl., 17th Annu. Rep. J. B. Lyon, N.Y. State Conserv. Dep., Albany.

O'Brien and Gere Consulting Engineers. (1965). "Joint Municipal Survey Committee, Greater Ithaca Regional Area: Comprehensive Sewerage Study." Syracuse, New York.

Oglesby, R. T., and Allee, D. J., eds. (1969). "Ecology of Cayuga Lake and the Proposed Bell Station (Nuclear Powered)," Publ. No. 27. Cornell Univ. Water Resour. and Mar. Sci. Cent., Ithaca, New York.

Oglesby, R. T., and Schaffner, W. R. (1977). The response of lakes to phosphorus. *In* "Nitrogen and Phosphorus—Food Production, Wastes and the Environment" (K. S. Porter, ed.), Chapter 2. Ann Arbor Sci. Publ., Ann Arbor, Michigan (in press).

Oglesby, R. T., Hamilton, L. S., Mills, E. L., and Willing, P. (1973). "Owasco Lake and its Watershed" Tech. Rep. No. 70. Cornell Univ. Water Resour. and Mar. Sci. Cent., Ithaca, New York.

Oglesby, R. T., Vogel, A., Peverly, J. H., and Johnson, R. (1976). Changes in submerged plants at the south end of Cayuga Lake following Tropical Storm Agnes. Hydrobiologia **48:**251-255.

Oglesby, R. T., Schaffner, W. R., and Mills, E. L. (1975). "Nitrogen, Phosphorus and Eutrophication in the Finger Lakes" Tech. Rep. No. 94. Cornell Univ. Water Resour. and Mar. Sci. Cent., Ithaca, New York.

Olsson, E., Karlgren, L., and Tullander, V. (1968). "Household Waste Water," Byggforskningens Rapp. No. 24. Natl. Swed. Inst. Bldg. Ros., Stockholm.

Peterson, B. J. (1971). The role of zooplankton in the phosphorus cycle of Cayuga Lake. Ph.D. Thesis, Cornell University, Ithaca, New York.

Peterson, B. J., Barlow, J. P., and Savage, A. E. (1973). "Experimental Studies on Phytoplankton Succession in Cayuga Lake," Tech. Rep. No. 71. Cornell Univ. Water Resour. and Mar. Sci. Cent., Ithaca, New York.

Peterson, B. J., Barlow, J. P., and Savage, A. E. (1974). The physiological state with respect to phosphorus of Cayuga Lake phytoplankton. *Limnol. Oceanogr.* **19**, 396–408.

Raffieux, P. (1671–1672). The Jesuit reflections and allied documents. **56**, 48–52.

Rickard, L. V., and Fisher, D. W. (1970). "Geological Map of New York. Finger Lakes Sheet." New York State Museum and Science Service, Albany, New York.

Sakamoto, M. (1967). Primary production by phytoplankton community in some Japanese lakes and its dependence on lake depth. *Arch. Hydrobiol.* **62**, 28.

Sass, D. B., ed. (1972). "An Atlas of Some Parameters of the Sedimentation, Minerology, Pesticide Retention and Topography of the Bottom of the Southern Portion of Seneca Lake (Schuyler County), New York," Environ. Stud. Contrib. No. 2 (aquatics). Alfred University Environmental Studies Program and the Aquatics Institute of the College Center of the Finger Lakes, Montour Falls, New York.

Sawyer, C. N. (1947). Fertilization of lakes by agricultural and urban drainage. *J. N. Engl. Waterworks Assoc.* **61**, 109–127.

Schaffner, W. R., and Oglesby, R. T. In press. Phosphorus loadings to lakes and some of their responses. Part I. A new calculation of phosphorus loading and its application to thirteen New York Lakes. *Limnol. Oceanogr.*

Shelton, R. L., Hardy, E. E., and Mead, C. P. (1968). "Classification, New York State Land Use and Natural Resources Inventory." Center for Aerial Photographic Studies, Cornell University, Ithaca, New York.

Singley, G. W. (1973). Distribution of heat and temperature in Cayuga Lake. Report prepared for New York State Electric and Gas Corporation by NUS Corp., Rockville, Maryland.

Stevens, T. H., and Kalter, R. J. (1970). "Technological Externalities, Outdoor Recreation, and the Regional Economic Impact of Cayuga Lake," Agricultural Economics Research 317. Dep. Agric. Econ., Cornell University, Ithaca, New York.

Stout N. J. (1958). "Atlas of forestry in New York," Bull. No. 41. State University College of Forestry, Syracuse, New York.

Sundaram, T. R., Easterbrook, C. C., Piech, K. R., and Rudinger, G. (1969). "An Investigation of the Physical Effects of Thermal Discharges into Cayuga Lake (Analytical Study)," CAL No. VT-2616-0-2. Cornell Aeronaut. Lab., Inc., Buffalo, New York.

Thompson, D. Q. (1972). Trees in history. *Cornell Plantations* **28**, 39–42.

U.S. Bureau of the Census. (1970a). "Census of Population. General Population Characteristics," Final Rep. PC(1)—B 34, New York. US Govt. Printing Office, Washington, D.C.

U.S. Bureau of the Census. (1970b). "Census of Housing. Block Statistics for Selected Areas of New York State," Final Rep. (HC(3)—163. US Govt. Printing Office, Washington, D.C.

U.S. Dept. of the Interior. (1971). "Water Resources Data for New York. Part I. Surface Water Records." U.S. Geol. Surv., Albany, New York.

Van Wazer, J. R. (1961). "Phosphorus and its Compounds." Wiley (Interscience), New York.

Vogel, A. (1973). Changes in the submerged aqatic flora at the south end of Cayuga Lake between 1929 and 1970. M.S. Thesis, Cornell University, Ithaca, New York.

Vollenweider, R. A. (1968). "The Scientific Basis of Lake and Stream Eutrophication, with Particular Reference to Phosphorus and Nitrogen as Eutrophication Factors," Tech. Rep. DAS/CSI/68. OECD, Paris. 27:1–182.

Vollenweider, R. A., Munawar, M., and Stadelmann, P. (1974). A comparative review of phytoplankton and primary production in the Laurentian Great Lakes. *J. Fish. Res. Board Can.* **31**:739–762.

von Engeln, O. D. (1961). "The Finger Lakes Region: Its Origin and Nature." Cornell Univ. Press, Ithaca, New York.

Wagner, F. E. (1927). Chemical investigations of the Oswego watershed. *In* "A Biological Survey of the Oswego River System" Suppl., 17th Annu. Rep., Chapter V. N.Y. State Conserv. Dep., Albany.

Webster, D. A., Bentley, W. G., and Galligan, J. P. (1959). Management of the lake trout fishery of Cayuga Lake, New York, with special reference to the role of hatchery fish. *N.Y., Agric. Exp. Stn., Ithaca, Mem.* **357**, 1–83.

Weigand, K. M., and Eames, A. J. (1925). The flora of the Cayuga Lake basin, New York. *N.Y., Agric. Exp. Stn., Ithaca, Mem.* **92**, 1–491.

Whitford, N. E. (1906). "History of the Canal System of the State of New York," Vol. 1, Suppl., Annu. Rep. State Eng. Surv., N.Y. State.

Wright, T. D. (1969a). Climatology. *In* "Ecology of Cayuga Lake and the Proposed Bell Station (Nuclear Powered)" (R. T. Oglesby and D. J. Allee, eds.), Publ. No. 27, Chapter IV. Cornell Univ. Water Resour. and Mar. Sci. Cent., Ithaca, New York.

Wright, T. D. (1969b). Hydrology and flushing characteristics. *In* "Ecology of Cayuga Lake and the Proposed Bell Station (Nuclear Powered)" (R. T. Oglesby and D. J. Allee, eds.), Publ. No. 27, Chapter V. Cornell Univ. Water Resour. and Mar. Sci. Cent., Ithaca, New York.

Wright, T. D. (1969c). Heat budget and water temperature. *In* "Ecology of Cayuga Lake and the Proposed Bell Station (Nuclear Powered)" (R. T. Oglesby and D. J. Allee, eds.), Publ. No. 27, Chapter VI. Cornell Univ. Water Resour. and Mar. Sci. Cent., Ithaca, New York.

Wright, T. D. (1969d). Currents and internal waves. *In* "Ecology of Cayuga Lake and the Proposed Bell Station (Nuclear Powered)" (R. T. Oglesby and D. J. Allee, eds.), Publ. No. 27, Chapter VII. Cornell Univ. Water Resour. and Mar. Sci. Cent., Ithaca, New York.

Wright, T. D. (1969e). Conductivity. *In* "Ecology of Cayuga Lake and the Proposed Bell Station (Nuclear Powered)" (R. T. Oglesby and D. J. Allee, eds.), Publ. No. 27, Chapter VIII. Cornell Univ. Water Resour. and Mar. Sci. Cent., Ithaca, New York.

Wright, T. D. (1969f). Transparency. *In* "Ecology of Cayuga Lake and the Proposed Bell Station (Nuclear Powered)" (R. T. Oglesby and D. J. Allee, eds.), Publ. No. 27, Chapter IX. Cornell Univ. Water Resour. and Mar. Sci. Cent., Ithaca, New York.

Wright, T. D. (1969g). Chemical limnology. Hypolimnetic oxygen. *In* "Ecology of Cayuga Lake and the Proposed Bell Station (Nuclear Powered)" (R. T. Oglesby and D. J. Allee, eds.), Publ. No. 27, Chapters X and XI. Cornell Univ. Water Resour. and Mar. Sci. Cent., Ithaca, New York.

Wright, T. D. (1969h). Ionic composition and trace elements. *In* "Ecology of Cayuga Lake and the Proposed Bell Station (Nuclear Powered)" (R. T. Oglesby and D. J. Allee, eds.), Publ. No. 27, Chapter XII. Cornell Univ. Water Resour. and Mar. Sci. Cent., Ithaca, New York.

Wright, T. D. (1969i). Plant nutrients. *In* "Ecology of Cayuga Lake and the Proposed Bell Station (Nuclear Powered)" (R. T. Oglesby and D. J. Allee, eds.), Publ. No. 27, Chapter XIII. Cornell Univ. Water Resour. and Mar. Sci. Cent., Ithaca, New York.

Wright, T. D. (1969j). Alkalinity and pH. *In* "Ecology of Cayuga Lake and the Proposed Bell Station (Nuclear Powered)" (R. T. Oglesby and D. J. Allee, eds.), Publ. No. 27, Chapter XIV. Cornell Univ. Water Resour. and Mar. Sci. Cent., Ithaca, New York.

Wright, T. D. (1969k). Plant pigments (chlorophyll *a* and phaeophytin). *In* "Ecology of

Cayuga Lake and the Proposed Bell Station (Nuclear Powered)" (R. T. Oglesby and D. J. Allee, eds.), Publ. No. 27, Chapter XV. Cornell Univ. Water Resour. and Mar. Sci. Cent., Ithaca, New York.

Youngs, W. D. (1969). Fish and other biota. *In* "Ecology of Cayuga Lake and the Proposed Bell Station (Nuclear Powered)" (R. T. Oglesby and D. J. Allee, eds.), Publ. No. 27, Chapter XVIII. Cornell Univ. Water Resour. and Mar. Sci. Cent., Ithaca, New York.

Youngs, W. D. (1972). An estimation of lamprey-induced mortality in a lake trout population. Ph.D. Thesis, Cornell University, Ithaca, New York.

Youngs, W. D., and Oglesby, R. T. (1972). Cayuga Lake: Effects of exploitations and introductions on the salmonid community. *J. Fish. Res. Board Can.* **29**, 787–794.

The Limnology of Conesus Lake

Herman S. Forest, Jean Q. Wade, and Tracy F. Maxwell

INTRODUCTION

Conesus Lake in Livingston County is the westernmost of typical Finger Lakes. The Finger Lakes have a similar origin in deeply scoured glacial valleys with thick unconsolidated sediments overlying the bedrock. The lake is 12.6 km long, and slightly over 1 km wide in most places. The shape is almost cylindrical, with the long axis tilting slightly toward the northeast. The waist is constricted by stream deltas at Long Point and McPherson's Point, separating north and south basins. Typically for Finger Lakes, the deepest portion (about 20 m) is in the south basin.

Four Towns (Geneseo, Livonia, Conesus, and Groveland) share the shoreline. The Village of Livonia, 3 km east, is the largest urban concentration in the watershed, but the hamlet of Lakeville surrounds the outlet of the lake, and virtually the entire shoreline is developed as residential area. In the watershed as a whole, about half the area is in active agriculture and a third is forested. The portion allotted to these uses has remained essentially constant for a century or more.

The municipal water supply is withdrawn by the Villages of Geneseo and Avon and distributed elsewhere as well. The lake is used both for summer and winter fishing, for boating, and for swimming. Its waters are generally rather clear in comparison to other fertile lakes, although heavy and prolonged blooms of blue-green algae do occur, and the crop of submerged aquatic vascular plants (macrophytes or "weeds") is one of the most luxuriant known in the region. Concern about water quality by residents led to establishment of inspection for disposal systems as early as 1925, and in 1969 bonds were voted to build New York's first complete perimeter sewer system for a major lake. Effluent from the Village of Livonia and all waste in the immediate vicinity of the lake was diverted from the lake basin in 1973 and treated in a new facility before discharge into Conesus Outlet Creek. Thus, Conesus Lake has become a large-scale laboratory experiment; here the altered ecosystem can be monitored toward understanding of long range changes and stability.

HISTORY

The Conesus Lake basin shares the rich history of western New York. Recently, a competent local history of Conesus Lake was published (Anderson, 1976). The first settlers spoke of Indian burial mounds when they occupied the Genesee Valley in the late eighteenth century. According to Ritchie (1965), these provided perhaps the first concrete evidence of a definite Hopewellian (southern Ohio) linkage for the New York burial mounds. The mounds place the time of the first Indian settlement in this area around Christian Era 310 ± 100. The Senecas later dominated western New York after a long succession of predecessors had either drifted into Canada or had been dispersed.

Etienne Brule, a French scout, was the first European to explore the Genesee country on or about the year 1610 (McIntosh, 1877). According to Doty (1925), there was a Seneca village of the pre-Revolutionary War era located near Lakeville, at the outlet of Conesus Lake. Although the Indian villages had disappeared by the early nineteenth century, the Indians were frequent visitors until at least 1860. Indian families still live nearby in 1976.

The first major intrusion by white men was by American revolutionary forces. In the spring of 1779, General John Sullivan, in command of 5000 men from the American army, marched from Pennsylvania into New York under the order from General Washington to destroy Fort Niagara, the stronghold of the British who were supplying the Indians in their attacks on settlers. The Sullivan expedition eventually returned to Pennsylvania, but the lush valley where grass grew 7–8 ft high and the groves of beautiful trees and open parklike places were a pleasure to the eye which could not be forgotten, and some soldiers returned to build homesteads.

The Sullivan expedition was important because it opened up western New York to the white settlers who could move in without fear of the Indians. Settlement began soon after the end of the Revolutionary War, in 1790, when the present towns, villages, and hamlets began to take shape. In September 1797, a treaty was signed at Geneseo with the Senecas, who still claimed the valley for their own. The Indians ceded all lands in this county (Livingston) to the whites except for some small reservations (French, 1860).

The necessity for clearing land was turned to a profitable industry. Ashes scraped off an acre after a good burn were worth $4.00 to $8.00. The making of potash began in 1796 and it became a staple product during early settlement. The ashes from elm, beech, and maple returned almost enough cash to pay for clearing. Curiously, the addition of larger amounts of potash fertilizer to farms is a significant trend in the 1970's.

Even though the nation's "bread basket" moved across the Mississippi before the Civil War, 55 rail cars of wheat are still shipped yearly from the Village of Livonia to cereal mills. The Livonia, Avon, and Lakeville Railroad with steam-driven locomotive carries 12,000 people a year during the June through October tourist season. With diesel locomotives, it freights not only wheat but also tank cars of corn syrup and cane sugar to and from the Western New York Syrup Corp., an industry established near Lakeville in 1971. Few significant processing industries have survived long in the basin, and no large-scale ones have ever existed there. According to the Gazetteer of New York State by French (1860), the Town of Conesus contained 13,455.5 acres of improved land, and 6889.5 acres of unimproved land. The Livonia Township had 19,444.2 improved land and 3882.5 unimproved. Thus, farming was already at a maximum in extent over a century ago. Winter and spring wheat formed the bulk of the crops sent to the commercial market in the mid-nineteenth century.

Lumbering in New York, as distinguished from land clearing, was initiated in 1867, but a half-million staves and some square timber were exported as early as 1819. Cooperage still flourished in Livonia in 1877. By 1887 the black walnut groves were greatly diminished. Indeed, the general peak of lumbering in New York was between 1880 and 1890. In its wake, very few uncut woods were left, even in the steep ravines leading into the lake from the south end and southern sides. Lumbering continued only as a farm woodlot enterprise. A unique industry operated from 1950 to 1965 high in the watershed valley between Webster's Crossing and Springwater. Here, an extensive peat deposit was stripped to an estimated extreme depth of 2–3 m. The pits were abandoned, but a small remnant remains of the rare bog forest which once covered the site.

Ice cutting on the lake, with storage at Lakeville, was conducted on a small but brisk commercial scale from 1909 until 1950. Ice cutting left a significant indirect mark on the lake since a road to the ice house on Fisherman's Point (later, Sand Point), near the outlet, contributed to the formation of a marsh there. The point had begun as an artificial peninsula on which a railroad siding had been laid to the hotels and steamer docks at the end. The marsh enclosure occurred 20 to 25 years later. Into the marsh discharged the untreated waste from a milk processing plant located in Lakeville for almost 70 years before 1969. It was noted as a conspicuous polluter in 1926. In addition to these extended activities, others have come and gone rapidly; for example, mass chicken production was attempted near Conesus in the late 1960's and persisted to 1975, with sporadic operation and a series of owners.

The two most significant activities for over 150 years have been farming and residential–recreational use. Although approximately half of the area is still agricultural, land in cultivation has decreased steadily in the second half of the twentieth century while total productivity has increased. Summer cottages near the shore were built in the nineteenth century, 35–40 having been counted in 1872, and a similar number at Lakeville. At least ten excursion steamers ran the lake from 1888 to 1900, and one fairly large motor launch was in service from Long Point as late as 1968.

The automobile era was largely responsible for surrounding the lake with private lots and residences. Building occurred chiefly between 1920 and 1950, when over 1100 houses could be counted. Twenty-five years later the number exceeded 1800. There was more building both near the shores and on lots without lake frontage. During the last two decades there has been a major shift to year-round residence, with recreation as a secondary consideration. The initial impetus was a housing shortage immediately following the end of World War II. The pattern of rural residence and automobile commuting over some distance became generally accepted even when the housing supply improved. The era of the trailer, with an ever increasing number of mobile homes up to the present time, follows this pattern. The lake perimeter residents commute to Geneseo, Avon, Rochester, and elsewhere.

The lake has supplied municipal water for Avon since 1888, and withdrawal for Geneseo began in the late 1890's. Lakeville and part of the lake perimeter subsequently were added to the Avon intake. An inspector for private disposal systems was at work in 1925, and the watershed sanitary inspection district was established about 1940. However, the rapid development of the shoreline which followed convinced residents that a comprehensive collection and treatment of wastewater was needed to protect the lake's water quality. In 1969, a county district from the four towns of Geneseo, Groveland, Conesus, Livonia, and Livonia Village voted for the required bond issue, and an advanced treatment plant was built on the Conesus Outlet, about 4 km north of the lake.* An interceptor sewer now surrounds Conesus Lake, receiving wastewater from Livonia Village and the near-shore residence areas. The system, which began operation in 1970–1973, is unique in New York. No other major residential lake wears such a protective ring. Figure 1 is a map of the watershed showing principal streams, cultural features, and other locations.

---

* A popular account of the episode was written by Brown (1970).

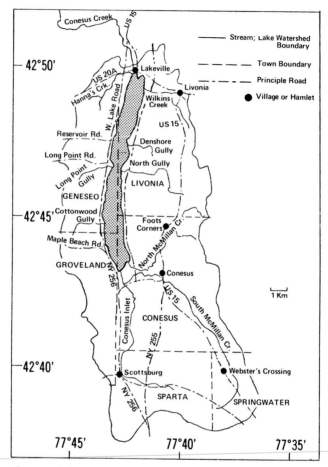

Fig. 1. Conesus Lake watershed showing principal streams, cultural features, and locations. [On the U.S.G.S.Livonia quadrangle map (1942) Densmore Gully was printed as "Denshore" and subsequent reports have repeated the error.]

DRAINAGE BASIN

Geology-Morphology-Soils

The area now containing the Conesus Lake basin area was an inland sea during the early Paleozoic era. Only Devonian rocks are now present, but there probably occurred deposition during Cambrian, Devonian, and Silurian periods as well, since these appear in the general region. Accumulation of sediments and later compaction into various rock strata occurred. The strata now dip to the south at 1° or 17 m/km. Figure 2 is a map of bed-

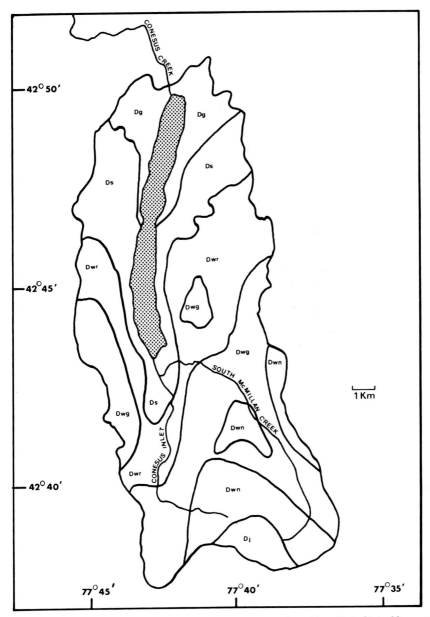

Fig. 2. Bed rock geology of the Conesus Lake basin from New York State Museum and Science Service Map and Chart Series, No. 15, 1971. All Middle Devonian, listed youngest to oldest. (1) Dj, (Java group) Hanover shale; (2) Dwn, Nunda formation; (3) Dwg, West Hill formation—Gardeau formation, Roricks Glen—Shale, upper Beers Hill shale, Grimes siltstone; (4) Dwr, West Falls group—lower Beers Hill, Dunn Hill shale, Millport shale, and Moreland shale; (5) Ds, Sonyea group—Cashaqua shale; (6) Dg, West River shale—Genesee group and Tully limestone.

rock geology. In the late Paleozoic, only a small portion of southern New York was submerged enough to leave strata, and during the Mesozoic, there was a general uplift and erosion. Present features of the bedrock morphology were formed only during the last 60 million years, the Cenozoic era. Glaciation within the last million years has been the major influence in shaping surface features, leaving only a thin cover of unconsolidated glacial sediments of the Wisconsonian age. The bed of Conesus Lake, however, is a very thick layer of unconsolidated sediments, and both inlet and outlet valleys are similarly filled to constitute a sizable underground reservoir and drainage route for water.

During the last advance and retreat of the Ontario lobe of the Laurentian ice sheet, the Conesus basin, a former stream valley, was scoured and deepened. The material picked up by the advancing glacier was deposited as the ice mass began to melt and retreat, and glacial outwash deposits were added to the older glacial sediments. Fairchild (1928) is the standard source for the general geology of the region, and von Engeln (1961) gives a popular account which is highly informative though not authoritative of the origin of the Finger Lakes. The detail of local glacial geology has never been mapped adequately.

Typically, the area is composed of sediment-filled U-shaped glaciated river valleys. The Conesus basin, however, is fairly flat to the north, becoming steeper and higher toward the middle of the lake. Till and outwash deposits cover the north, but bedrock emerges southward at the higher ground. Relief alongside the lake reaches 360 m, and the highest elevation of over 550 m is located at the southern edge of the basin, along the axis of the divide between the headwaters of Conesus Inlet and South McMillan Creek.

The soils of the basin are extremely variable and contain more than 75% of the soil types identified in the survey of Livingston County [U.S. Department of Agriculture (USDA), 1956]; Stout (1970) wrote the following summary:

> In the northern third of the Conesus watershed, deep, well-drained, high to medium lime soils from glacial till on nearly level to gently sloping topography predominate. In the middle third, there is intermingling of deep, well-to-poorly drained, medium to low lime soils on hilly or moderately steep topography. At the southern end are found generally deep, poorly drained, low lime soils on nearly level to moderately steep topography. The soils included strongly acid types and hardpan.

The source of lime is probably the Onondaga limestone, which crops out north of the lake. The soil survey (USDA, 1956) classifies the soils according to the use and management of the various soils groups on the basis of their relative suitability for crops and pasture. Five classes and 24 management groups are classified. For this writing only the five major classes need be considered. The best soils are classified as Class 1 soils, the poorest Class 5, with Classes 1, 2, and 3 suited for agriculture. Class 4 soil is poor for crops, but usable for pasture. Class 5 soil is poor for crops and pasture and is recommended for forest.

The largest portion of soils in the Conesus basin were judged as Classes 1–3. Class 4 soils, mainly the Allis silt loam types, are located on or near U.S. Route 15, south of Conesus and east of Scottsburg. Class 5 soils are the undifferentiated alluvial soils (Ab), steep ledgey land, with 25–60% slope (Sp), steep Lansing, Ontario, and Honeoye soils, with 30–45% slope (So), and steep Wooster, Valois, and Bath soils, with 25–40% slope (St). Since only 15% of New York is covered with prime agricultural soils, the Conesus basin has a high proportion. Yet most of the areas classed as highly viability farmland are located at the sides of the watershed, in its northern half: they are the edges of outside agricultural areas. Figure 3 is a simplified soil map of the watershed.

Population

The population is concentrated in villages, hamlets, and around the perimeter of the lake. Stout (1970) noted that only 2.1% of the drainage basin was used for residential development. The Child *et al.* (1971) determination was 2.3% or 3.57 km². Most of the total, 2.93 km², is adjacent to the lake. Villages and hamlets include Livonia, Conesus, Scottsburg, South Livonia, Webster's Crossing, Union Corners, and Lakeville.

Boyle (1974) counted over 1800 residences connected to the Conesus Lake sewer district, and estimated that up to 10% more had been constructed in the 2 years since the district map had been prepared. The estimate may be high, but only 3 years previously (1970), Stout had counted about 1600. Additional variation of over 100 residences can occur due to the particular fringe areas chosen for inclusion.

A mail survey conducted by Stout revealed a surprisingly low annual average occupancy among lake residences, 1.78 persons/residence when weighted for all months. Residents and officials believe that there are from 2500 to 3000 permanent residents near the lake, with the number increasing to approximately 6000 during the summer months. Boyle (1974) also accepted local estimates that population near the lake may swell to upward of 10,000 individuals during weekends and holidays.

The total population of the drainage basin is difficult to determine because it overlaps parts of six townships; Conesus, Geneseo, Groveland, Livonia, Springwater, and Sparta. Except for Conesus and Livonia, the principal population centers of the towns are outside the drainage basin. According to the 1970 census, the total population of the townships, except-

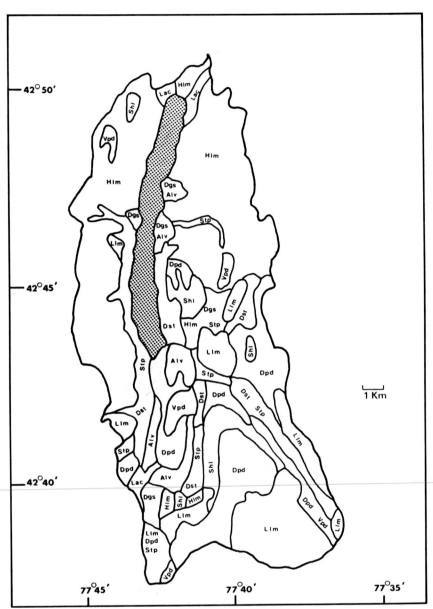

Fig. 3. Soils of the Conesus Lake basin. Categories adapted from color groupings in Soil Survey, Livingston County, New York (U.S. Department of Agriculture, 1956). Hlm, medium- and high lime soils from glacial till; Llm, low lime soils from glacial till and outwash; Lac, Soils from fine-textured lacustrine deposits; Dgs, Deep well-drained soils from gravel and sand deposits; Dpd, Deep poorly drained soils; Dst, Deep well- to poorly drained soils on hilly or moderately steep topography; Vpf, Deep very poorly drained soils; Alv, Deep well- and moderately well-drained soils from recent alluvium; Shl, Shallow well- to poorly drained soils; Stp, Steep soils.

ing Springwater, was 18,276 individuals [U.S. Department of Commerce (USDC), 1972a]. In the watershed, Mills (1975) used 4180 individuals to compute phosphorus loading, and Stewart and Markello (1974) accepted a population of 5697.

Natural and Cultural Characteristics—Land Use

Both Stout (1970) and Child et al. (1971) used the same raw data obtained from aerial photography (1967–1969) to obtain gross summaries of land uses. Child meaningfully divided the watershed into an "Inlet" portion (south of the lake) and "North" (adjacent to the lake). Both authors agree that about 50% is active agricultural land, another 10% inactive, and that forest covers about 30%; Conesus Lake itself is 8% of the watershed (here, 7% is accepted). All other uses, including the critical residential use, cover 2–3% of the area. Figure 4 is a land use map of the watershed.

There are disparities among almost all measurements of geography and usage, but most of these are not of great magnitude. There is a sizable spread in measurements of the total areas themselves. Part of the differences can be accounted for by choice of boundaries. The greatest addition is the inclusion of the watershed of Conesus Outlet Creek, which adds a third to the total (e.g., Berg, 1963). Excluding this area, the figures accepted by most authors are close: New York State Department of Health (NYSDH) (1961) basin survey (corrected for location below outlet), 180 km^2; Stout, 185 km^2; Stewart and Markello (1974) 180.52 km^2. The estimate of Child is rather low, 158.45 km^2, and the estimate of Mills (1975) is quite high, 231 km^2. The high figure is probably derived from the 89 square miles listed by Berg, but this actually includes the watershed of the outlet stream. The figure of 180.5 km^2 is accepted here for the lake and its input streams.

Land Use Changes

Residential. The study by Stout (1970) was specifically directed at land use changes from 1930 to 1970. It utilized photographs of the Federal Agricultural Stabilization and Soil Conservation Service from 1938, 1942, 1954, and 1963, and also the County census of agriculture (Bond, 1947), and Bratton (1954, 1964). The Agricultural Census (USDC, 1973) was not available at the time, and 1974 data were published afterward. The State's annual agricultural statistical report by the New York Department of Agriculture and Markets (NYSDAM, 1974) contains no local data. Stout also conducted a survey by interview of 35 farms totaling 40 km^2, or about one-quarter of the total watershed. She assembled data on crops, fertilizer, and herbicide use, and change in practice. In a mail survey, occupancy of

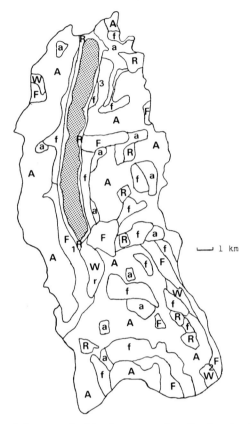

Fig. 4. Conesus Lake watershed land use map. A, active farmland; a, inactive farmland; F, forest over 30'; f, forest predominantly under 30'; R, residential or public; r, wildlife refuge; W, wetland. Notable natural communities: 1, mesic ravine; 2, bog forest at edge of peat excavations; 3, upland arbor vitae.

residences on the perimeter of the lake was determined for 1969. Land use data from the aerial survey for 185 cells of 1 km² each were plotted by LUNR (land use–natural resources) categories on a 1:24,000 scale to correspond with USGS quadrangle maps, 7.5-min series. Thus, a thorough investigation of land use has been accomplished for the watershed. The investigator noted the limitations of the different sources with sensitivity. The most notable change over time was the increased number of residences around the lake. This constituted virtually the total addition to population in the watershed. When weighted for increased occupancy, following World War II, the population during 1930–1970 increased fivefold to its present level of more than 5000. The trend toward occupancy of former farm houses (as well as lake cottages) as residences was noted by Stout, but the amount is a small part of the total population. Residence construction has continued since 1970, an increase of 10% being indicated by 1975 with very few cottages abandoned. Mobile homes have been used commonly since 1950. These were placed densely along the filled shore at the lake outlet, distributed on vacant lakeshore lots, and scattered throughout the watershed, individually or in concentrations, such as at Webster's Crossing. The Village of Livonia has had sewers and a wastewater treatment plant since 1937, but other watershed residences used individual disposal systems until the recent construction of the Conesus Lake sewer district. The Livonia system was diverted to the new facility on Conesus Creek on October 23, 1972, and within a year all individual disposal systems around the lake had been connected to the new perimeter sewer.

Agricultural. Although residential and recreational use may be the chief environmental factors influencing the condition of the lake, the major area of the watershed is agricultural, and agricultural land use can be documented in some detail. In comparison to residential use, the level of use for agriculture has been extraordinarily stable. In 1969, the average farm had been held by the owning family for 40 years, and principal products of today are recognizable from 1860. Even the amount of agricultural land has not changed markedly. Stout (1970) noted the difficulty in demonstrating any sizable changes in area, although farms had been consolidated into larger holdings by 1970. Boyle (1974) found that a balance was still being maintained between areas put into and taken out of cultivation. County agricultural data (NYSDAM, 1974) substantiate the conclusions drawn from the basin. Consolidation has reversed slightly: more farms with smaller acreage are indicated—the direction is contrary to the state trend. Meanwhile, the increase in farm value in Livingston County significantly exceeds the state gain. Apparently, here there are buyers who find farming attractive as investments or as a style of living.

French (1860) listed half of the Town of Conesus as "improved," which would not be far from the portion in residence and agriculture today. Almost 75,000 bushels of wheat flowed from its fields, and there was hay, potatoes, apples, tobacco in some quantity, corn for silage, and some eating corn. Dairy farming, in terms of butter and cheese rather than fresh milk, was certainly important. In a rough economic equivalency, dairying might have constituted 25% of the total produce. In the years which followed, dairy farming became the principal agricultural activity, but it has begun to wane during the past 20 years. The dairies at Webster's Crossing and Lakeville have been abandoned. Wheat has remained important, and corn and oats are also significant grains. There are crops of buckwheat from time to time, but data are lacking. Cultivation of potatoes has virtually disappeared, although potatoes were grown in numbers by two farms in 1969. In addition to the variety of grains, as has been noted, dry beans are a significant addition to the nineteenth century crops. The area has also shared the general shift in hay content from grass to legumes, as improved pasture practices grew.

Stout concluded that specialization was increasing, and holdings were being consolidated by renting as well as purchase. With the decline of "family" farming and increased "business" orientation, added use of commercial fertilizer was a notable trend. Particularly, the use of high-analysis fertilizer, such as N10-P20-K20, had increased. Stout calculated applications by crop, area, and totals for nitrogen, phosphorus, and potassium. The trend continued in New York until 1974 (latest available statistics) with the rise in the application of potash being quite sharp. However, 1974 was probably the breakpoint when added fertilizer could no longer be translated into cash profit, according to Livingston County's cooperative agricultural extension agent. Stout also documented that chemical herbicides were applied according with general practice in the state. The use of 2,4-D has dropped to a low level since 1970, however. The second most significant shift during the past 5 years has been the introduction of biological control for alfalfa weevil with a corresponding reduction of chemical control.

Stability of Forest Area

Forests cover the second largest area in the watershed. What is obvious from a field vegetational survey is surprisingly difficult to demonstrate quantitatively: secondary succession of cleared areas toward woodland is occurring in several areas. The increases which are visually significant probably amount to no more than 5% of the watershed, and they are simply lost in the large-scale analyses. Stout (1970) compared overlays of aerial photographs from 1938–1963, and could not measure a significant increase in forests. Planimeter measurement of 1954 photographs gave 27.9% forest. However, the margin of error is about 5%, consequently the figure approaches the 1969 total of 28.4% (Stout, 1970) or 34.7% (summaries of Child *et al.*, 1971). Since Stout and Child used essentially the same computerized aerial survey, the difference in their figures might be explained by the choice of slightly different watershed boundaries.

Floral Composition

Examination in detail does reveal distinct patterns in the wooded area. The summaries of Child *et al.* (1971) show the portion of forest in the Inlet portion as roughly double that of the North. The proportion of active agriculture is much more even, but somewhat higher in the North. Stout (1970) calculated that the proportions of brushland and mature forest were approximately equal, with perhaps 3% in plantations. Ground surveys undertaken in 1975 revealed no managed plantations, and most can be considered now as brushland of limited duration. The relationship between agriculture and forest has been affected by a series of federal programs which subsidized the withholding of land from cultivation. Livingston County contained the highest portion of conserved agricultural acreage of any New York county in 1960, actually 14% of its total. The prime lands in the County were cultivated again when the programs were abandoned a decade or so later, but some areas in the Conesus basin were not profitable, and they have simply been left to secondary succession. The pattern of woodlands in the basin is both complex and intriguing. No study of the principal communities, except for the wet woods at the lake inlet, had been conducted before 1975. A survey of communities and both terrestrial and aquatic successional patterns was undertaken to obtain information for this writing, and results will be reported elsewhere (Forest *et al*, 1977a).

DeLaubenfels (1966a,b) mapped two forest zones in Livingston County, the oak–northern hardwood, and the elm–red maple–northern hardwood. The map scale is too small for precise placement, but Conesus Lake seems to be located in the former zone near the latter, which generally covers the Lake Ontario plain. The fundamental environmental factor recognized as separating the two is soil drainage. While these categories may have some validity in distinguishing state-wide forest zones, they are of no value in application to the limited area of the Conesus watershed.

The survey of 1975 revealed that an oak–sugar maple mixture is almost universal in mature woods on the highest plateaus and slopes of the watershed. Oaks include *Quercus alba* L.(white), *Q. velutina* Lam. (black), and *Q. rubra* L. (red). *Quercus rubra* is more prevalent in mesic soils, the two others inhabiting drier areas. Hickories [*Carya ovata* (Mill.) K. Koch (shagbark) and *C. ovalis* (Wang.) Sarg. (false shagbark)] are also present. The ravines entering the valley or the lake from about 300 m elevation downward contain oak woods, with some hickory, on the more exposed, generally higher and steeper slopes, and grade to sugar maple below the crests and into the bottoms, where basswood appears. Locally, beech (*Fagus grandifolia* Ehrh.) accompanies the maple. The more moist and perhaps less disturbed ravines grade into the hardwood–conifer forest, with hemlock [*Tsuga canadensis* (L.) Carr.] most common, and white pine (*Pinus strobus* L.) secondary. Such mesic ravines are restricted to the inlet portion of the watershed except for an extension adjacent to the southwest shore of the lake, and some reaches of North McMillan Creek on the east side of the lake. The inlet area, extending southward at least 4 km, contains an exemplary forest dominated by swamp maple, with some ash (*Fraxinus americana* L.) and other species. The swamp maple has been variously referred to as either red maple (*Acer rubrum* L.) or silver maple (*A. saccharinum* L.), but it is a hybrid population of the two species. Swamp maple is widely distributed in wet sites throughout the glaciated northeast and southward along the Appalachian Mountains. Kelly (1973) assigned the provisional name of *Acer x freemanii* E. Murray to the species, but its common name of swamp maple serves to distinguish it satisfactorily. Silver maple does not grow in New York except where introduced. Red maple is ecologically distinct as a tree of drier hillsides or more acid bogs but it intergrades with swamp maple. Swamp maple is common locally wherever soil is saturated well into the growing season. Elsewhere in western New York, the species lines the banks of rivers, and it was probably a major component of the original lakeside woods around Conesus. Basswood (*Tilia americana* L.) and walnut [*Juglans nigra* L. (black) and *J. cinerea* L. (white)] also must have been important there, in addition to maple, elm, and ash. Most of the natural vegetation near the lake has been replaced by stone, lawns, planted sugar maples, or various exotic plantings, but evidence suggests the original composition. Fortuitously, a unique stand of patriarchal burr oaks (*Q. macrocarpa* Michx.) has been allowed to persist on the stream delta which pinches the middle of the lake from the west side. Long Point has been an amusement park, bathing beach, and resort area throughout the twentieth century under their canopy. McPherson's Point, on the opposite side, may once have been similar, but it now has only cottonwoods and a few swamp maples.

Three highly extraordinary natural communities were identified in the 1975 field studies. High in the valley of South McMillan Creek, near Springwater, are the abandoned peat pits which have been discussed elsewhere. At their northeast corner is a small undisturbed bog wood. The species include not only larch [*Larix laricina* (DuRoi) Koch], white pine, and hemlock, but also balsam fir [*Abies balsama* (L.) Mill.], which is a "north woods" species generally associated with the high altitudes in the Adirondacks and Tug Hill Upland. The balsam fir persists in an area where the summer temperatures would ordinarily be considered too warm for survival. Red maple, *Sphagnum* moss, and some herbaceous plants also add to the uniqueness of the community. The soil conditions are evidently ideal. The second community is also a remarkable relict of wet, cold, postglacial times. It is located adjoining VanZandt Road, only 0.4 km east of the lake and 3 km from the outlet, where the country is gently sloping and mostly open. This is a virtually pure stand of arbor vitae (*Thuja occidentalis* L.), maintained by cold water running in braided streams at or near the surface. There are only a few series of cold-water bogs in western New York, where this species grows often with larch. To find a pure undisturbed stand on a hillside, south of its usual range in the region, was startling. The third notable woodland community is a ravine system almost adjacent to the southwest corner of the lake. This is a mesic community with the mixture of deciduous and evergreen trees. There are very few large trees, indicating that the area was heavily logged within 75–100 years, its stage being young maturity. The variety of tree species is highly unusual, with good specimens of flowering dogwood (*Cornus florida* L.), Sassafras (*S. albidum* (Nutt.), tulip (*Liriodendron tulipifera* L.), and American chestnut [*Castanea dentata* (Marsh.) Berkh.]; the chestnut being about 15 cm. in diameter and bearing viable seed. Other chestnut sites occur southward to the edge of the watershed. Several species of ferns grow here, and at least four species of *Lycopodium* (club moss, running cedar, ground pine) can be found intermingled. No other site such as this is known in the state. Near the stream in the main ravine grows a variety of bryophytes, including a number of unusual species (Specimens in herbarium, SUNY College at Geneseo).

The largest area of wet succession developed adjacent to the inlet. The swamp woods characterized by maple and ash have already been identified as extending some distance along the inlet to the south. To the east of the inlet a "textbook" transition of communities developed, from submerged aquatic plants, through a floating-leaf zone, emergents, to saturated muck with the liverwort (*Riccia fluitans* L.), forget-me-not (*Myosotis scorpiodes* L.) cattails (*Typha* sp.), sensitive fern (*Onoclea sensibilis* L.), and a "weedy" zone into the wet woodland. Much of the wetland was destroyed between 1951 and 1967 in the development of the DaCola Shores residential area. After 1967, the state exercised regulatory jurisdiction and defended the shallow water and immediate shore area below the legal mean high water mark (819.57 ft or 249.8 m). However, it was possible to cut timber, fill, or build channels. Consequently, no undisturbed area remains, although there is about 30–40 m of unfilled frontage. The destruction adjacent to the inlet stream was complete on the west side with the construction of a system of channels or lagoons between filled areas. Five years after the action began, the area has been developed no further. Typical shallow water or pond succession is occurring in the lagoons, with heavy blooms of algae, and establishment of aquatic vascular plants.

Land Use and Animal Communities—Recreational Prospects

Although agriculture and residence may continue as the most well-defined land uses, recreational use poses more problems, and suggests the need for far more information. The fishery of the watershed will be considered subsequently in its due importance, but birds and mammals also constitute recreational resources. Of the latter, little is known. Waterfowl have received some attention by local residents and under state game management and refuge programs, but no study had been made of upland birds or mammals. Such studies were undertaken in 1975 and will be reported elsewhere (Forest *et al.*, 1977b). These limited investigations indicate that the variety of plant communities is complemented by animal variety. In comparison with almost any other basin of a Finger Lake, the watershed of Conesus may be demonstrated as being one of the most productive of amount and kind.

The most recent bird count, conducted by the Rochester Birding Association on December 28, 1975, illustrates the richness of Conesus biota. In a region including both Hemlock and Honeoye Lakes, the Conesus report included 76 species of the 85 total for the whole region. Among waterfowl was a record high number (900) of canvasbacks; *Aythya valisineria* (Wilson); common loons; *Gayia immer* (Brunnich); great blue herons; *Ardea herodias* (L.); whistling swans; *Olor columbianus* (Ord.); Canada geese, *Branta canadensis* (L.); ring-necked ducks; *Aythya collaris* (Donovan); buffleheads, *Bucephala albeola* (L.); oldsquaws; *Clangula hyemalis* (L.); ruddy ducks; *Oxyura jamaicensis* (Gmelin); common mergansers, *Mergus merganser* (L.); and killdeer, *Charadrius vociferus* L.

At the time of the statistical summaries on land use by Stout (1970) and by Child *et al.* (1971), very little public land was held in the basin, and indeed, there was very little public access to the lake itself. Even now, useful rights-of-way along streams are posted as private, and the single public boat launching site is now clearly inadequate for access to all areas. There are also some points of informal entry, however. A recent purchase by the state has greatly increased the public land holdings. The entire swampy wooded area along the inlet creek, south of the lake, is now managed by the New York State Department of Environmental Conservation. The total, 8 km², is about 5% of the watershed.

Future predictions or even projections for land use in the basin are too conjectural for inclusion here. It is highly probable that an express highway (Interstate 390, or the Genesee Expressway) will be built parallel to the lake, perhaps 5 km westward. Fears have been expressed that residential and recreational use of the lake and its shores will come under even heavier pressure than at present. Boyle (1974) discussed such developments in a study prepared for Livingston County on a background of present land use and the regulatory laws of the individual towns. There are unfortunately little substantial data of any kind to consider in trying to project the amount, kind, and location of developments. Of course, events in Conesus basin are somewhat subject to economic conditions in the Rochester region, New York, and the nation.

Socioeconomic Status

The United States government census provide the only substantial and readily available statistics on socioeconomic criteria for Livingston County, but their precise application to the Conesus Lake basin is not possible because political subdivisions do not conform to natural boundaries. Some generalities are suggested which can be judged against the experience of residents, realtors, community leaders, professional planners, and ecological investigators. In the census (USDC, 1972b), Livingston County upholds the general impression of being above average economically for the state. Its median family income of $10,520 was well below that of Monroe County (metropolitan Rochester), which is close to the highest in the state at $12,423. However, Livingston at least equaled the metropolitan counties in the Buffalo–Niagara area, was above adjoining Genesee County (with the City of Batavia) to the northwest, and was substantially above Cortland County at $9142, and two counties with similar populations in the depressed Mohawk Valley: Montgomery at $9006 and Fulton at $8653. The poverty levels reported for all these counties are inversely related to family income, except that Genesee County was a bit lower than Livingston (Table 124 of the census reference).

Within the basin, Livonia possesses several of the attributes of a small agricultural trade center, including grain storage facilities, and an active group of shareholders in AGWAY (the state-wide, farm-based cooperative). Also, a dairy farmer was mayor for several years. Lakeville, at the lake outlet, had the milk processing plant at one time, but it became much more of a vacation and day recreation service center, as well as a travel stop. Instability in enterprises, such as steamboat excursions and boat liveries, is evident from neglect, abandonment, and new building. However, two industries are still located there, a foundry and a recently established syrup preparation plant.

A small group of older residences has persisted, and the growth of cottages and three mobile home parks have merged into the hamlet. Of the other centers, Conesus Village is static with some attrition and a fringe of modest to substandard growth; it is now essentially a residential area. Scottsburg is smaller and similar. Webster's Crossing was aptly characterized by a field assistant as "Appalachia." Conesus basin does have poverty pockets.

The present socioeconomic structure of the residents can be understood to some extent in historical context. A century ago there were about 35 houses by the lake and a similar number in Lakeville, which was still agricultural in character with the interest in travel service, but not in residential vacationing to a major extent. The lake houses belonged to wealthy land owners of Livingston County and Rochester friends. Indeed, some of the large holdings of that time have not been completely dispersed. Conesus has long since ceased to be a lake for the wealthy, but status is suggested in the spacious lots, ample houses, and "no trespassing" signs which remain at Eagle Point on the west shore. Some groupings elsewhere are above average, but relatively few individual residences are impressive as they stand among mediocre dwellings. In appearance, Conesus is a middle or lower-middle class lake, even though market values are quite high.

After World War II, the summer cottages could be sold at inflated prices on the hungry housing market, and finally, mobile homes were available to be put almost anywhere a patch of ground could be found. Also, by cash or credit, most found a way to buy more than one boat, with power and speed increasing until the 1970's. The pressures to buy and own tend to warp general statistics of what can be afforded.

Although the market price of houses around the lake is substantially above the area level (realtors estimate 50% above), no shortage of buyers is apparent. The brisk rental business of cottages to winter occupants from the College does not appear to be declining because of increased purchases for permanent residence. Some houses are even being purchased as rental investments. Summer rentals are relatively few in number, at approximately $150–$200 a week. Conspicuously, new housing is being built in the second or third tier, back of the saturated perimeter of the lake. Lake frontage land is now worth at least $150 a foot, and $250 appears to be more typical. Most older residents have been able to hold their property. There are many professionals (lawyers, doctors, dentists) around the lake, but the less affluent have managed as well. The growth of the State University College at Geneseo and the flight from urban Rochester by skilled craftsmen and others has significantly augmented the pool of those who could afford the increased prices. Their personal incomes during recent times kept pace with the rise of land values.

One of the attractions of the residential community is that it is *not* a neighborhood or a community except in a limited sense. The Conesus Lake Association (incorporated in 1932) has rather effectively provided the limited social structure which most residents appear to have desired. It resembles the urban neighborhood associations which began to form a generation later in performing certain quasi-governmental functions. The Fourth of July celebration which it organizes is characteristic of other lakes, but its initiative on a number of substantial community problems is impressive. The Association can be credited as the vehicle which obtained an effective sanitary inspector before other lakes, also fire hydrants, safety regulations, and lake level control. It stood against filling of shore wetlands and for the perimeter sewer. The Association considered a control program for aquatic "weeds" in 1968, and accepted a recommendation against action.

Scattered throughout the watershed, dwellings range from old mobile homes to remodeled farmhouses and contemporary designs for the successful businessman or Rochester technician. There is space, scenery, and a minimum of interference.

Interference, in terms of undesired personal contact, regulation of activity, and costs of services, had inexorably increased. Paradoxically, the resistance to orderly planning and regulation has increased both the need for restrictions and the difficulty in adopting or enforcing them. Foremost in the fears of future interference is the Genesee Expressway (Interstate 390) which is planned for a north–south route in the area between Conesus Lake and the Genesee River to the west. The prospect was discussed by Boyle (1974), in a student planning study, who concluded that the maximum expected growth could be accommodated by rational planning and stringent land use regulation. It was quite evident to Boyle and to any objective observer that use of the environment would have to be rationed, or the "tragedy of the commons" would be played out to its end once more. Concrete evidence of the present mood is observable in terms of chained roads and fenced property, which impedes both research and recreation. Outsiders are generally blamed for overuse or abuse of the lake, even though there is extremely little public access space. Trash, beer cans, and too many motorboats are the reasons generally cited. Even the most remote highlands report nuisances.

Since socioeconomic status is a relative measure, it is meaningful to compare Conesus with other nearby lakes which have residential populations. Canandaigua, for the most part, is distinctly more "high" or "high-middle" class, while Honeoye is noticeably lower on the socioeconomic scale, and Silver Lake is a bit lower than Conesus.

CLIMATE AND HYDROLOGY

Regional Climate

Livingston County, with the entire Conesus Lake watershed, lies in Region IV of the six climatic zones recognized by Carter (1966) for New York. Region IV extends generally across the Erie–Ontario drainage area, except for the Genesee Valley above the Letchworth area and the Thousand Island portion of Lake Ontario. The Susquehanna River Valley, including lower portions of its principal tributaries, is also a part of Region IV, with adjacent uplands assigned to Region III. Both were characterized by cold snowy winters, but summers were cool and wet for Region III and warm and dry for Region IV. Other characteristics of Region IV included (1) mean temperature for January, $-4°C$; (2) frost-free season 150–180 days; (3) potential evapotranspiration 58–69 cm; (4) growing degree months 60°C (converted from 140°F); (5) precipitation 63.5–102 cm. Elsewhere (Eschner, 1966) a 38 cm annual average runoff was indicated.

Region IV was considered by Carter as one of the driest areas in the state. It can be added that Livingston County is apparently the driest of all counties. Although the general characterization can be applied reasonably well to the Conesus Lake watershed, it fails to recognize significant variation within the area.

No official meterological observation stations are located within the Conesus Lake watershed. The Geneseo Village water supply intake does maintain daily records of water level, temperature, and precipitation.

Both Geneseo and Avon–Lakeville intakes record the amount of water withdrawn. The three nearest meterological stations with extended temperature records are at Hemlock Lake, 7 km east of the middle of Conesus Lake, Groveland, 7 km to the southwest, and Mount Morris, 13 km west of the southern end.

Topoclimatology

In the Conesus Lake watershed of 180.5 km², elevation ranges from 249.3 to about 550 m, the mean elevation of the general upland region being approximately 396 m. Lougeay (1975) has identified a strong correlation between temperature and elevation in this area of western New York, with a mean annual temperature decrease of 0.52°C per 100 m of increased elevation. The average annual temperature of Hemlock Lake valley is 8.5°C, and if they are comparable, the annual mean temperature for Conesus drainage would be 7.8°C, since it is a bit higher in mean elevation. The inaccuracy of standard precipitation measurements has been well documented. An undercatch of from 15 to 65% is not uncommon depending upon wind exposure and the portion of snowfall. Lougeay (1976) estimated the undercatch for this region to be at least 10%. Again, assuming similarity between Conesus and Hemlock valleys, the annual measured precipitation value of 77.85 cm should be increased to 85.0 cm. This value seems more representative of the Conesus Lake watershed. In recent years precipitation has been about 6% above the long-term average.

R. Lougeay (personal communication) has concluded that approximately half of the precipitation is discharged from the watershed as stream runoff. During 1969–1974 discharge from Conesus Lake averaged 47.0 cm. The amount represents 52% of the adjusted precipitation or 90.19 cm for these years. If evapotranspiration accounts for the remaining 48% output from the watershed, then the annual average would be 43.29 cm in the years 1969–1974. The open water surface of Conesus Lake covers about 13 km² or 7% of the watershed. The available data are not adequate to determine actual evapotranspiration, since the evaporation rate at Geneva for the 6 warm months is far higher: 76 cm for 1973 and 1974. It has been noted that the pan data approximate the lake rather than land surface. Furthermore, there is a third route of output: groundwater flow. W. J. Brennan (personal communication) suggests that the subsurface sediments may carry a substantial portion of water. Such flows have been calculated for a nearby valley of similar origin to the Conesus valley.

Radiant Energy Budget

Climatic radiation balance is strongly influenced by the local microclimatic setting and surface ground cover. Thus, the radiative energy balance for the Conesus Lake watershed would be an intricate mosaic of microenvironmental settings. Only limited generalizations can be drawn from the data now available.

Solar insolation has been recorded at Geneseo, 8 km west of Conesus Lake, for 1973 and 1975 (Table 1.). Long-term insolation records are available for Ithaca, New York, 105 km to the southeast (Phillips and McCulloch, 1972). Mean daily incoming solar energy for the region ranges from approximately 520 cal cm²/day in July to 85 cal cm²/day in December, with an average daily insolation value of approximately 320 cal cm²/day for the year. Net radiation for the region varies from −25 cal cm²/day in December and January to almost 300 cal cm²/day in June.

Precipitation

It is difficult to assess basin precipitation accurately from rain gauges outside the watershed. The annual precipitation records among the three most proximate stations varied as much as 14%, with the station at Hemlock Lake highest in recent years (1968–1974). The range was between 59.4 cm (Mount Morris, 1971) and 101.7 cm (Groveland, 1972). However, 1972 was the year of unusual rains associated with Hurricane Agnes, so a more typical high value is 87.4 cm (Hemlock, 1970). This level was reached only seven times at this station since 1930, and the 1972 level was exceeded once. During this time an approximately equal number of years exceeded and were lower than 80 cm. For comparison, the information supplied by the USGS for the National Eutrophication Survey of Conesus estimated the mean annual precipitation at 78.7 cm and the sampling year level at 83.8

Herman S. Forest, Jean Q. Wade, and Tracy F. Maxwell

TABLE 1

Daily Values (1975) of Incoming Solar Radiation (cal cm^{-2} day^{-1}) 77°48′W 42°48′N[a]

Day of month	Jan	Feb	Mar	Apr	May	Jun	Jul	Aug	Sep	Oct	Nov	Dec
1	—	200	217	—	100	—	—	515	182	134	68	86
2	—	—	141	—	538	—	—	440	261	146	47	79
3	—	214	198	—	484	—	—	446	432	328	41	94
4	—	123	253	—	47	—	—	—	255	317	155	115
5	—	90	206	—	199	—	—	530	322	267	209	138
6	—	47	78	—	65	—	—	102	307	222	165	16
7	—	169	62	—	591	—	—	—	394	305	121	152
8	—	178	259	—	547	—	—	549	228	310	95	93
9	—	179	—	—	582	—	—	568	377	58	163	32
10	67	185	—	—	577	—	—	302	372	182	97	39
11	123	162	—	—	534	—	—	378	288	114	181	121
12	90	171	—	—	212	—	—	509	197	43	43	75
13	99	192	—	—	—	—	—	225	288	108	149	76
14	82	103	—	—	—	—	—	496	353	169	47	76
15	94	—	—	—	—	—	—	255	346	172	165	20
16	136	—	—	—	—	—	—	122	24	111	133	107
17	97	55	—	—	—	—	508	435	326	106	171	72
18	34	—	—	—	—	—	476	501	70	23	110	92
19	28	93	—	150	—	—	316	474	204	146	163	77
20	189	154	—	59	—	—	348	428	116	58	162	45
21	126	270	—	230	—	—	498	212	139	192	86	63
22	70	229	—	217	—	—	549	47	114	243	67	80
23	164	43	—	302	—	—	541	324	247	217	100	116
24	81	37	—	31	—	—	250	—	88	219	126	164
25	33	131	—	44	—	—	511	—	—	129	91	65
26	86	149	—	195	—	—	593	—	90	163	53	44
27	152	221	—	247	—	—	414	—	91	239	49	83
28	118	189	—	394	—	—	—	—	355	230	45	95
29	58		—	504	—	—	506	—	342	58	75	121
30	82		—	534	—	—	549	—	259	112	119	58
31	178		—		—		514	193	—	201	—	85

[a] Recorded at Geneseo, New York 8 km west of Conesus Lake. Sensor, Matrix solameter model MK 1-GM. Recorder, Weathermeasure potentiometric recorder model EPR-2T. Planimetered by staff. Data by R. Lougeay, Dept. of Geography, SUNY College, Geneseo.

cm. The source is not cited, but the figures are close to those values measured at Hemlock Lake. Therefore, a convenient metric estimate of typical precipitation in the Conesus Lake watershed from those data is 80 cm. Extremes in a 50-year period range approximately from 55 cm to 110 cm.

Temperature

Mean annual temperature at Hemlock from 1969 through 1974 (8.19°C) showed monthly means ranging from −9.3°C (January, 1970) to 22°C (August, 1973). At Mount Morris, the range was within a degree in both directions; thus, the figures may be taken as representative. There is also close agreement that February is slightly colder than January, with an average mean of −5.4°C for the 6 years, and July, at 21°C, is a half degree or so warmer than August.

Lake Evaporation

No systematic record of evaporation is available closer than Geneva, which is 60 km eastward, at the outlet of Seneca Lake. These evaporation pan data would approximate the potential evaporation from the open lake surface. The highest monthly means, for July 1970–1974, are from 17.5 to 21.1 cm with April and September rates almost equal at 54% of July evaporation. October evaporation is 35% of July level, and no record is available for November through March.

Lake Level

Forest and Mills (1971) compared the fluctuations of annual precipitation (Mount Morris and Hemlock Stations), withdrawal of water (Geneseo and Avon–Lakeville intakes),* and lake level, but they discerned no relationship between water usage and lake level. Subsequent data have been added, and are presented here in Fig. 5. The two USGS topographic maps which include Conesus Lake differ in the assigned water level; the northern Livonia map (1942) gives 817 ft (249.0 m) and the Conesus map (1951) 818 ft (249.3 m). Records at the Geneseo intake since 1930 indicate a mean level about equal to the lower figure until 1953, and almost 0.5 m higher since. In reference, the legal level set for state jurisdiction was the "mean high water level" of 819.57 ft (249.8 m). The level has been regulated to some extent, but the control was irregular and impossible to document in detail. There is no

_____

* The paper was printed with an incorrect graph, but correct data were available.

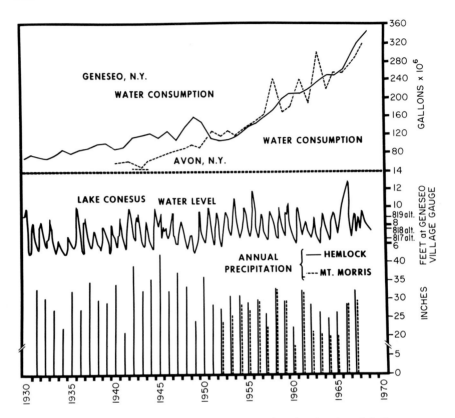

Fig. 5. Conesus Lake level, water withdrawal, and regional precipitation.

evidence that the lake level was held consistently higher during the past 20 years. Paradoxically, the precipitation (at Hemlock Lake) has been generally lower during the higher mean water level years. The number of years when 80 cm was exceeded was almost 50% higher during the low level years, and the amount of excess was much greater. Estimates of annual evapotranspiration are not available for these years, however. Without local measurements of wind speed, temperature, and atmospheric humidity, it is difficult to assess all of the variables which determine lake levels (S. Walker, 1975).

Water Supply

The withdrawal of water for urban supply has tripled in 20 years. The amount now approaches 2.7×10^6 m³ a year, which is about 2% of the lake volume. Alternatively it is also 6.4% of the average yearly outflow from the lake. There is rather regular annual oscillation between a spring high and a late fall low. Over a much briefer extent of time, heavy rains or drought may strongly affect the lake level at any season. Unfortunately, no available climatic data blended over extended periods correlate at all with the observed lake level.

Stream Flow

The Upper Genesee River drainage basin study (NYSHD, 1961) was conducted in part to establish state "best use" classifications. Except for the Livonia reservoir area of North McMillan Creek, all tributary streams were recommended for the lowest "D" classification with best use for agriculture. All streams except for Conesus Inlet and Wilkins Creek are intermittent and automatically classified as D, although water might be of high quality during times of flow.

Tributary streams were gauged in April and July 1969 (H. S. Forest, T. F. Maxwell, E. L. Mills, and G. J. Stout, unpublished data). Although all streams flowed in the spring and some in summer, it was concluded that the Inlet and North McMillan Creek provided between 50 and 70% of the total water input.

The National Eutrophication Survey of Conesus Lake (USEPA, 1974)* included flows and chemical composition of the major entering streams, and most of the principal secondary streams. The inlet streams, including South McMillan Creek, drains 41% of the watershed. North McMillan is second in watershed area with 11%, and five gauged streams had similar-sized drainage areas (3–5%); Cottonwood Point Creek, in the same size range, was not gauged. Direct drainage and minor watercourses constitute 23%.† The USEPA listed mean flows in direct relationship to their subwatershed areas except that the miscellaneous discharges are somewhat more (30%). This 7% difference was drawn from the five secondary streams (Table 2.). The total input balances precisely the outflow listed for Conesus Creek, even though no provision was made for evaporation or water withdrawal from the lake for municipal supply. A much more reserved conclusion can be accepted, however; the largest, and southernmost streams contribute at least 50% of the lake water. There is evidence that the amount is still higher, since vegetation patterns and constant stream flow indicate that rainfall may be higher in the southern watershed. Topoclimatological evidence cited in *Climate and Hydrology* supports the ecological observa-

* The USEPA report contains some significant errors. Its map of Conesus Inlet and South McMillan Creeks is quite wrong, and it has called North McMillan Creek "Davis Creek."

† This is the EPA figure. The subtraction of Cottonwood Point Creek drainage leaves a remainder of about 20%.

TABLE 2

Drainage Area and Mean Flow for Tributaries of Conesus Lake[a]

Name	Drainage area (km²)	Mean flow (m³/sec)
Long Point Gully	5.44	0.065
North Gully Creek	6.73	0.082
Densmore	6.22	0.076
Wilkins Creek	4.92	0.062
Hanna's Creek	6.99	0.084
Conesus Inlet	74.33	0.90
Davis Creek[b]	20.20	0.25
Minor tributaries and immediate drainage[c]	42.73	0.65
	167.57	2.17(9)
Outlet–Conesus Creek (this area includes the lake)	180.5	2.17

[a] Source: USEPA National Eutrophication Survey Report, Working Paper No. 156. All measures and total converted from English units.

[b] Davis Creek is North McMillan Creek on all known maps and in practice by residents.

[c] The watershed of Cottonwood Point Creek which enters the west shore of the lake about at the midpoint of the south basin has an area of 6–7 km². Consequently, the drainage by the miscellaneous sources should be reduced about 15%–20% of the total instead of 23.6%.

tions. Also, evapotranspiration is probably relatively lower in the Inlet portion of the basin.

Flow at Conesus Outlet was calculated for 15 years (1919–1934) with results reported in the Upper Genesee River Drainage Basin Study (NYSDH, 1961) and in more detail by Gilbert and Kammerer (1965). These figures were used by Stewart and Markello (1974) with a correction added for municipal water being withdrawn from the lake in 1969. However, withdrawal during the gauging period was less than 10% of the 1969 total. A more suitable correction for the period would be subtraction of 3%, since the gauging station was 2.4 km downstream from the outlet of the lake. The average discharge before correction calculated by Stewart and Markello was 43.1 × 10⁶ m³/year. The correction for the excluded watershed reduces the total to 42 × 10⁶ m³/year. These differences are insignificant since accuracy within such a range of variation is not possible. The 15-year program also recorded miminum average discharges for periods from 1 to 274 days and a consecutive 7-day minimum average for 10 years. The four lowest figures in these data are less than 0.02 m³/sec. Duration of daily flows was also recorded. Other information included maximum (Dec. 1, 1928), 17.75 m³/sec; maximum mean daily (Dec. 2, 1928), 16.75 m³/sec; minimum mean daily (Dec. 18, 1932), 0.01 m³/sec; and average (15 years), 1.37 m³/sec.

These figures do not contain the minor correction for the watershed size. Also present, withdrawal of water is equivalent to 0.057 m³/sec. The only other extended flow measurements were obtained for the National Eutrophication Survey (USEPA, 1974) from November, 1972 through October, 1973. The mean flow was determined as 2.17 m³/sec or almost double the 1919–1934 average. This deviation seems suspiciously large, particularly since mean precipitation calculated for 1924–1934 was about 73 cm, only 9% below the amount recorded for the 1973 water year. Other data from the more recent survey conform more to expectations from the 15-year record (1919–1934) and to measurements conducted in 1969. A maximum mean flow of 16.75 m³/sec was recorded on March 4, 1973, and 13.3 m³/sec for the month of April, 1973, the highest month. December, 1972 was the second highest month. The lowest month was October with November, September, and August, following in that order. The lowest daily discharge in 15 years as well as the highest discharge of the outlet stream were both in December. Most streams are considered intermittent from July through October, but the time and duration of the dry period varies. Wilkins Creek flowed constantly with effluent from the Livonia sewage treatment plant. The increment of the effluent was not reflected in the USEPA data, although diversion occurred during the year of collection. Finally, the flow has been controlled sporadically for many years by a weir at Lakeville.

Stream Chemistry

There is little information on the chemistry of streams tributary to Conesus Lake. In 1959 data were collected on Conesus Outlet, Wilkins Creek, and South McMillan Creek, as well as the lake itself, for the basin survey (NYSDH, 1961).

Hardness, alkalinity, and chlorides were the only chemical measurements which need to be considered here. At that time, the quantities in the lake (four measurements) were nearest to those of its outlet stream, as could be expected. However, a dairy discharged waste into the outlet stream at that time. Wilkins Creek, with effluent from the Livonia Village sewage treatment plant, was 5–10 times higher in chlorides, approximately twice as high in hardness, and 50% higher in alkalinity. South McMillan Creek, the longest tributary which joins with Conesus Inlet near its mouth (one measurement), was four times higher in alkalinity, approximately double in hardness, and intermediate between the lake and Wilkins Creek in chloride level (Table 3).

The only other sources of stream chemistry are the National Eutrophication Survey (USEPA, 1974) and a few analyses for phosphate in the summer of 1968 and the following winter [Forest and Mills, 1971, and unpublished, but part included in Stout, 1970]. The data of Forest and Mills agree in the ranking among streams. The levels of total phosphorus and nitrogen for seven tributaries are summarized in Table 4. As expected, Wilkins Creek was very high and the Inlet was decidedly higher than most other tributary streams. Long Point Creek was higher than the creek at Cottonwood Point. In general, winter levels were far higher than summer, and they were much more nearly uniform (data in Stout, 1970).

LIMNOLOGY

Historical Review

The limnology of Conesus Lake has been viewed historically from various physical and biological aspects. All investigators have approached it in hopes of understanding the structure and dynamics of the lake as a unit. The pioneering study of 1910 by Birge and Juday (1914) was made at a time when physical limnology and morphometry were tabulated as the critical description of lakes. Biological limnology, particularly the search for organisms which accurately reflected the condition of the lake as a whole, followed, although the understanding has been sought to some extent for the practical application to fish management. More recently, the flow of

TABLE 3

Comparison of Water Chemistry in Conesus Lake with Three Tributary Streams from Upper Genesee River Basin Survey, 1961[a]

	Hardness (mg/liter)	Chlorides (mg/liter)	Alkalinity (mg/liter)
Conesus Lake (average of 4 measurements, 8/19)	111.0	11.25	100.0
Conesus Outlet (average of 10 measurements, 8/10 and 9/10)	123.2	16.3	100.3
Wilkins Creek (average of 4 measurements, 8/3 and 9/14)	180.0	98.0	149.75
South McMillan Creek (single measurement, 9/23)	184.0	51.0	382.0

[a] Data collected August and September 1950; September 1959.

TABLE 4

Summary of Nitrogen and Phosphorus Levels for Conesus Lake, Tributary Streams, and Outlet Stream

Location[b]	Total N (mg/liter)		Total P[a] (mg/liter)	N/P ratio
	NO_2 and NO_3	NH_3		
Conesus Lake				
14 measurements—north May, July, Oct 1972	0.076	0.066	0.025	
11 measurements—south May, Jul, Oct	0.079	0.042	0.017	
Conesus Outlet Creek				
13 measurements Nov 1972, Oct 1973	0.165	0.051	0.031	
Hanna's Creek				
10 measurements Nov 1972–Jul 1973	1.10	0.10	0.098	24/1
Wilkins Creek				
12 measurements Nov 1972–May 1973	1.02	0.055	0.065	7/1
Long Point Gully				
9 measurements Nov 1972–May 1973	1.42	0.057	0.067	11/1
North Gully Creek				
12 measurements Nov 1972, Oct 1973	0.70	0.048	0.046	31/1
Densmore Gully				
7 measurements Jan–May 1973	0.92	0.038	0.043	41/1
Davis Creek (North McMillan)				
9 measurements Nov 1972–Jul 1973	0.38	0.038	0.025	52/1
Conesus Inlet				
12 measurements Nov 1972–Oct 1973	0.23	0.1	0.079	21/1

Location[c]	Date	Total P (mg/liter)
Inlet		
North McMillan Creek	6/27/68	0.058
South McMillan Creek	6/27/68	0.062
	7/4/68	0.045
Conesus Inlet Creek		
a. Upstream from swamp	7/4/68	0.055
b. Near lake, at bridge	6/13/68	0.185
	6/20/68	0.156

(Continued)

TABLE 4 *(Continued)*

Location[c]	Date	Total P (mg/liter)
	6/27/68	0.156
	7/4/68	0.101
Lake		
a. 30.5 m N of inlet	7/4/68	0.114
b. 30.5 m N of inlet	8/1/68	0.071
c. 152.4 m N of inlet	8/1/68	0.045
d. 304.8 m N of inlet	8/1/68	0.045
e. 457.2 m N of inlet	8/1/68	0.022
Mid-lake		
Reservoir Road (Geneseo pumping station.)	6/6/68	0.042
	6/13/68	0.016
	6/20/68	0.033
	6/27/68	0.036
Streams before entry:		
Cottonwood Pt. (west side, southern half)	6/27/68	0.019
Long Point (west side, waist of lake)	6/6/68	0.055
	6/27/68	0.062
Wilkins Creek (east side about 1.3 km from outlet	6/6/68	1.49
carrying effluent from Livonia Village disposal	6/13/68	1.53
plant)	6/20/68	>1.63
	6/27/68	0.521
Outlet Stream	6/13/68	0.035
	6/20/68	0.078
	6/27/68	0.058

[a] All measures converted from phosphate (PO_4).
[b] From Appendix B,C, and p. 11, National Eutrophication Survey, 1974.
[c] From Forest and Mills (1971) (Table II).

materials and energy in the watershed and lake are being examined to achieve mathematical summaries which will not only describe the lake and enable comparison with others, but actually provide equations in which management decisions can be tried or modeled.

Although limnology has advanced enormously as a science since 1910, and a considerable accumulation of data is now available, the understanding of Conesus Lake is far from satisfactory. It is particularly instructive to compare Conesus with other lakes and bays of the Finger Lakes region, since individuality, rather than uniformity, is characteristic of the group. All investigators agree that Conesus is a relatively productive lake. Nevertheless, it is particularly both a weedy and a clear lake. Although plankton production is high, the phytoplankton level is somewhat misleading when averaged over a full year. A peculiarity of Conesus is that there can be an enormous mobilization of nutrients and energy resources to produce extremely heavy "blooms" which are of several weeks duration.

Effective grazing primarily by a single crustacean, *Daphnia,* helps keep the open water rather clear most of the time. Conesus Lake contains a richly varied biological community which has proved remarkably stable during the years when a large quantity of both septage and effluent from a municipal wastewater treatment plant entered the lake. Since both sources have been effectively eliminated as of 1973, Conesus Lake presents a highly unique "experiment" for observing limnological changes.

Following Birge and Juday's work on the Finger Lakes (1914, 1921), the principal reports on Conesus Lake have been as follows: Muenscher (1927), aquatic plants and zooplankton; Berg (1963), Conesus was included in a tabular summary of the physical limnology of regional lakes; Greeson and Robinson (1970) and Greeson and Williams (1970a; 1970b), geographic information; Savard and Bodine (1971), algae; Forest and Mills (1971), rooted aquatics and general aspects of the environment including pollution and wetland destruction, The National Eutrophication Survey (USEPA, 1974), attempted nutrient budgeting and the computation of loadings; Stewart and Markello (1974), included the first modern limnological analysis; Mills (1975), phytoplankton; Chamberlain (1975), zooplankton; and Forest (1976), rooted aquatics. The last four studies are comparative in nature, including at least three other regional lakes. Oglesby *et al.* (1975) have provided a comparison of 11 Finger Lakes, including Conesus, with respect to nitrogen, phosphorus, and indeed, most of the accepted limnological criteria. An alternate computation of phosphorus loading was offered by Oglesby and Schaffner (1975).

Location, Dimensions, and Retention

Unfortunately, unresolved disparities remain in the data. Some are minor, but others are material enough to affect critical computations such as nutrient loading and trophic state. These differences may be resolved either by identifying accurate measures or simply accepting an arbitrary figure, but resolution is not yet possible in some cases. The values accepted here for location and dimensions are listed in Table 5.

The maximum depth is unresolved. Stewart and Markello (1974) made most careful measurements, 62 ft or almost 19 m. The earlier (1939–1940) large-scale map based on 4000 reference points indicated 66 ft at the bottom.* Partly because the contour and volume measurements depend on this depth, 20.2 m is accepted here (Oglesby *et al.,* 1975). The disparity is also

---

* This map was made by the Conesus Lake Sportsman's Club. The original is held by the New York State Department of Conservation, Region 8, Avon, New York, and copies on 22 fitted 8½ × 11 sheets are filed at the Environmental Resource Center and will be deposited in the Milne Library, State University College at Geneseo, New York.

TABLE 5

Location and Dimensions for Conesus Lake, Livingston County, New York

Location
Arbitrarily: Latitude[a] 42°47'   N
                              mid-lake Long and McPherson Pts.
             Longitude[a] 77°43'   W
             Elevation (249.3 m)   (818 ft)

Parameter	Accepted measurements
Drainage area (d)	180.5 km² (including lake)
Surface area (s)	12.9 km²
d/s	14
Length	12.6 km
Width, maximum	1.34 km
Width, mean	1.06 km
Shoreline	29.6 km
Depth (maximum)	20.2 m
Depth (mean)	11.5 m
Volume	156.8 × 10⁶ m³
Hydraulic retention time	3.2 years

[a] USGS location is at Geneseo Village pumping station, latitude 42°47'39", longitude 77°43'15".

reflected in determinations of mean depth. Stewart and Markello (1974) give two computations, and the greater is accepted here. The most serious unresolved difference is in volume. Together with differences in watershed size, volume effects determination of hydraulic retention time. Stewart and Markello calculated a volume of 117.1×10^6 m³ by using seiche timing and 148.5×10^6 m³ from the bathymetric map. Mills (1975) and our own determinations are slightly higher. The high figure was accepted here (157×10^6 m³).

Published figures for hydraulic retention time range from 1.4 years (this was Mills' calculation; he used the large volume figure, which was also accepted for the National Eutrophication Survey) to 3.2 years (Stewart and Markello from their higher alternative volume). Stewart and Markello's accepted figure for modeling was 2.6 years, which was derived from their lower alternative volume, and they expressed the conclusion that the correct figure probably lay between the two. The National Eutrophication Survey's figure was 1.7 years.

An independent calculation has been made dividing the lake volume by the total calculated annual discharge. The lower discharge figure of 43×10^6 m³/year (NYSHD, 1961; Gilbert and Kammerer, 1965) gives a retention time of 3.64 years, and the higher figure of 68.3×10^6 m³/year (USEPA, 1974), 2.6 years. The earlier data were based on 15 years data;

the latter limited to a single year. With some allowance for the lake not being at full volume and for possible higher runoff, the most plausible of previous calculations is the highest, 3.2 years (Stewart and Markello, 1974). On the other hand, if two plausible estimates for runoff are used, the retention time would be less. The disparity between estimated runoff and measured outflow is *prima facie* evidence of underground flow, perhaps a third of the surface water (see tabulation below).

Runoff (cm/year)	Rentention time (years) (runoff/lake volume)
40	2.26
50	1.8

A correction of 4% has been subtracted for the higher evaporation from the lake surface, but none for withdrawal by municipalities.

Area, contour, and volume of depths from 0 to 20 m were determined by R. T. Oglesby (unpublished, 1974), and independent calculations of the area and volume are given for comparison (Table 6). Since the computations were in English units, some irregularities have been introduced by conversion. The general morphometry of the bottom is reasonably consistent, although there are disparities between the two determinations. About one-sixth of the lake is shallow, less than 3 m, and gentle slopes between 9 and 12 m and 15 and 18 m each cover one-quarter of the bottom. Between them the slope is steeper and the layer covers only one-eighth of the bottom. About one-fifth of the bottom is in the zone of rooted plant growth, but not all of it is, in fact, suitable for their growth. Figure 6 is a simplified bathymetric map of Conesus Lake.

PHYSICAL LIMNOLOGY

Temperature Regime

It is generally recognized, e.g., Stewart and Markello (1974) and Mills (1975), that Conesus is a dimictic lake with seasonal temperature changes typical of north temperate lakes. K. M. Stewart (personal communication) has most helpfully provided a 6-year record of summer stratification. The onset is defined as a maintained difference of at least 2°C between top and bottom, and a complete breakdown as a difference of no more than 1°C, which disappears completely.

During 1967–1969 and 1972–1974, the onset of stratification occurred during the first and second weeks of May for 2 years each, and the fourth

TABLE 6

Depth, Volume, and Area in Conesus Lake

Depth (m)[a]	%	Area (km²)	Contour length (km)	%	Volume (m³ × 10⁶)
0	100	13.67	29.07	39.0	61.14
5	79.3	10.84	26.16	32.5	50.92
10	69.8	9.54	26.20	22.7	35.55
15	36.1	4.93	9.62	5.8	9.22
20	0.4	0.06	1.35		
20.2		bottom			

Depth (m)[b]	Area (%) Cumulative	Area (%) Layer	Area (km²)	Volume (m³ × 10⁶)
0–3.05	100	17.2	2.18	26.97
3.05–6.1	82.8	5.9	0.75	9.25
6.1–9.15	76.9	4.5	0.57	7.06
9.15–12.20	72.4	25.1	3.19	39.36
12.20–15.25	47.3	13.6	1.73	21.3
15.25–18.30	33.7	26.0	3.32	40.76
18.30–20.13	7.9	7.5	0.96	11.76
20.13–	0.2	0.2	0.02	0.3

[a] Measurements by R. T. Oglesby (unpublished) based on Conesus Lake map (scale 1:21,760).

[b] Measurements by P. G. Savard (unpublished) recomputed to metric scale from 1:4800 scale map (1939–1940). Volume based on Oglesby's total of 156.8 m³ × 10⁶.

TABLE 7

Summer Stratification in Conesus Lake[a]

Year	Onset	Complete breakdown
1967	4th week of May	mid-Oct
1968	1st week of May	4th week of Oct
1969	mid-May	4th week of Oct
1972	1st week of May	3rd week of Oct
1973	4th week of April	end of Oct
1974	2nd week of May	2nd week of Oct

[a] Data provided by K. M. Stewart (unpublished).

week of April and the fourth week in May during a single year. Breakdown occurred between the second week and the end of October. There was no relationship between time variations for the two events (Table 7.).

Mills recorded the initial week of thermal stratification as June 6, 1972 and June 13, 1973 and also observed autumnal overturn during the week of September 25 both years. The epilimnion becomes thicker in the course of a summer, but there is only a scanty record of the change. Information from the reports indicates a depth of 5–7 m in July, yet observations recorded in

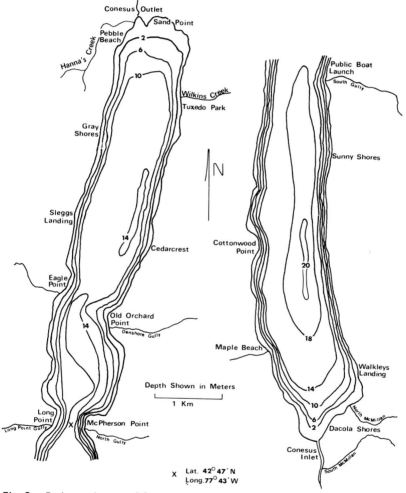

Fig. 6. Bathymetric map of Conesus Lake with shore points and streams. "Denshore" of maps since 1942 is actually "Densmore."

college course studies during 1970 and 1971 indicate 8 to 9 m in July, 11 m in early September, and 13 m later (H. L. Huddle, personal communication). Chamberlain (1975) considered the mean thickness of the epilimnion to be 8 m, which is generally consistent with other information, but perhaps a little thin. Thermal profiles for Conesus Lake have been determined in detail by Stewart and Markello (1974) in 1969, and temperature data were integrated for layers by Mills (1975) for 1972 1973 (Fig. 7).

Present temperatures of Conesus Lake may be compared with those recorded in 1910 by Birge and Juday (1914) and with current temperatures in two nearby Finger Lakes. Chamberlain (1975) compared surface and bottom temperatures in 1910, 1972, and 1973, but no trend is apparent (see tabulation below).

August 1910 (°C)		August 1972 (°C)		August 1973 (°C)	
Surface	Bottom	Surface	Bottom	Surface	Bottom
21.8	12.5	21.0	11.5	23.0	11.0

Average temperature of the hypolimnion and epilimnion were compared by Mills (1975) for Conesus and Hemlock Lakes, while data from Stewart and Markello (1974) provide a comparison with Honeoye Lake (Table 8). Hemlock Lake is deeper than Conesus by about 27% in extreme depth and 15% in mean depth, while Honeoye is only about half as deep as Conesus Lake.

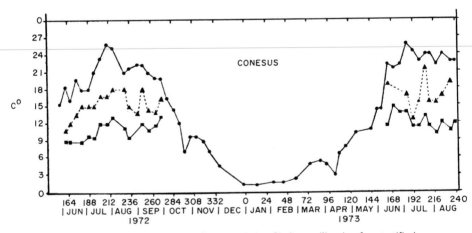

Fig. 7. Temperature variation in Conesus Lake. Circles, epilimnion for stratified season and total water column for rest of year; triangles, metalimnion during stratified season; squares, hypolimnion during stratified season. From Mills (1975).

TABLE 8

Comparison of Temperatures during Summer Stratification for Conesus, Hemlock, and Honeoye Lakes

	Date	Epilimnion (°C)	Hypolimnion (°C)
Conesus average			
Mills (1975)	1972	21.1	11.4
	1973	23.3	12.4
Stewart and Markello (1974)	Jul 23, 1969	Approx. 25.2	Approx. 11.0
Hemlock average			
Mills (1975)	1972	21.1	8.1
	1973	23.0	10.1
Honeoye			
Stewart and Markello (1974)	Jul 23, 1969	Approx. 26.0	Approx. 22.0

The maximum temperature at the surface does not differ greatly among the three lakes. Its time of occurrence in Conesus is July rather than August, Mills having recorded maxima of 25.5°C in early July 1972 and late July 1973. The temperature in the epilimnion is the same in Hemlock and Conesus Lakes, and the temperature difference in the hypolimnion is much smaller than the depth difference. Indeed, fluctuations within the two lakes from year to year exceed the difference between them.

Formation of a complete ice cover occurs in late December and usually persists until middle or late March. Thickness is rarely less than 30 cm, and in or during January 1977 the ice actually measured over 90 cm thick. Stewart (1972) recorded wavy patterns of isotherms during inverse stratification. He attributed the pattern to continued input from lateral streams, and the conclusion is supported from a consideration of stream location and knowledge that winter stream flow does occur.

Seiche Period and Wind

Little information is available on short-range water movements. Stewart and Markello reported a seiche oscillation period along the north–south axis of the lake of 45 min. Variations and turbulence in the general northward flow of water depend greatly on whether lateral streams are running, but it is quite clear from biological evidence, which will be discussed subsequently, that nutrients in stream discharges are skewed northward. The effect of prevailing southwesterly winds is sometimes pronounced, particularly in the northern portion of the lake where the relief is lower.

Spherical colonies of planktonic algae may accumulate on the leeward shores in a layer several centimeters thick, and detached vascular aquatic plants form considerably bigger windrows.

Clarity

The clarity of Conesus Lake is high in relation to other Finger Lakes, and to other productive lakes. The characteristic has been substantiated by all investigators who have compared the lakes, from Birge and Juday to present day, and it has been maintained in the face of an increase of tenfold or more in human waste and activity.

The clarity is less than Skaneateles, which is quite unproductive, but overlaps the ranges of both Canandaigua and Canadice Lakes, which are far less productive. Hemlock Lake's clarity is less, although productivity is generally judged slightly lower (e.g., Mills, 1975; Oglesby et al., 1975; K. M. Stewart, personal communication).

Sampling through space and time produces a range of Secchi disk readings, yet the pattern of data is consistent and meaningful. Birge and Juday's single measurement was 6.3 m (August, 1910). This was during the characteristic late summer period of high clarity, the time of maximum standing crop of submerged vegetation. Water temperature in the epilimnion at this time is within 1° to 3°C of the maximum (about 22° to 25°C) and dissolved oxygen around 9 mg/liter.

Secchi disk values for the lake are summarized in Table 9. The yearly means calculated by Mills were 4.7 m (1972) and 5.1 m (1973), but the latter did not include the fall months. The lowest reading in 1972 had been in October during a very heavy bloom of blue–green algae. Stewart and Markello's mean of monthly means was slightly higher, 5.3 m, with the first 3 months of the year repeated. Although clarity is high typically in the summer, the highest value ever recorded (10 m) was obtained under ice in January, 1969 (K. M. Stewart, with H. S. Forest and students). Turbidity from flooding following Hurricane Agnes (June 1972) reduced clarity less than the bloom of blue–green algae in autumn of the same year.

Interestingly, clarity improved rapidly after the Agnes flooding, and by late August had attained 6 m, one of the five highest readings recorded. On the other hand, transparency decreased during August 1973 from a high initial peak, but began to increase again late in the month. The peculiar pattern of 1973 is also suggested in the graphs of Hemlock, Owasco, and Skaneateles Lakes (Mills, 1975), indicating perhaps that unusual weather conditions were responsible, Godfrey's higher readings (P. J. Godfrey, 1971, unpublished) were in March (5.5 m) and June (6.5 m) with a steady decline to September (2.5 m) 1971. The pattern is markedly different from the more

TABLE 9

Secchi Disk Values for Conesus Lake

Source	Value (m)
Birge and Juday (Aug 1910)	6.3
Stewart and Markello (Jan 1969–Mar 1970)	
(2–5 measurements/month)	
Mean of monthly means (approx.)	5.3
Standard deviation of monthly means	3.8–8.1
Extremes	1.2–10

Mills (May 1972–Aug 1973) (52 readings)	Summer (m)			Autumn overturn (m)			Winter and spring overturn (m)		
Year	Mean	Min	Max	Mean	Min	Max	Mean	Min	Max
1972	4.7	1.8	7.0	2.4	1.5	3.9	3.3	2.0	4.5
1973	5.1	3.6	7.0						

Source	Month	Value (m)
P. J. Godfrey (1971) (unpublished)	Jun	6.5
	Jul	5.5
	Aug	5.8
	Sep	—
	Oct	3.3

substantial records, yet it is not altogether implausible. Blooms of blue–green algae visible to the naked eye have been observed to begin in June and last through the summer. Transparency would decrease if the phytoplankton crop increased.

CHEMICAL LIMNOLOGY

Dissolved Oxygen

Sources of data include Birge and Juday (1914), P. J. Godfrey, (1971, unpublished), the USEPA National Eutrophication Survey (USEPA, 1974), Stewart and Markello (1974), and Mills (1975). Birge and Juday found no dissolved oxygen near the bottom at an unknown depth in mid-August 1910. Godfrey sampled near the bottom in deep water at an unspecified depth. Stewart and Markello prepared profile diagrams for dissolved oxygen from data collected January 28, April 29, July 23, and October 23, 1969.

A series of 52 samplings from June 1972 through August 1973 was plotted by Mills (1975) and is included in Oglesby et al. (1975). The data for the epilimnion (during stratification) and total water column were integrated (the "top" for Table 10). Mills second plot of bottom values may be compared in Fig. 8.

The general pattern of dissolved oxygen is reasonably consistent among all sources of data. The highest levels occur in winter near the time of ice formation, under ice, and perhaps around the time of ice breaking. Some degree of depletion occurs between the two winter peaks. From a high of 12–14 mg/liter, the level may fall to about 10 mg/liter in upper waters and 6 mg/liter at the bottom. This decrease is not consistent, and is probably affected in time and amount by the inflow of lateral streams. Dissolved oxygen at spring mixing, varying from 8 to 12 mg/liter, is generally higher than the level of fall mixing, about 8 mg/liter.

All observers have noted the severe depletion of oxygen in the lower waters of Conesus Lake in the late summer (Aug–Sep). This pattern is not unique to Conesus, but has been recorded in Hemlock Lake (Mills, 1975; Oglesby et al., 1975), Honeoye Lake, and even in Canadice Lake (during October) by Stewart and Markello (1974). The lowest level in Conesus and shallower lakes approaches zero, while the deeper ones (Canadice and Hemlock) may retain 2 mg/liter at the bottom. This difference is not large, and, if Birge and Juday (1914) were correct, Hemlock Lake may have no oxygen at the bottom during some years. Given the range of variation from year to year for oxygen and other chemicals, it is not necessary to presume that the oxygen level at the bottom of Hemlock Lake has risen since 1910, or that the level in Conesus Lake has changed appreciably in 65 years.

Stewart and Markello (1974) remarked that the anoxic condition appeared important in the release of phosphorus from the sediments. Their data demonstrate a very high peak of phosphorus near the bottom in August and September. Yet, a study of the bottom contour of Conesus Lake further

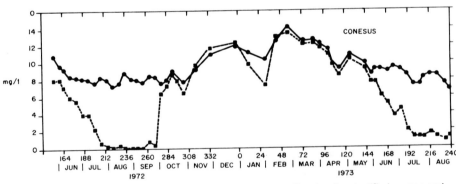

Fig. 8. Dissolved oxygen in Conesus Lake. Circles, epilimnion for stratified season and total water column for rest of year; squares, bottom. From Mills (1975).

decreases the importance and uniqueness of oxygen depletion. An extremely small portion of the lake bottom becomes anoxic, and depletion is not complete every year. To a depth of 16 m or more, there may be 4 mg/liter of dissolved oxygen. Less than 8% of the bottom is in the deepest 2 m of the lake (whether the extreme depth is taken as 19 or 20.2 m), and less than 5% may be completely anoxic.

Phosphorus

The preliminary studies of Forest and Mills (part in Forest and Mills, 1971; part in Stout, 1970; and part unpublished) established some correct general relationships about the phosphorus regime. Forest and Mills concluded:

1. The total phosphate level was within the range expected of open waters of a natural mature lake, and the level approximately doubled after extended moderate rains. Also as expected, the level is considerably higher in winter, but data on the seasonal flux were not published.
2. The small streams draining farmland contributed little additional phosphorus, at least during ordinary flow periods. At such times the input streams, including South McMillan Creek (which extends almost to Springwater), are about equal to the outlet stream in phosphorus level.
3. Conesus Inlet was conspicuously higher in phosphorus (about quadruple), and the source accounts for half or more of the water in the lake.
4. The marsh through which the inlet stream flows is a major source of phosphorus in times of ordinary flow. Streams entering the marsh were the same as those entering the lake directly.
5. The elevated level of phosphorus extends only a short way from the inlet, perhaps 150 m. This area of stream influence corresponds approximately to the enriched stream mouth community of rooted aquatic plants.
6. The phosphorus concentration in Wilkins Creek (which contained effluent from the Livonia Village treatment plant) was five to ten times or more higher than the open lake level. The National Eutrophication Survey (USEPA, 1974) judged the contribution from two to nine times higher than other streams. Stewart and Markello later calculated that the annual weighted mean concentration of the lake was exceeded over 100-fold. Although the total flow of Wilkins Creek was small, they considered that its phosphorus contribution must be a sizable portion of the total. The Eutrophication Survey assigned 8% of the total input to Wilkins Creek, although the effluent was diverted shortly after the testing year began. Forest and Mills attributed 8% of the total to Wilkins and other waste inflows between its mouth and the lake outlet.

TABLE 10

Summary of Dissolved Oxygen (DO) Levels for Conesus Lake

Source	Date	Location	Depth (m)	DO (mg/liter)	Notes
Birge and Juday (1914)	1910 Mid-Aug	?	?	0	Near bottom in deep water
Stewart and Markello (1974)	1969 Jan 28	South basin	0	14+	Deepest part of lake
			3–12	13.5	
			15	10	About 19 m
			18	6+	
	Apr 29		0–15	11.5	Amounts estimated from graphic profiles
			18	11	
	July 23		0–6	8	
			10	4	
			12–16	3	
			18+	1	
	Oct 28		0–15.5	8+	
			18	9	
P. J. Godfrey (unpublished)	(1971) Jan	South basin (?)		7.9	
	Mar			8.4	Bottom
	Jun			4.4	
	Jul			1.8	
	Aug			.6	
	Sep			.3	
	Oct			1.7	
USEPA (1974)	1972–1973 May 27	Station no. 363901	0	12.0	Adjacent to Wilkins Creek
			3.05	12.0	North basin
			9.14	14.0	
	Jul 27		0	—	
			1.22	8.6	
			4.57	6.2	
			7.62	3.8	
			9.75	3.4	

	Depth (m)		Comment
Oct 13	0	—	
	1.22	8.1	
	4.57	7.4	
	7.62	7.7	
	10.67	8.0	
	12.80	8.3	Near the inlet, south basin
May 27	0	11.4	Station no. 363902
	3.05	12.4	
July 27	0	—	
	1.22	9.6	
	4.57	8.4	
	10.67	3.6	
	14.02	3.0	
Oct 13	0	—	
	1.22	8.9	
	4.57	8.6	
	7.92	9.3	

Oglesby *et al.* (1975)
Mills (1975); selected
data estimated from
graph

1972–1973			
Jun	top	10.5	Deep portion of south basin
	bottom	8	
Jul	top	8	Top is epilimnion or total water
	bottom	4	column during times of
Aug–Sep	top	8	mixing
	bottom	near 0	
Oct	both	8	
Late Dec	both	12	
Early Feb	top	10	
	bottom	7.5	
Late Feb	both	13–14	Under ice
May	both	9.5	Under ice
Jun	top	9.0	
	bottom	4.0	
Late Jul–Aug	top	8.0	
	bottom	1.5	
Late Aug	top	7	
	bottom	1–1.5	

165

Forest and Mills discussed the restriction of Wilkins Creek's influence to the northern, or even northeastern, part of the lake, and unpublished data (Mills) showed that the high level of phosphate was perceptible only a few meters away from the mouth. Consequently, simple percentage calculations, whether 16 or 8%, probably overstate the creek's influence on the lake water as a whole. There was a great influence on rooted vegetation, however, as subsequent studies have demonstrated. The other tributaries will be discussed subsequently.

7. Forest and Mills estimated that at least 48% of the phosphorus input was retained, and later data on winter phosphorus levels would have raised the estimate to about 60%. The estimate by the USEPA (1974) was 57%, based on input measurements and allowances for precipitation, direct septage, and immediate drainage.

8. One feature of the phosphorus regime which has been added from subsequent studies is the occurrence of a very high spike of phosphorus at extreme depths from July through October, related to oxygen depletion. This pattern was expected and thoroughly confirmed in 1969 observations reported by Stewart and Markello. They also noted the dissipation of the high phosphate water as the metalimnion increased in depth during the summer.

9. The second characteristic of the phosphorus regime has been the identification of autumn (Sep–Nov) as the time of highest levels for both total phosphorus and soluble-reactive phosphorus. Oglesby et al. (1975) have suggested that decomposition of rooted aquatic vegetation during this season may release large quantities of phosphorus which quickly become part of rapidly expanding populations of blue–green algae. This possibility is noteworthy, because higher phosphorus levels in the fall are generally attributed to the transport from anaerobic depths and the bottom during overturn.

Data were collected from the open waters by various investigators from 1969 to 1973, but there are considerable differences in the sampling and analysis. A synthesis of the information is attempted cautiously. A redrawing (Fig. 9) has been made for total phosphorus from Stewart and Markello (1974), and a plot of the Mills data for soluble-reactive phosphorus (Mills, 1975; Oglesby et al., 1975) is included in Fig. 10. Table 11 summarizes data from available sources from 1969 to 1973.

Unfortunately, few reconciliations are possible among data of the different investigators. Only Stewart and Markello sampled the deepest portion of the lake and demonstrated the high phosphorus level there. At this site an elevated level extended up to a depth of 12 m or higher for a brief

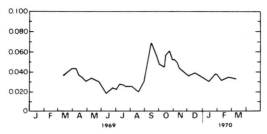

Fig. 9. Total phosphorus (mg/liter) in Conesus Lake (weighted mean). Annual mean, 0.037 mg/liter. Redrawn and modified from Stewart and Markello (1974).

period in September. However, there is little evidence of a gradient by depth elsewhere in the lake. The USEPA data for May 1973 do show a gradient between the surface and 9 m, but the result is discounted both because of the scanty data and lack of a rational explanation. During most of the year, despite a very wide range of measured levels, the total phosphorus of a single water column varies little. The only data comparing the two ends of the lake (USEPA, 1974) reveal little difference. Both sites were located adjacent to enriching streams but may have been out of their range of influence.

In comparing total phosphorus reported for the upper waters by Stewart and Markello (1974) and USEPA (1974), the May level was approximately 0.020 mg/liter and the July level probably lower, perhaps half that amount, in the photosynthetic zone. The fall levels do not compare well, due to variation in overturn time or unknown factors. The fall level is over twice that of May, but not far above the level at spring mixing. Stewart and Markello (1974) calculated a fall overturn concentration of 0.052 mg/liter and a spring mixing of 0.043 mg/liter. The seasonally integrated values of Mills (1975) are much lower.

The seasonal flux of total phosphorus (Stewart and Markello, 1975) and soluble-reactive phosphorus (Mills, 1975; Oglesby *et al.*, 1975) may be compared. Total phosphorus descended from about 0.040 mg/liter in mid-winter to about half as much in the summer, then rose sharply in September to about 0.060 mg/liter, fell, and rose again to about the high level in October, and descended gradually to the mid-winter level. Soluble-reactive phosphorus concentration was irregular, and some oscillation is suggested. The winter level, ignoring two sharp sags, was about 0.006 mg/liter; the summer level went from a trace amount to 0.003 mg/liter, disregarding the high peak in June 1972 which was associated with flooding following Hurricane Agnes. There is a fall peak in soluble-reactive phosphorus which coincides well with the fall peak in total phosphorus noted by Stewart and

168

TABLE 11

Phosphorus Levels for Conesus Lake

Source	Date	Location	Depth	Species and amount total (mg/liter)	Notes
Stewart and Markello (1974)	1969–1970	South basin deep point		0.010 (Pt.)	Individual values and estimates from graphic data
	Mar		0	0.010 (Pt.)	
			3–12	0.025	
			15	0.05	
			18	0.075	
	End of Apr–Late Jun		0–16	±0.020	
			18	0.075	
	Late Jul		0–6	<0.010	
			9–16	±0.025	
			18	0.100	
	Mid-Sep		0–9	<0.010	
			12	0.080	
			15	0.230	
			18	0.840	
	Late Oct		0–18	±0.040	
	Late Dec		0–16	±0.035	
			18	0.040	
	Late Feb		0–16	±0.035	
			18	0.050	
	Annual range		0–18	0.005–0.840	Actual figures from Stewart and Markello (1974)
	Annual mean		0–18	0.037	
	Spring mixing			0.043	
	Fall overturn			0.052	
P. J. Godfrey (1971) (unpublished)	1971 late Jul		epilimnion	0.006	
			hypolimnion	0.020	

USEPA	1972	Depth	Species and amount (mg/liter)		
			Total P	Soluble P	Notes
Station no. 363901 off Wilkins Creek North basin	May 27	0	0.013	0.006	
		3.0	0.020	0.013	
		9.1	0.061	0.020	
	Jul 27	1.2	0.013	0.015	
		4.5	0.011	0.008	
		7.6	0.011	0.008	
		9.7	0.014	0.009	
	Oct 13	1.2	0.013	0.015	
		4.5	0.034	0.025	
		7.6	0.036	0.029	
		10.6	0.036	0.026	
		12.7	0.030	0.024	
			0.029	0.022	
			0.028	0.022	
Station no. 363902 near inlet South basin	May 27	0	0.016	0.008	
		3.0	0.020	0.010	
	Jul 27	0	0.011	0.009	
		1.2	0.012	0.009	
		4.5	0.011	0.008	
		10.6	0.010	0.009	
		13.9	0.011	0.009	
	Oct 13	0	0.023	0.012	
		1.2	0.022	0.013	
		4.5	0.024	0.013	
		7.9	0.029	0.029	

(Continued)

TABLE 11 (Continued)

Source	Location	1972–1973	Depth	Soluble-reactive phosphorus	Notes
Mills (1975),[a] Oglesby et al. (1975)	South basin	May–Jun	Upper 10 m	0.027	
		Late Jun		0.014	
		Early Jul		0.012	
		Mid-Sep		0.028	
		Late Oct		0.006	
		Mid-Nov		0.012	
		Nov–Dec		0.001	
		End of Jan		0.006	
		Feb–Apr		—	
		Mid-May			
		End of Aug		0–0.002	
Mills (1975)		Annual range			Figures from table and text
		May–Dec 1972		Trace–0.191	
		Jan–Aug 1973		Trace–0.310	
		Seasonal integrations[b]			
		Summer 1972		0.004	
		Summer 1973		0.0008	
		Autumn overturn 1972		0.015	
		Winter 1973		0.012	
		Spring mix 1973		0.003	
		Annual range 1973		0.008–0.171	(total P)
		Jan–Aug			
		Winter and spring mix 1973 (seasonal integration)		0.012	(total P)
		Summer 1973 (seasonal integration)		0.012	(total P)

[a] Data chiefly collected by Mills was shared with Oglesby and Schaffer. The annual mean for total phosphorus is derived as 0.67 mg/m². Mills (1975) actually used potential phosphorus: total/mean depth (11.5 m). The soluble or available phosphorus (Oglesby et al., 1975) cannot be compared with other sources.

[b] The seasonal integrated average in 1973 was much lower (0.8 mg/m³) than 1972 (4.4). The 1972 average includes a very high peak

Markello. The highest level of 0.028 mg/liter for soluble-reactive phosphorus was reached in October 1973. The phosphorus regime of Conesus Lake will be discussed subsequently under the subject of the lake ecosystem and its trophic status.

Nitrogen

Nitrate (NO_3) was recorded among chemical parameters by Berg (1963) as 0.9 mg/liter (0.239 mg/liter N), and in 1971 P. J. Godfrey (unpublished) obtained levels of 0.19 mg/liter (0.051 mg/liter N) in the epilimnion and 0.03 mg/liter (0.008 mg/liter N) in the hypolimnion. The National Eutrophication Survey of 1972–1973 (USEPA, 1974) included depth profiles at two stations sampled in May, July, and October. Analyses included nitrate plus nitrate nitrogen, and ammonia–nitrogen. A pattern is suggested by the data for NO_2 plus NO_3 nitrogen as the concentration decreased with depth, except for the fall sampling when the column was either uniform or nitrogen increased with depth. However, the data are limited in amount and rather irregular. The minimum value for both nitrogen classes was 0.02 mg/liter N and the maximum was 0.27 mg/liter N for nitrate–nitrite and 0.15 mg/liter N for ammonia. Most values were far lower than the maxima.

Stewart and Markello (1974) plotted nitrate levels at seven depths from the surface to 18 m over 16 months to March 1970 (see Fig. 13). A higher winter concentration was characteristic of the nitrogen regime. This pattern has been noted in other lakes and is generally attributed to biological assimilation in summer and increased nitrification in winter. The more comprehensive data did demonstrate a summer stratification, the concentration generally increasing with depth to 15 m. However, the oxygen deficiency reduced the nitrate before autumnal overturn. During ice cover, nitrate was quite high near the surface, but otherwise, the concentration generally increased with the depth.

Mills (1975; also Oglesby et al., 1975) collected data from May 1972 through August 1973 and prepared a seasonal flux of nitrate nitrogen for the upper 10 m at Conesus Lake. Mills' seasonal pattern (Fig. 11) and data summaries are complementary to those of Stewart and Markello (Table 12).

Mills' seasonal plot of nitrate nitrogen in the upper 10 m of water confirms the generally higher winter levels, showing a maximum of over 0.4 mg/liter in late February with ice still on the lake. Warm season levels were quite low, only traces during most of the summer of 1972 and early fall.

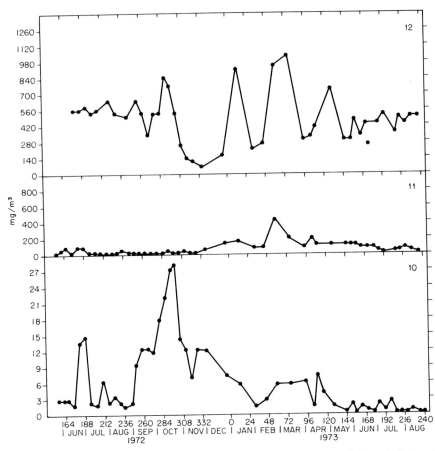

Fig. 10. Soluble-reactive phosphorus in Conesus Lake (upper 10 m). From Oglesby *et al.* (1975).

Fig. 11. Nitrate nitrogen in Conesus Lake (upper 10 m). From Oglesby *et al.* (1975).

Fig. 12. Soluble-reactive silicon in Conesus Lake (upper 10 m). From Oglesby *et al.* (1975).

Other Chemical Components

Mills (1975) found dissolved silicon levels generally reaching a minimum during the summer and maximum in fall and winter in four New York Lakes, including Conesus. Conesus was highest among the group in concentration of silicon in the hypolimnion. Mills' 16-month plot for soluble-reactive silicon (SRS) showed a series of peaks and valleys with few plateaus. The highest peaks were October, January, March, and May; the lowest valleys were December, February, April, and June (Fig. 12). Consequently it is not possible to relate silicon concentration to lake dynamics,

TABLE 12

Summaries of Nitrate Nitrogen Levels for Conesus Lake

	NO_3-N values (mg/liter) from Stewart and Markello (1974) 1969 data	NO_3-N values (mg/liter) from Mills (1975)	
		1972 data	1973 data
Annual range	0.02–0.741	Trace–0.18	0.019–0.607
Spring mixing	0.136		0.145
Fall overturn	0.051	0.046	
Summer	0.053^a	0.048	0.081
Winter	0.160^a		0.199

^{a} Estimated from graph for upper 10 m.

while nitrogen concentration, for example, does reflect other variables (Fig. 13).

Mills and previous investigators have also tested the waters of Conesus for other chemical factors, which have been assembled here in Table 13. In comparison with other Finger Lakes, Conesus is generally high in all dissolved ions. Comparison cannot be made with lakes intercepting salt beds such as Seneca Lake or waters which are subject to input from a very large and urbanized watershed such as Irondequoit Bay. The ionic content of Conesus Lake totals about 750 mE, with a slight preponderance of cations. This balance is typical of all Finger Lakes except Seneca Lake (Oglesby *et al.*, 1975). Calcium, magnesium, and sodium comprise most of the cation content in a ratio of approximately 35:25:10. Well over half of the anion content is bicarbonate with chloride second in concentration.

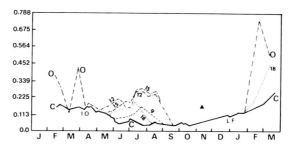

Fig. 13. Nitrate nitrogen (mg/liter) in Conesus Lake (various depths). I.O., ice out; L. F., lake freezes; triangle, outlet; C, consolidated 0.3, 6, 9, 12, 15, 18 in depth except for conspicuous deviations where shown. Redrawn and modified from Stewart and Markello (1974).

TABLE 13

Some Chemical Measurements for Conesus Lake[a]

	pH	Cond. (micro-mhos)	Alk. (as CaCO₃)	Ca	Mg	Na + K (mg/liter)	Fe	HCO₃	SO₄	Cl
NYSDH (1961)										11.25
Before 1963 (Berg, 1963)	7.7	309	108.2	40	11	9.4	0.05	132	31	13.0
1971 (Godfrey)			100							
Epilimnion	8.4		99.8	44			0.06			27.1
Hypolimnion	7.7		107.9	53			0.06			27.4
1972 (USEPA)	8.1	339	118							
1973 (Mills)[b]	8.2	330	118	41	13.2	14.8 (Na, 12.2; K, 2.6)	0.03	133.7	27.8	29.5

[a] Chloride has doubled and sodium has probably increased 50%. Such increases elsewhere have been due to road salting where mine waste or natural solution are not contributors.

[b] Mean concentrations of combined surface and bottom samples collected April 22, July 17, and August 28, 1973. Mills considered the pH range generally as 8.0–8.5, with levels ranging down to 7.5 either in winter or in the hypolimnion during summer stratification. He also reported the mean concentration of total dissolved solids as 201 mg/liter.

BIOLOGICAL LIMNOLOGY

Attached Algae and Plankton

Two reports of the algal populations of Conesus Lake have been pre-pared: Savard and Bodine (1971) and Mills (1975). The first, a floristic study, was based chiefly on summer and fall collections (1968). Included were quantitative studies of planktonic colonies during the summer and of diatoms made approximately biweekly in 1967 by K. M. Stewart. The combined list includes 101 species in 66 genera of microscopic algae and Charophytes.

The most conspicuous attached microscopic algae are filamentous green algae, although early Spring collecting by Mills later also revealed the yellow–green genus *Tribonema* which is common in this season. Also, the diatom genus *Melosira* forms large filamentous masses from time to time, but not regularly. Oglesby *et al.* (1975) noted the coincidence of the unusual *Melosira–Aphanizomenon* bloom in the fall of 1972 when soluble-reactive phosphorus was high. Soluble-reactive silicon peaked 2 weeks earlier than the phosphorus, and the height of the bloom followed the phosphorus peak by 1 month. Large masses of the green algal genus, *Spirogyra,* consistently are present in shallow waters, in spring and well into the summer, wherever there is some attachment. A conspicuous fall crop may also occur. Four species were identified by Savard and Bodine [*S. ellipsospora* Trans., *S. maxima* (Hass.) Wittr., *S. protecta* Wood, and *S. varians* (Hass.) Kuetz.]. The list would undoubtedly be much longer if a special study were conducted, since zygospores are required for identification to species level.

The green alga *Cladophora glomerata* (L.) Kuetz. is of particular interest because of its conspicuously heavy crop in Conesus and the lower Great Lakes. Furthermore, its fertilization by phosphorus was demonstrated for Lake Huron (Neil and Owen, 1964). *Cladophora* is common in hard water streams of western New York, wherever there is attachment, flow, and light. It also grows to a depth of 15 m in Lake Ontario (collections by H. S. Forest, July, 1971), but the huge crops which may wash up on beaches develop in much shallower water, no lower than 5 m deep. *Cladophora* is not abundant in the Finger Lakes and Ontario bays, but large crops occur in two waters known to be high in nutrients: Irondequoit Bay and Conesus Lake where it was observed by Muenscher (1927) in 1926. It grows only in very shallow water, 1–1.3 m in Conesus and 0.5 m or less in Irondequoit Bay. In Conesus Lake, an unusually heavy growth developed every summer at the mouth of Wilkins Creek, before effluent from Livonia was diverted from it in 1972. Subsequently, the crop declined, but it was again substantial in 1976. Within 15 m of the creek mouth, *Cladophora* festooned the rooted aquatics, and actually attained a comparable wet weight. On July 25, 1971 quadrats collected at 0.75–2.5 m averaged 522 gm dry weight ($\frac{1}{10}$ fresh weight)/m². This quantity is found only in fertile communities of submerged vascular plants. The observation is notable because it is unusual to have the opportunity to measure a crop of attached algae by methods used for macrophytes. Even where masses appear impressive, they weigh very little.

Muenscher (1927) recorded the presence of the large, erect, branching green algae *Chara* and *Nitella*, although the second identification may have been in error, since a species of *Chara* which is structurally similar to *Nitella* is found in the lake. Muenscher's herbarium specimens at Cornell University included three species of *Chara* but no *Nitella*. *Nitella* does occur in a nearby fertile lake, Honeoye, so it could have disappeared from Conesus during the following 40 years. The species identified by Savard and Bodine (1971) were *Chara Braunii* Gmelin (*C. fragilis* Desv. of Muenscher's specimens), *C. coronata* N. Br., and *C. vulgaris* L. (*C. foetida* A. Br. of Muenscher's specimens). *Chara* is not as conspicuous as vascular plants in Conesus Lake, but plants may occur in any part. Thick beds may be found in the northwestern quarter of the lake, and at Old Orchard Point on the east side of the north basin. The depth is generally 1.5 m or less. Charophytes are a much more conspicuous part of the submerged plant community around the infertile shores of Canandaigua Lake, but also in Honeoye Lake, which is generally placed higher in trophic rank than Conesus Lake and is less transparent.

Savard and Bodine collected both planktonic and attached or simply "settled" algae. They called attention to the resemblance of some colonies of the planktonic form *Gleotrichia echinulata* to *Calothrix parietina*, a nonplanktonic species. Subsequently, the problem was pursued under H. S. Forest by T. F. Maxwell (1972) and K. R. Khan (1974). Gelatinous colonial forms of 5 cm or more in diameter were found to develop quite abundantly, almost covering certain vascular aquatics, particularly *Myriophyllum*. Noncolonial and small colonial forms were found in the muddy bottoms in winter. Between 1969 and 1974, the developmental morphology of *Calothrix parietina* (Naeg.) Thur. was investigated both in Conesus Lake and laboratory cultures. All of the field forms, including the conspicuous bloom plankton, were related to the single species and reported by Maxwell (1972) and Khan (1974). Other blue–green algae customarily have been reported under different names for planktonic and nonplanktonic forms, and a limited discussion of the problem will follow after planktonic algae have been considered.

Mills collected extensive data on the planktonic algae of the lake, sampling regularly on 25 occasions through 1973, and almost daily during the

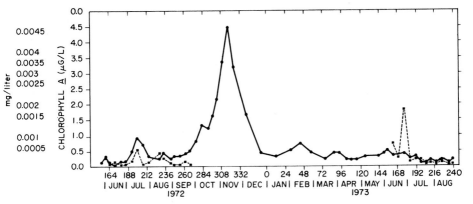

Fig. 14. Chlorophyll *a* in Conesus Lake. Circles, epilimnion for stratified season and total water column for rest of year; squares, hypolimnion for stratified season. From Oglesby *et al.* (1975) with scale change.

summer of 1972. Thus, both short-range variation and seasonal fluctuation could be documented. Data were assembled to show distribution of (1) species, seasonally, taxonomically, spatially (depth); (2) size categories; (3) density, cell count, biomass, life form; (4) content of chlorophyll *a*, phaeophytin, and carotinoids for volume of water. Oglesby *et al.* (1975) included Mills' graphic summaries for chlorophyll *a* and phytoplankton biomass integrated for the upper 10 m of water. These are included here in Figs. 14 and 15, respectively. The information was utilized in estimating the trophic ranking of Conesus with other lakes, both in the Finger Lakes group and elsewhere. It is unquestionably the most exhaustive investigation of planktonic algae undertaken for New York lakes. Mills ranking summaries may be found in his paper on Oneida Lake (Mills *et al.,* in Volume II, this treatise).

A total of 159 phytoplankton and incidentally suspended species were identified by Mills in Conesus Lake. The greatest number of species, characteristically present during the summer and fall, belonged to the Chlorophyta. *Oocystis lacustris* Chod., *Dimorphoccus lunatus* A. Braun, and desmids such as *Cosmarium reniforme* Ralfs., *Cosmarium pyramidatum* Breb., and *Staurastrum natator* var. *crassum* W. and G. S. West typified this pattern, while *Ankistrodesmus falcatus* (Corda) Ralfs and *Sphaerocystis schroeteri* Chod. were generally present year-round. Phytoflagellates belonging to the Cryptophyta and Chrysophyceae appeared frequently throughout the year. The Cryptophyta were represented by *Cryptomonas erosa* Ehr., *Cryptomonas pusilla* Bachm., and *Cryptomonas ovata* Ehr., while the Chrysophyceae consisted of flagellates, nonmotile colonial forms, and a filamentous species: *Chromulina ovalis* Pasch.,

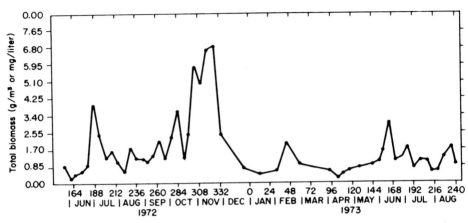

Fig. 15.   Phytoplankton biomass in Conesus Lake (upper 10 m). From Oglesby *et al.* (1975).

Dinobryon sertularia Ehr., *Ochromonas* sp., *Cladomonas fruiticulosa* Stein, and *Tribonema minus* (Wille) Hazen. The two last species were cold tolerant, occurring during the winter months. *Ceratium hirundinella* (O.F.M.) Schr., a large pyrrophyte or dinoflagellate, was observed only during the summer months, while a second important member of this group, *Peridinium cinctum* (Muell.) Ehr., was seen year-round. The predominant diatoms were species of *Cyclotella* sp., *Navicula* sp., *Asterionella formosa* Hass., *Fragilaria crotonensis* Kitt., *Fragilaria virescens* Ralfs, *Melosira granulata* (Ehr.) Ralfs, *Stephanodiscus astraea* (Ehr.) Grun., and *Tabellaria fenestrata* (Lyngb.) Kuetz. The number of species of blue–greens was greater than in Hemlock, Owasco, or Skaneateles Lakes, although the total species diversity was about equal to Owasco and below Hemlock.* *Anabaena flos-aquae* (Lyngb.) Breb., *Aphanizomenon flos-aquae* (L.) Ralfs, *Lyngbya limnetica* Lemm., and *Microcystis aeruginosa* Kuetz. were the most important forms. *Merismopedia trolleri* Bach. was present during the summer months and tended to show increased abundance when decreased dissolved oxygen concentrations developed in the hypolimnion. In general, the number of species composing the phytoplankton community exhibited seasonal maxima during the summer and minima during the winter. A pronounced decrease in the number of species occurred after Hurricane Agnes (June, 1972).

The National Eutrophication Survey (USEPA, 1974) sampled Conesus Lake for phytoplankton three times during 1972, in the same period when

---

* See the subsequent discussion of nomenclature for species of blue–green algae, No. 4 of following list.

Mills sampled at weekly or biweekly intervals. The EPA report identified five dominant genera and "other algae" on each occasion with counts of individuals as number/ml. The chrysophyte genus *Dinobryon* dominated the flora on May 5, with 982 of the 1477 total. On July 27 two diatoms dominated the flora: *Fragilaria* with 588/ml and *Cyclotella* with 149/ml out of the 1193 total. On October 13, the blue–green genus *Anabaena* dominated with 1642 of the total 3434 algae/ml.

The decline in numbers from May to July appears contradictory to Mills' data for biomass (Fig. 12), which indicated a slight decline to the first of June, but a dramatic temporary increase following the Hurricane Agnes flooding in late June. The greater number observed in October corresponds to the fall bloom of blue–green algae observed by Mills. Further correlation between the two studies does not seem meaningful. The green alga, *Schroederia*, was reported among the five dominants by the USEPA in May and July, but the genus did not once appear in Mills' 28 collections (May–Dec), or in Mills' collections of 1973, nor had the species been listed in 1968 (Savard and Bodine, 1971). Most of the other dominant genera were present consistently through the entire period of Mills' sampling in 1972. In terms of numbers, the fall population increase is of modest proportions, only double. Mills did show lower amounts of phytoplankton for his mid-October collections, although the general level was much higher from late September to December.

Before considering Mills' extensive data summaries, it is appropriate to list the limitations of sampling and identification in order to determine the reservations which must be made in accepting derived formulations.

1. Approximately 25 species of Savard and Bodine's list were attached or other nonplanktonic forms. Most of these were green algae.

2. Twelve species or genera of green algae were identified in both studies. Only two of these appeared constantly in the 1973 sampling: *Ankistrodesmus falcatus* (Corda) Ralfs and *Pediastrum boryanum* (Turp.) Menegh.

3. Fourteen diatoms were on both lists, which totaled 78 identifications. Of these, however, there was good coincidence for most of the commonest species: *Asterionella formosa* Hass., *Cocconeis placentula* Ehr., *Fragilaria crotonensis* Kitt., *Melosira granulata* (Ehr.) Ralfs, *Meridion circulare* (Grev.) Agardh., *Stephanodiscus astraea* (Ehr.) Grun., *Synedra cyclopum* Brut., and *Tabellaria fenestrata* (Lyngb.) Kuetz. Of these, the first four were the most frequent and abundant planktonic diatoms in 1967, and the fifth was absent only during the summer.

4. The blue–green algae require careful consideration because their taxonomy is difficult, and extensive revisions are occurring at family, genus,

and species levels. Mills' generalization that more species of blue–greens occur in Conesus than in the other lakes studied is probably correct, but the total number of species is certain to be reduced in taxonomic review. The species discussed here have been verified over a 7-year period.

Anabaena flos-aquae (Lyngb.) Breb. is unquestionably common in the lake, and may be found at any time of the year. Summer blooms are characteristic, but blooms under ice are not beyond expectation.

Aphanizomenon flos-aquae (L.) Ralfs. The species was not found in 1967 or 1968, but bloomed in fantastic abundance during October of 1970, reaching the consistency of green paint when concentrated by current or wind. Mills found it throughout 1973, but fall occurrence seems more typical. It is much less consistent than *Anabaena* both seasonally and perennially. Recently, Drouet (1976) has assigned this form to *Calothrix parietina* (Naeg.) Thur., and reduced *Aphanizomenon* to synonymy. While it may seem startling to limnological investigaters that such different forms could be ecological and developmental phases of the same species, both T. F. Maxwell (1972) and K. R. Khan (1974), using material from Conesus Lake, indicated plausibility for Drouet's taxonomic revision.

Anacystis cyanea (Kuetz.) D.&D. Mills and most limnological literature refer to the species as *Microcystis aeruginosa* Kuetz. Although it was constant (all 25 samplings) during 1973, it is not generally common or consistent. It is a characteristic species for the lake, but never dominates its plankton as the species does in certain other, waters such as Irondequoit Bay.

Merismopedia species have been reported in plankton from time to time in lakes of the region. Mills included two species, found only in the fall, *M. tenuissima* Lemm. and *M. trolleri* Bach. In their revision of the coccoid blue–green algae, Drouet and Daily (1956) found the name with priority for the genus to be *Agmenellum,* and they reduced most of the myriad of species names to synonymy. (*Merismopedia tenuissima* Lemm. is one of the synonyms for *Agmenellum quadruplicatum* Breb.) Most important in regard to plankton, none of the genus develops gas vacuoles like the characteristic planktonic blue–greens. Certain bacteria are similar in form and do contain granules which might be interpreted as gas vacuoles, and specimens of these bacteria have been the basis for some of the described species of *Merismopedia*. The earliest name of the commonly confused bacterial species is *Erythroconis littoralis* Oerst., but it is generally known as *Thiopedia rosea* Winogradsky.

Calothrix parietina (Naeg.) Thur. is listed in limnological literature as *Gleotrichia echinulata* (J. E. Smith) Richt., which is its colonial planktonic form. It occurred with impressive bloom in 1968 and 1969, extending from summer to fall. Maxwell collected it frequently in 1970 and 1971,

Khan from 1970 to 1974. Yet, the species was not noted in Mills' 25 collections during 1973. The species must, however, be considered as most characteristic for the lake. Indeed, it seems to have been in bloom condition when Muenscher surveyed the lake in early September, 1926 (reported as *Rivularia*).

The problem of reconciling planktonic and nonplanktonic forms for the Oscillatoriaceae was largely accomplished by Drouet (1968), but the task of transmitting the information to limnologists and others who must identify algae in field investigations has not begun. It is not possible to reconcile confidently Mills' identifications with the revised taxonomy without studying actual specimens. Yet, the treatment given here is based on a knowledge of both the taxonomy of the family and specimens which have been collected in Conesus Lake, or might be present there.

Schizothrix calcicola (Ag.) Gom. This small species is probably the most ubiquitous species of algae on earth. It is highly polymorphic and may develop the gas vacuoles of planktonic forms. Mills' *Phormidium tenue* (Menegh.) Gom. is probably this species.

Microcoleus vaginatus (Vauch.) Gom. This species is the second commonest of the Oscillatoria family, and a long list of "species," varying in size and sheath, have been described for filaments with a tapered end and a button, or cone tip. Forms with and without gas vacuoles can be found. Mills' *Oscillatoria prolifica* (Grev.) Gom. unquestionably belongs here.

Microcoleus lyngbyaceous Kuetz. This species occurs in several forms, one of them the well-known plankter called *Oscillatoria rubescens* DC.

The list is minimal, and could be extended to include most of the common nonmarine species of blue–green algae if a serious collecting was undertaken for a variety of sites in and adjacent to Conesus Lake. However, extension of the discussion is not appropriate at this time.

5. The dinoflagellate or pyrrophyte, *Ceratium hirundinella* (O.F.M.) Schr., has been found in Conesus Lake consistently, although the population apparently drops from time to time. Mills did not find it in winter collections, but it has appeared in winter and constantly in warmer seasons during other years.

6. Mills alone found and identified a number of Cryptophytes and flagellated Chrysophytes. A number of species were constant components of the plankton, and may be characteristic of the lake.

In summary, even though the algal flora of Conesus has been sampled more extensively and studied more skillfully than the flora of any other lake in the region, the knowledge of it must be considered quite limited. A characteristic flora consisting of a few species can be described. The lake is rich in species, but the preponderance of green algae and diatoms can be expected in almost any mature lake in the temperate zone. The most characteristic species, however, are the groups of common diatoms and blue–green algae which have been designated previously, and one or two dinoflagellates. Conspicuous blooms of blue–green algae have been noted, but the green colonial flagellate, *Pandorina,* was conspicuous in the spring of 1968, and other species, diatoms and *Ceratium,* probably develop high populations irregularly. Aside from the relatively constant species, the flora fluctuates widely and cannot be defined in precise terms for a long list of species. Indeed, the number tends to obliterate perspective.

Mills' data are so extensive that it is somewhat difficult to truncate within the limits of this article, but some significant examples have been selected. Biomass of algae, by major taxonomic group and season, is summarized in Table 14. The summer abundance of blue–green algae is demonstrated quantitatively in the data. Green algae were represented by the most species, but rather surprisingly, Cryptophytes as well as diatoms attained about equal the mass of the green algae, which are represented by most species. The summer study was conducted partly for the methodologic determination that weekly collection was frequent enough. In the seasonal

TABLE 14

Biomass of Algae in Conesus Lake by Taxonomic Group and Season[a]

		Daily mean	Weekly mean
Integrated means by day and week summer 1972 (gm/m³/ day fresh weight) by taxonomic group	Dinoflagellates	1.70	2.01
	Pyrrophyta	0.18	0.34
	Cryptophyta	0.23	0.31
	Yellow–greens Chrysophyta		
	Chrysophyceae	0.05	0.07
	Diatoms		
	Bacillariophyceae	0.18	0.22
	Blue–greens		
	Cyanophyta	0.80	0.79
	Greens		
	Chlorophyta	0.23	0.20
Seasonal integrated averages of total biomass (gm/m³ fresh weight)	Summer stratification 1972	1.39	
	Autumnal overturn 1972	3.42	
	Winter 1972	0.90	
	Spring overturn 1973	0.65	
	Summer stratification 1973	1.44	

[a] From Mills (1975). Data from May 1972 through Aug 1973.

averages, the very high mass for autumn of 1972 reflected the bloom of blue–green algae which is discussed elsewhere. The summer levels were only 50% above the winter, and the low spring level probably includes simple dilution. The seasonal flux of total biomass in the upper 10 m of water is shown in Fig. 12.

The algae were also tabulated into size categories: netplankton (70 μm), large (20–70 μm) and small (10–20 μm) nannoplankton, and ultraplankton (less than 10 μm). The entire plant, whether colony, filament, or single cell, was tabulated as a unit. Conesus was characterized by netplankton from October through June. During the summer of 1972 ultraplankton shared size dominance with netplankton, and both small nannoplankton and ultraplankton increased during June through August period of 1973. These data reflect precisely the same characteristics noted for the species distribution. The common species are netplankton, and there is a considerable degree of fluctuation in the flora with respect to other species. The Cryptomonad flagellates increased the representation of small size categories, and accounted for a spring peak in 1973.

The vertical distribution of selected phytoplankton species was plotted for an entire year. It seemed striking that many species were either uniformly distributed or reached maximum values in the upper 5 to 10 m. Nevertheless, as Mills stated, this distribution is to be expected if some species are characteristic of the stratified season. There is not enough understanding of this aspect of algal dynamics to utilize the data further at this time. In an unpublished study, T. F. Maxwell found winter diatoms concentrated somewhat below the surface, but still above the 5-m depth. Thus, a strategy is indicated by which plankton utilize environmental resources.

Mills plotted surface plankton counts for the four lakes investigated against chlorophyll a, for 1972 and 1973, with a correlation of 0.78 (n = 172), but there was considerable scatter in the data. He plotted both chlorophyll a and chlorophyll a + phaeophytin in mg/m^3 from May 1972 through August 1973 (Fig. 14). The two lines are close to each other, and the pattern simply reflects the pattern of abundance which has already been discussed. The extremely high fall peak is again apparent. Seasonal means and annual ranges were determined for both pigment groups and also for carotenoids.

Other Microbiology

Data have been reported only for total coliform bacteria. Expectedly, surveys in mid-lake have always recorded low numbers. The single substantial study of shallow waters and adjacent streams was completed in 1968 and reported by Forest and Mills (1971). A study for the Village of Avon (H. S.

TABLE 15

Occurrence of Zooplankton Species in Conesus Lake

Zooplankton species	1910[a]	1965[b]	1972–1973[c]
Calanoid copepods			
Diaptomus minutus Lillj.	x		x
D. sicilis Forb.		x	x
Cyclopod copepods			
Cyclops sp.	x		
C. bicuspidatus Forb.[d]		x	x
C. vernalis Fisch.			x
Mesocyclops edax Forb.		x	x
Cladocera			
Bosmina (Eubosminia) longirostris Mull.			x
Daphnia hyalina Ley.[e]	x		
Daphnia pulex de Geer	x	x	x
Leptodora kindtii (Focke)		x	x
Protozoans			
Difflugia lebes Ehr.			x
Glaucoma scintillans Ehr.			x
Rotifers			
Asplanchna priodonta Gosse			x
Ascomorpha saltans DeB.			x
Conochilus unicornis Rous.			x
Kellicottia longispina (Kell.)			x
Keratella cochlearis Gosse			x
K. hiemalis Carl.			x
K. quadrata (Mull.)			x
Monostyla quadridentata Harr.			x
Polyarthra spp.			x
P. euryptera Wier			x
P. vulgaris Carol.			x
Triochocera multicrinis Jenn.			x

[a] Birge and Juday (1914).
[b] Hall and Waterman (1967).
[c] Chamberlain (1975).
[d] This is the author according to Chamberlain, but Pennak (1953) give Claus as author.
[e] *Daphnia hyalina* Leydig 1860 is considered a separate, but primarily European, species by current taxonomists. It is sometimes confused with other species. Based on the descriptions available to Birge and Juday, the most probable identity of Conesus specimens is *D. pulex*. However, in other Finger Lakes, the name would have been applied to *D. galeata mendotae*. Note by H. D. Chamberlain June 6, 1976.

Forest and P. G. Savard, unpublished, 1969) showed only natural levels near the water supply intake in the Pebble Beach area during 1968 and 1969. The counts of 50–500 total coliform/100 ml were typical of deeper waters of the lake. The tests at Hanna's Creek, Sand Point swamp, and the short beach between where mobile homes were parked (Fig. 6) did strongly indicate fecal contamination. Boyle (1974) reported very few counts obtained from the Livingston County Health Department near Hanna's Creek. He concluded that the level of coliform bacteria was substantially reduced after the Conesus Lake sewer system intercepted septic systems around the perimeter of the lake in 1973.

Zooplankton

Birge and Juday (1914) reported their sampling for zooplankton of August 25, 1910, and these data have been used to compare with more recent samplings by Hall and Waterman (1967) and by Chamberlain (1975). The two earlier studies identified five and three species of microcrustacea. The only species in common was *Daphnia pulex* de Geer and both lists included *Diaptomus* and *Cyclops*. Chamberlain studied the zooplankton of Conesus as well as Skaneateles, Owasco, and Hemlock Lakes. This investigation was coordinated in time and scope with that of Mills (1975) on phytoplankton. Like the Mills study, it is one of the most comprehensive studies of its kind on New York lakes. The remaining published studies are those of Muenscher (1927), who noted a few bottom forms, and R. Walker (1975), who sampled the aquatic invertebrate populations of Conesus Outlet Creek in 1974, and the lake and its tributaries in 1975. Species found in zooplankton are listed in Table 15, and a taxonomic note for the table was provided by H. D. Chamberlain (personal communication). Other identified invertebrates, including some planktonic forms, are assembled in Table 16; snails are listed in (Table 17).

Chamberlain collected from mid-lake in weekly or biweekly intervals from May 19, 1972 to August 28, 1973, by vertical haul of a tow net. During stratification in 1973, samples were taken from the three thermal layers, the lowest near the bottom. The types of data compiled by Chamberlain included (1) individuals in m^3 and m^2 by sex and length categories; (2) grouping into larger taxonomic categories by area and volume of water; (3) living or wet weight for each species by length category, deriving a biomass estimate for each sampling day; (4) dry weight derivations based on published and original data; (5) relationships with seasonal and other environmental variables.

To summarize briefly, 21 species of zooplankton were found. The diversity was the least of the four lakes studied: Skaneateles, Owasco, Hem-

TABLE 16

Checklist of Some Invertebrates for Conesus Lake

Invertebrate	Station no.[a]	Source[b]
Protozoa		
Coleps sp.	3	J
Dileptus americanus Kahl	3	J
Dileptus sp.	3	J
Multicilia lacustria Laut.	3	J
Paramecium busaria Ehr.	3	J
P. caudatum Ehr.	3	J
Stentor coeruleus Ehr	3	J
Vorticella sp.	3	J
Coelenterata		
Hydrozoa		
Hydra sp.	3	R and J
	8	R
Turbellaria		
Dugesia tigrina (Girard)	3	R and J
Rotatoria		
Bdelloidea sp.	3	J
Kellicottia longispina (Kellicott)	3	J
Annelida		
Oligochaeta		
Aeolosoma beddardi Mich.	8	R
A. hemprichi Ehr.	3	J
Chaetogaster limnaei K. von Baer	3	J
Lumbriculus inconstans (Frank Smith)	8	R
Stylaria sp.	3	R
Tubifex tubifex (Muller)	3	R
	8	R
Hirudinea		
Dina sp.	3	R and J
Helobdella sp.	3	R
Macrobdella decora (Say)	3	R
Bryozoa		
Fredericella sultana (Blumenbach)	3	J
Arthropoda		
Crustaceae		
Cladocera (water fleas)		
Daphnia pulex (deGeer)	3	R and J
Graptoleberis testudinaria (Fischer)	8	R
Anostraca (fairy shrimp)	3	J
Conchostraca (clam shrimp)	3	J
Copepoda		
Orthocyclops sp.	3	R
Cyclopidae sp.	3	J

TABLE 16 (Continued)

Invertebrate	Station no.[a]	Source[b]
Ostracoda		
Cypridonsis sp.	3	R and J
Potamocypris sp.	8	R
Isopoda (aquatic sow bugs)		
Lirceus lineatus (Say)	3	R
	8	R
Amphipoda (scuds)		
Gammarus fasciatus Say	3	R
	8	
G. linnaeus Smith	3	R and J
Decapoda (crayfish)		
Astacidae sp.	3	J
Cambarus bartoni (Fabr.)	3	J
Orconectes virilis (Hagen)	3	R
Arachnoidea		
Hydracarina (water mites)		
Diplodontus despiciens Mull.	3	R
Hydrochoreutes ungulatus (Koch)	3	R
Hygrobatea sp.	3	R
Lebertia sp.	3	R
Limnesia sp.	3	R
Megapus sp.	3	R
	8	R
Insecta		
Collembola (springtails)		
Isotomurus sp.	3	R
Ephemeroptera (mayflies)		
Caenis sp.	3	R
	8	R
Tricorythodes sp.	3	R
Odonata (dragonflies and damselflies)		
Anomalagrion sp.	3	R
Argia sp.	3	R
Ischnura sp.	3	R
	8	R
Aeschnidae sp.	3	J
Enallagma sp.	3	J
Hemiptera (bugs)		
Gerris sp.	3	R
Limnogonus sp.	3	R
Plea sp.	3	R
Megaloptera (alderflies, dobsonflies and fishflies)		
Sialis sp.	3	R
Trichoptera (caddis flies)		
Helicopsychidae sp.	3	J
Leptoceridae sp.	3	J

(Continued)

TABLE 16 *(Continued)*

Invertebrate	Station no.[a]	Source[b]
Coleoptera (beetles)		
Agabus sp.	3	R
Dubiraphia sp.	3	R
Endalus sp.	3	R
Galerucella sp.	3	R
Suphisellus sp.	3	R
Hydroporus sp.	3	J
Diptera (flies, mosquitoes, and midges)		
Anatopynia (Psectrotanypus) sp.	3	R
Atherix variegata Walker	3	R and J
	8	R
Bezzia sp.	3	R
Chaoborus sp.	9	R
Chironomus (Camptochironomus) sp.	9	R
C. (Dicrotendipes) sp.	3	R
C. (Endochironomus) sp.	3	R
C. (Microtendipes) sp.	3	R
	8	R
C. (Polypedilum) sp.	3	R
	8	R
C. (Pseudochironomus) sp.	8	R
C. (Tribelos) sp.	3	R
Culicidae sp.	3	J
Pentaneura sp.	3	R
	8	R
Procladius (Psilotanypus) sp.	3	R
	9	R
Mollusca		
Pelecypoda		
Anodonta sp.	3	R
Sphaerium sp.	9 and outlet stream	R
Gastropoda (see Table 17)		

[a] Description of stations: station no. 3—240 Pebble Beach Road on Conesus Lake. The shore vegetation consisted of emergents, such as *Pontederia* sp. and *Scirpus* sp. The submerged plants included *Chara* sp. as well as the various commoner vascular species found in the lake. The bottom was gravel near the shore with silt and sand extending outward. Samples were collected at 0- to 1-m depths along this undisturbed section of the shoreline. Station no. 8—2801 Old Orchard Point. The beach at this station was heavily used by cottagers, and most aquatic vegetational development was retarded. The gravel bottom supported *Chara* sp. and occasional floating rafts of vegetation were seen, depending on the wind direction and velocity. Samples were collected from depths of 0–1.5 m. Station no. 9—near south end of the lake, opposite Walkley's Landing. Samples were taken from a depth of 13.7 m, with a silty muck bottom.

[b] J, data collected by J. Wade; R, data collected by R. Walker.

TABLE 17

Checklist of Snails for Conesus Lake and Vicinity

Family	Genus, species	Collection location	Identification source[a]
Physidae	*Physa elliptica* Lea	Stream, lake	Walker
	P. gyrina Say	Lake	Wade, RBB[a]
	P. integra Haldeman	Lake	Wade, RBB
	P. sayii Tappan		RBB
Lymaeidae	*Lymnaea stagnalis* (Linnaeus)[b]	Lake	Wade
	Lymnaea catascopium Say		RBB
	Fossaria modicella Say		RBB
Planorbidae	*Helisoma trivolvis* (Say)	Lake, stream	Wade, Walker, RBB
	Helisoma campanulata (Say)	Lake	Wade
	Gyraulus parvus (Say)	Stream, Lake	Wade, Walker
	Promenetus exacuous (Say)[b]	Stream, Lake	Walker, Wade
Ancylidae	*Laevapex fuscus* (Adams)[b]	Stream, lake	Walker, Wade
Viviparidae	*Viviparus georgianus* (Lea)	Stream, lake	Wade, Walker
	Campeloma decisa (Say)	Lake	Wade, RBB
Pleuroceridae	*Pleurocera acuta* Raf.	Lake	Wade
	Goniobasis livescens (Menke)	Lake	Wade, Walker, RBB
Hydrobiidae	*Amnicola limnosa* (Say)	Stream, lake	Wade, Walker
Valvatidae	*Valvata tricarinata* (Say)	Lake	Wade, Walker

[a] From J. Q. Wade (Wade and Vasey, 1975 and unpublished); R. A. Walker (1975); and RBB, Robertson et al. (1948).
[b] Unverified.

lock, and Conesus. Hemlock was highest in diversity at 29 species. The Conesus list contains about one-third of the total species (61) found by Chamberlain in all 11 Finger Lakes. Comparison among the lakes must be made with strong reservations. Interestingly, Conesus produced the fewest number of species in all three of the published lists, among the four lakes. Their trophic rank ranges from fertile (Conesus and Hemlock) to highly infertile (Skaneateles). If this pattern is not accidental, it is in marked contrast to the richer variety of many other life forms in Conesus Lake. On the other hand, Conesus exhibited the highest mean number of species/day (11.20 in 1973), while Skaneateles was the lowest (5.45 in 1972). Indeed, by this measure the lakes conform to their generally accepted trophic rank. In ascending order they are: Skaneateles, Owasco, Conesus, and Hemlock, the last two exchanging third and fourth rank during the two years.

In terms of zooplankton population abundance, Conesus exceeded Hemlock by most criteria. The lakes again usually fell into order. The greatest number of individuals in each Sedwick–Rafter counting cell was 380 for Conesus Lake and 421 for Hemlock, but Conesus attained highest values in numerical abundance by lake surface area (8,005,968 individuals/m^2) and by lake volume (500,378 individuals/m^3, July 1973). The species occurred generally throughout the year, except that about half of the kinds of rotifer were found only in warmer weather. The most abundant species in both Conesus and Hemlock lakes was the rotifer *Conichilus unicornis* Rous. It reached an abundance of 408,948 individuals/m^3 in late July 1973. *Daphnia pulex* (cladoceran) was the only species found on every sampling date (and also, as noted, the only species reported in 1910, 1965, and 1972–1973). Its density peaked earlier, in June and July, but also increased in December 1972. Two other species generally present were the rotifer *Kellicottia longispina* Kell. and the copepod *Cyclops bicuspidatus* Forb. Chamberlain meticulously traced the annual population flux of these and other species.

When the species were grouped, Conesus was revealed as dominated by rotifers, followed by crustaceans.* The position at Conesus in relation to the other lakes was usually intermediate with respect to calanoid copepods and nauplii, but the other lakes were not constant in rank. Conspicuously, the greatest abundance developed in the epilimnion, with a noticeable peak in the metalimnion and no distinct seasonal trend in the hypolimnion. Added individuals, not primarily added species, account for the very high summer peaks in volumetric abundance.

In 1972, the high surface runoff resulting from Hurricane Agnes (about June 20) apparently disrupted the rise in population level of zooplankton. A

---

* In terms of biomass, the cladocerans are dominant, followed by cyclopoid copepods, and rotifers rarely are significant (Fig. 16).

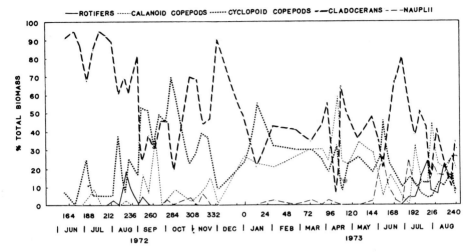

Fig. 16. Zooplankton biomass by taxonomic groups in Conesus Lake. From Chamberlain (1975).

August. Therefore, 1973 is the only available year for data which may be typical. In that year the peak extended from June through August, reaching the highest level in July (Fig. 17). The pattern in Hemlock Lake was generally similar, but the acceleration began earlier, and there was a sharp decline in late June between two high peaks. In contrast, Skaneateles had no significant summer increase, and Owasco only a modest rise. A possible interpretation of this difference in the two pairs of lakes will be suggested subsequently, in the discussion of submerged vascular plants. Chamberlain's determination of biomass again showed Conesus as the leading producer among the four lakes, with a high value of 857.6 mg/m³. The mean total biomass for 1972–1973 was 145 mg/m³, almost three times as high as the other lakes. In terms of unit surface area, the biomass was calculated as equivalent to 1100 mg/m², less than Skaneateles and Owasco Lakes (which are far deeper) but twice the value for Hemlock Lake (slightly deeper). Chamberlain also related the total zooplankton productivity to the photoplankton mass, particularly to the high crops of blue–green algae. By almost any criterion of productivity, Conesus Lake is impressive.

It seems improbable that an estimation of long range trends could be attempted from the data available, yet a comparison was made. Chamberlain found that the zooplankton in Conesus Lake had changed no more than in the other lakes. It may be noted again that during this period the human waste received by Conesus increased enormously in relation to the others, particularly Hemlock Lake. Although the number of rotifers may

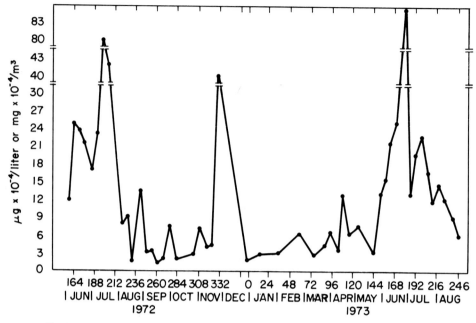

Fig. 17. Zooplankton population level in Conesus Lake. From Chamberlain (1975).

have increased considerably in Conesus Lake, their total biomass is small. Species richness appears to have increased in all of the lakes. The amount of increase in standing crop of Conesus was judged to be intermediate between Hemlock Lake (by far the highest) and the other two lakes.

It is not possible to do full justice here to all of Chamberlain's work, which included a geometric portrayal of "relative living space," and considerations of niche, nutritional and temperature relationships, and other approaches to analysis of the zooplankton community. Among his conclusions, he emphasized that Conesus had the greatest species dominance, the most predictable species, the largest species biomass, and largest mean biomass. These conclusions parallel those which could be derived as to the trophic status of Conesus Lake from phytoplankton data. The total number of species in Hemlock Lake, however, was greater, again parallel to the phytoplankton! Conesus is strongly dominated by a single large cladoceran, *Daphnia pulex,* with copepods in second rank. This is in general conformity to the pattern of productive lakes and in contrast to the dominance of calanoid copepods in less productive lakes. The very large *Daphnia* may, however, reflect specifically the absence of alewives, *Alosa pseudoharengus* (Wilson), from Conesus Lake, as Chamberlain suggested.

Other Invertebrates

Muenscher (1927) included a few notes on insect larvae in his report of vegetation in Conesus Lake. All were found in mud, in almost all samples at depths from 3.3 to 14.5 m. They were not found in deeper or shallower waters nor in gravelly or sandy bottoms. *Chironomus* was the only genus found except along transect V, located near the inlet of the lake, where *Chaoborus (Sayomia)* was found at 12.1 m.

R. Walker (1975) sampled nine sites in the Conesus drainage basin during 1974 and 1975. His stations were located on Conesus Creek (no. 1, no. 2), shallow (0–1.5 m) shore locations in the lake (no. 3 and no. 8), at 13.7-m depth near the inlet (no. 9), and the remainder were in streams tributary to the lake. Stream stations included a cold water stream (no. 4), a warm water stream (no. 7), and Conesus Inlet (no. 5, no. 6). Bottom condition and vegetation were described for each site.

None of the sites were in mid-lake so that the sampling was directed primarily at nonplanktonic forms. However, some crustaceans from the lake stations should be noted in relation to Chamberlain's study. *Daphnia pulex* de Geer reliably appeared in the deep water station, another *Daphnia* only in the inlet stream. *Orthocyclops modestus* Herr., *Ostracoda* sp., *Potamocypris* sp., and *Gammarus fasciatus* Say do not appear on Chamberlain's consolidated list from 11 Finger Lakes. *Gammarus,* which was found in both shallow lake station and in five stream stations as well, has been observed frequently at the edge of the lake, sometimes in very dense concentrations. It seems to be restricted to the shallows, and appears to dominate these areas even more conspicuously than *Daphnia* dominates deeper waters.

In all, 175 invertebrate species were identified, the fauna being primarily littoral and benthic. Both large and small invertebrates were included. Habitat notes provide for some ecological correlations.

Categories include Coelenterata, Hydrozoa, *Hydra* sp.; Turbellaria, *Dugesia tigrina* (Girard); Nematoda (single collection from outlet stream); Annelida (Oligochaeta and Hirudinea); Arthropoda; and Mollusca.

Arthropoda

In addition to the group already noted, an aquatic isopod (*Lirceus lineatus* Say) and two species of crayfish were identified. Insects predictably provided most of the species on the list. Nine orders were represented, with Diptera showing the greatest variety.

Mollusca

Walker found one clam (*Anodonta* sp.) in the lake, and the other (*Sphaerium* sp.) in the inlet and outlet streams. Also, further studies have been conducted on gastropods. Robertson *et al.* (1948) included a few snails collected on the west side of Conesus Lake in their survey of the gastropods of western New York. J. Q. Wade began to study the biology of the snails in 1971. The investigation is still in progress, but a preliminary report has been published (Wade and Vasey, 1975). Wade also collected and identified 32 other invertebrate species from shallow waters at 240 West Lake Road. A consolidated list of invertebrates, excluding snails, by Wade and by Walker in the lake itself can be found in Table 16.

The following discussion of the snail populations of Conesus Lake is based on information provided by Wade. The environment seems highly favorable for snails because of the high calcium content of the waters and rich flora of vascular plants and attached algae in generally calm shallow waters. There is a general seasonal migration to burrows in deeper mud during the winter and emergence to the weedy shores about mid-May. However, Pleuroceridae have been found active in a mild November (1972), with the water temperature at 10°C.

Viviparus is the most common and abundant of the larger snails. *Viviparus georgianus* (Lea), the most common species, requires moderate oxygen levels (3–8 mg/liter) according to Harman and Berg (1971). Only three individuals of *Campeloma decisa* Say, which requires 8–12 mg/liter of dissolved oxygen, were found during 1971–1973, but several large specimens were collected during 1975. Pleuroceridae were commonly picked from cement breakwalls and iron structures where the wave action was at times strong. The greatest population of this species was found in the rapid waters of the outlet stream. Disk-shaped Planorbidae and the Physidae were found only in the shelter of weeds in shallow waters. Tiny snails were found free floating or attached to plants or other snails. In 1971 and 1972 collections were made at four widely separated sites, while the 1973 collections were

TABLE 18

Distribution of Individual Snails by Family in Conesus Lake, 1971–1973[a,b]

	1971	1972	1973
Individuals	321	219	365
Viviparidae	53.2%	42.0%	69.3%
Pleuroderidae	31.2	47.0	26.8
Planorbidae	8.0	4.0	3.8
Physidae	7.2	7.0	0.1

[a] From J. Q. Wade.
[b] Note: Tiny snails, including *Valvata* and *Amnicola*, were not tabulated.

restricted to a single site, 240 West Lake Road, in Grey Shores (for location, see Fig.6). This site, noted elsewhere, is in the area where *Chara* is abundant and a small marginal community (*Scirpis, Pontederia, Sagittaria*) has begun to develop. A summary of all known collections is presented in Table 17, and distribution of individuals by families in 1971–1973 in Table 18.

The snails were found to be intermediate hosts for certain flat worms. Approximately five host-specific trematode larvae have been tentatively identified, including the cercariae of the xiphidioceraria type, furcocysto-cercous, and pleurolophocercus types. The snail hosts were of the Viviparidae, Physidae, Planorbidae, and Pleuroceridae families.

Fish

General Information

The sources of information on fish are from the New York State Department of Environmental Conservation (NYSDEC) Region 8, Avon, New York, and the Biology Department, State University College at Geneseo. Data from files and oral statements have been supplied by E. N. Holmes and W. J. Abraham of the NYSDEC and R. M. Roecker of the State College.

The earliest source of fish biology for the region is the report by Greeley (1927). An annotated list of species with some locations was included, and there are a few specific references to Conesus Lake, but no comprehensive list was attempted. A short but valuable case history on Conesus Lake (Abraham, 1975) was completed concurrently with this writing. State fisheries management records begin in 1940, but there is no entry for some years.

Dr. Roecker began collection and identification in 1964. His data sheets record species identification, relative abundance, and environmental information. The data have been summarized only to the extent that a checklist has been prepared for Livingston County, revised to July 1974. With the inclusion of a few uncertainties, there are 59 species in the county list and 39 from Conesus and its inlet stream within about 1 km of the lake (Table 19). Half of the difference is in the number of species of minnows and simply reflects a choice of collecting sites. As a general observation, yellow perch, *Perca flavescens* (Mitch.), is the most numerous of sports fish. Other important game fish are the walleye, *Stizostedion vitreum vitreum* (Mitch.), the northern pike *Esox lucius* L., smallmouth bass, *Micropterus dolomieui* Lac., and largemouth bass, *M. salmoides* (Lac.). Other game fishes which contribute to the fishery and constitute significant elements in the ecosystem are the sunfish, pumpkinseed, *Lepomis gibbosus* (L), and

TABLE 19

Checklist of Fish from Conesus Lake and Conesus Inlet, 1964–1974[a]

Family and genus and species	Common name
Salmonidae	
Salmo gairdnerii Rich.	Rainbow trout
Salmo trutta L.	Brown trout
Salvelinus fontinalis (Mitch.)	Brook trout
Umbridae	
Umbra limi (Kirt.)	Central mud minnow
Esocidae	
Esox lucius L.	Northern pike
E. niger Les.	Chain pickerel
Cyprinidae	
Campostoma anomalum (Raf.)	Stoneroller
Cyprinus carpio L.	Carp
Hybognathus nuchalis Agassiz	Silvery minnow
Nocomis biguttata (Kirt.)	Hornyhead chub
Notemigonus crysoleucas (Mitch.)	Golden shiner
Notropis atherinoides (Raf.)	Emerald shiner
N. bifrenatus (Cope)	Bridled shiner
N. cornutus (Mitch.)	Common shiner
N. dorsalis (Agassiz)	Bigmouth shiner
N. heterodon (Cope)	Blackchin shiner
N. spilopterus (Cope)	Spotfin shiner
N. volucellus (Cope)	Northern mimic shiner
Pimephales notatus (Raf.)	Bluntnose minnow
P. promelas Raf.	Northern fathead minnow
Rhinichthys atratulus (Hermann)	Blacknose dace
Semotilus atromaculatus (Mitch.)	Creek chub
Catostomidae	
Catostomus commersonii (Lac.)	White sucker
Hypentelium nigricans (Les.)	Northern hogsucker
Moxostoma macrolepidotum (Les.)	Northern shorthead redhorse
Ictaluridae	
Ictalurus nebulosus (Les.)	Brown bullhead
Noturus gyrinus (Mitch.)	Tadpole madtom (rare)
Cyprinodontidae	
Fundulus diaphanus (Les.)	Banded killifish
Centrarchidae	
Ambloplites rupestris (Raf.)	Rock bass
Lepomis gibbosus (L.)	Pumpkinseed
L. macrochirus Raf.	Bluegill
Micropterus dolomieui Lac.	Small mouth bass
M. salmoides (Lac.)	Large mouth bass
Pomoxis nigromaculatus (Les.)	Black crappie
(no records, but known from larger streams of Genesee drainage)	

TABLE 19 *(Continued)*

Family and genus and species	Common name
Percidae	
Etheostoma blennioides Raf.	Greenside darter
E. flabellare Raf.	Fantail darter
E. nigrum Raf.	Johnny darter
Perca flavescens (Mitch.)	Yellow perch
Stizostedion vitreum vitreum (Mitch.)	Walleye
Atherinidae	
Labisdesthes sicculus (Cope)	Brook silverside

a From R. M. Roecker.

bluegill, *L. macrochirus* Raf., and also the brown bullhead, *Ictalurus nebulosus* (Les.). Other fish may also play significant roles in the ecosystem, but they have not received the attention and formal study accorded to sports fish. During spawning season suckers of very large size, mostly *Catostomus commersonii* (Lac.), have commonly been observed in numbers which fill the lower reaches of South McMillan Creek. In addition, during the dives made for study of submerged vegetation by H. S. Forest and associates, huge carp, *Cyprinus carpio* L., were observed on virtually every occasion over a period of 8 years.

Sports Fish Management

Salmonids Rainbow trout have been introduced systematically since 1969. The program strategy was based on the limnological finding that the middle layers of lake water were cool enough and sufficiently oxygenated to enable summer survival of "cold water" species. About 5000 brook trout were introduced in 1972 and 1973, and brown trout from 1971 to 1973 in similar numbers. The brook trout failed to survive and the brown trout continue as a few individuals found in the lake or, during the spawning season, in the McMillan Creeks.

The program for maintaining a population of rainbow trout has been somewhat successful, with a stocking of about 5000 yearlings from 1969 to 1973, and 15,000 in 1974–1975. Also, a number of hatchery surplus adults were placed in the lake in 1973. Individual fishermen have been well rewarded both in summer angling and ice fishing, although in summary the effort has produced only marginal return. The trout have moved actively to and from the McMillan Creeks during spawning season, as well as throughout the lake. Size of the introduced fish has proved important. Stocked yearlings must be at least 23 cm (9 in.) long to insure reasonable survival, since smaller individuals were likely eaten by northern pike. In future, stocking will be terminated. It is the judgment of fisheries biologists that only reproduction by an established population is satisfactory. Stocking simply has not produced a return which justified the cost and effort.

Walleye. The walleye population in Conesus Lake has been studied recently by DEC biologists, following reports for about 3 years that angling success declined in Conesus Lake. Indeed, professional fish biologists agree that the population has declined across the state for most waters. The 1975 study was completed in December with the analysis of 35 diaries from fishermen.

Fortunately, comparative data were available from trap netting conducted by the New York State Conservation Department from 1966 to 1968. The population at that time was estimated to be 12,000 adults, and the fishing experience was satisfactory. In the early spring (Mar–Apr) of 1975, 442 adults were trap netted, marked, and recapture attempted with the aid of electric boat shocking. Calculations based on the experiment indicated a population of about 6000 adults, a 50% decline. The complementary study of fishermen's diaries accounted for 219 fish, and produced a higher population estimate, 10,000. Such a difference is to be expected given the limited amount of data available, and the very nature of this study. As an interesting footnote to the diary study, it was observed that the skill and tenacity of fishermen vary enormously. Their catch varied from none to 87 walleyes per individual. Such differences are characteristic of most creel census investigations in the experience of state fisheries investigators.

The investigation provided valuable information both for basic ecology and application to management. Spawning areas for walleye have been identified: South McMillan Creek near Conesus Inlet, and the lake margin generally, particularly in gravelly areas. This discovery is most encouraging for the prospect of continuing the species at a high stable level, since natural spawning rather than stocking would supply replacements. The ability of the natural population to maintain the present population level will be subject to test in coming years, since stocking was suspended for 1974 and 1975 only. If stocked fish do contribute significantly, the population will sag.

The spawning area used by northern pike at Sand Point, adjacent to the outlet, was destroyed by land filling in 1966. This left only the small and disturbed area of wet shore near the inlet as a spawning ground in the lake. However, the inlet stream and South McMillan are spawning areas for this species and most of the spawning occurs in the marsh area at the DEC wildlife management area south of the lake. Field notes by R. M. Roecker in 1974 recorded his judgment that extensive shore and bank modifications of 5 years before had not permanently disturbed the general fish populations immediately upstream in Conesus Inlet in the state management area.

Nevertheless, South McMillan Creek itself is vulnerable to damage a short distance upstream before it enters the management area and joins Conesus Inlet. Its classification of "D" does not place it within the state's stream protection law.

A critical management proposal suggested by the study is that a meaningful size limit should be placed on walleye. Indeed, a state-wide size limit is now being considered. An adequate natural breeding population is strongly desired, because stocking of fry is unsuccessful, and stocking of fingerlings not feasible. Data indicate that the population in Conesus has declined because of overfishing. Evidence includes a surprisingly high reported return of fish tagged during the 1960's through 9 years. The exploitation rate (which excludes removal by natural causes) from 1966–1970 was 40–45%. The mean length reported in 1975 was 46 cm (18 in.). Even though very small fish are not retained by fishermen, the breeding population of males, at 45–50 cm and almost 1 kg in weight, is reduced somewhat. The females, which are larger, 60–62 cm long and almost 2 kg in weight, are probably reduced severely. A limit of 45 cm (17–18 in.) is indicated to restore the balance between production and harvest. In addition, some causes of population decline such as unfavorable weather, may lie outside management.

Study, Management, and Law

The study has been a most productive one, partly because base data were available. Yet, the attainment only emphasizes the handicaps under which fisheries management is attempted. There is no systematic and sustained investigative program which would provide either a historical continuity or an environmental perspective. The changes in seasonal dynamics, rooted aquatics, and zooplankton may well be reflected in the crop of game fish, but there is no way to find out about these relationships. Even direct information on fish production is limited. It is not possible to substantiate from published sources the generally expressed opinion that Conesus Lake is a highly productive lake for fish. Various reports exist which might provide a definitive comparison of growth rate or total crop in terms of fish weight/unit area with other lakes, but data are readily available. That fisheries management has succeeded at all under these conditions is due to personal knowledge and judgment of the state field staff.

The management problems designated in 1976 and the future by the regional DEC fisheries staff (Abraham, 1975; also personal communication) include the decline of walleye population, insufficiency of public access, possible conflict of interest between management of fish and land animals in the inlet and marsh areas south of Conesus Lake, and the lack of base line data to assess important changes which are observed in the field. In addition to these stated technical and policy problems, it seems clear that adequate legal means are required for implementing rational control. Past laws have been inadequate, and whether recent legislation proves to be more effective remains to be demonstrated in the future.

Aquatic Vascular Plants

Marginal Plants

The exemplary profile of successional communities at the southeastern corner of the lake has been discussed briefly in relation to the swamp maple community of terrestrial vegetation. Although shoreline alteration has largely devastated the vegetation, it is still possible to see the remarkable series: submerged, floating, emergent, woody emergent, saturated muck plants, dry and wet weedy area, and the swamp forest with maple (*Acer x freemanii* E. Murray) highly dominant, ash (mostly *Fraxinus pennsylvanica* Marsh.), and an occasional elm (*Ulmus americana* L.) present.

Since 1967 aquatic vegetation within the definition of Fassett (1957) has been collected and studied in the Conesus area (Forest *et al.*, 1977a) but little has been published. Supplementary reports to this article will be prepared for the grasses and sedges, and the general terrestrial flora will contain some marginal aquatics. Only two publications list aquatic plants. Muenscher (1927) surveyed the lake in early September 1926. Forest and Mills (1971) reported their survey of 1968 which used aquatic transects similar to Muenscher's. Their list of marginal plants was intended for comparison with the earlier study and does not represent a comprehensive flora (Table 20). Plants ranging in habit from floating leaved, emergent, shore, and inshore were included in Muenscher's list: *Typha, Sparganium, Peltandra, Nuphar, Nymphea,* and *Megalodonta beckii* (Torr.) Greene (a stick-tight). Four duckweeds were recorded, two species of *Lemna,* and *Spirodela* and *Wolffia.* Perhaps the most conspicuous omission, later found to be abundant near the inlet, was *Decodon verticillatus* (L.) Ell., the water willow. Also, Muenscher listed *Equisetum,* which seems to be only incidental to the marginal flora, but he omitted the sensitive fern *Onoclea* which is abundant and characteristic of wetlands.

With few exceptions, the species noticed by Muenscher can still be found, even though most of the wet shore has been destroyed by land filling. The last sizable area in the north, Sand Point Marsh, was filled between 1966 and 1968. A limited community of emergent plants is developing in the Gray Shores area on the western shore, where extensive *Chara* beds are located. There remains only the inlet area and a few patches near stream mouths.

TABLE 20

Occurrence of Aquatic Vascular Plants in Conesus Lake

Name[a]	1926[b]	1968[c]	After 1968
Sphenophyta (nonflowering)			
Equisetum fluviatile L. as E. limosum L.	x		x
Anthophyta (angiosperms or flowering plants)			
Marginal, emergent, and floating plants			
Megalodonta beckii (Torr.) Greene			x
as Bidens beckii Torr.	x		
Cephalanthus occidentalis L.		x	
Cicuta maculata L.		x	
Decodon verticillatus (L.) Ell.		x	
Lemna minor L.	x	x	
L. trisulca L.	x	x	
Nuphar variegatum Engelm.		x	
as Nymphozanthus variegatus (Engelm.) Fern.	x		
Nymphea odorata Ait.	x	x	
Peltandra virginica (L.) Kunth.	x	x	
Polygonum amphibium L.		x	
Sagittaria latifolia Willd.	x	x	
S. rigida Pursh		x	
as S. heterophylla Pursh	x		
Scirpus acutus Muhl.		x	
as S. occidentalis (Wats.) Chase	x		
S. americanus Muhl.	x	x	
S. atrovirens Willd.		x	
S. validus Muhl.	x	x	
Sparganium chlorocarpum Rydb.	x	x	
S. eurycarpum Engelm.	x	x	
Typha angustifolia L.	x	x	
T. latifolia L.	x	x	
Wolffia punctata Griseb.			x
Submerged plants			
Ceratophyllum demersum L.	x	x	
Elodea canadensis Michx.	x	x	
Heterantera dubia (Jacq.) MacM.	x	x	
Myriophyllum exalbescens Fern.	x	x	
and M. spicatum L.	x		x
Najas flexilis (Willd.) Rostk. and Schmidt	x	x	
Potamogeton amplifolius Tuckerm.	x	x	
P. angustifolius Berch. and Presl.		x	
P. crispus L.	x	x	
P. epihydrus Raf. as var. cayugensis (Wiegand) Benn.	x		
P. foliosus Raf.	x		
P. gramineus L.			x
as var. graminifolius Fries.	x		

(Continued)

TABLE 20 (Continued)

Name[a]	1926[b]	1968[c]	After 1968
P. natans L.	x		
P. nodosus Poir.			x
as P. americanus var. novaeboracensis			
(Morong) Benn.	x		
P. pectinatus L.	x	x	
P. pusillus L.			x
P. richardsonii (Benn.) Rydb.	x	x	
P. zosteriformis Fern.		x	
as P. compressus L.	x		
Ranunculus trichophyllus Chaix.		x	
as R. aquatilis var. capillaceus DC	x		
Utricularia vulgaris var. americana Gray	x	x	
Vallisneria americana Michx.	x	x	

[a] Names conform to revision of appendix in Fassett (1957).
[b] Data from Muenscher (1927).
[c] Data from Forest and Mills (1971).

Submerged Vascular Plants

Muenscher recorded the presence of submerged species and judged several of them to be "predominant." This varied flora included *Ceratophyllum, Elodea, Heteranthera, Myriophyllum, Najas, Utricularia,* and ten species of *Potamogeton*. A search of the Wiegand Herbarium at Cornell University located specimens for six species of *Potamogeton*, but they have not been reviewed taxonomically. Field investigations in 1968 found only six species, but specimens in the New York State Museum, Albany and subsequent field investigations will probably add to the modern list. The most recent review of the genus in New York (Ogden, 1974a) locates other species generally in the region, but does not identify collecting sites. Taxonomic review of specimens is needed to establish the number of species with some certainty. Nevertheless, it is concluded that the species diversity has not noticeably diminished since 1926. One additional submerged plant has been found, *Ranunculus*, the water buttercup.

The meaning of Muenscher's "predominant" must remain uncertain, but since it was applied to several species, the conclusion is inescapable that the mixed, varied community of 1926 was not greatly different from that found in 1968 or 1975. Muenscher noted that most of the weed beds contained several species with *Potamogeton* species predominating. He has indicated that *Chara* beds near the outlet may have been more extensive in 1926, and *Vallisneria* sometimes occurred in pure stands over extensive areas. The published data recorded from five transects are not sufficient to verify the general conclusions.

Beds of *Chara* may have been more extensive in 1926, but many *Myriophyllum* plants were also found, and *Ceratophyllum* was found at 5.5 m, along with a sparse growth of *Chara* and *Potamogeton*. The data do not indicate pure stands of *Vallisneria*, which probably were simply observed without sampling. *Vallisneria* is still quite conspicuous in shallow waters in some places, but close examination reveals a mixed community. *Myriophyllum* may have increased significantly since 1926, but *Ceratophyllum* and *Heteranthera* have probably increased too. Indeed, the four genera vary in abundance from place to place and from year to year. None have conspicuously displaced or replaced another.

Forest and Mills compared the earlier and current communities, and reported preliminary estimates of standing crop from quadrats along a longitudinal transect from shallow to deep water near the inlet, as well as surveys of other locations. The quantitative investigations have continued from time to time through August 1975. Samples were usually collected from randomly placed quadrats in defined communities and depths. The quadrats were defined communities and depths. The quadrats were defined by metal rings 0.2 m in area except in preliminary procedures. Each sample was weighed and the species proportioned and identified. The methodology, rationale, and some results of these investigations have been reported recently (Forest, 1976); however, the bulk of the data has not yet been compiled. The record for Conesus Lake, although far from adequate, is probably one of the most substantial in existence for a northern glacial lake. There are also some comparative data for five other Finger Lakes (Silver, Hemlock, Canadice, Honeoye, and Canandaigua) and two largely self-contained bays of Lake Ontario (Irondequoit and Sodus).

The species list of Conesus is a representative flora for the Finger Lakes region. Species which are most common in Conesus are typical of fertile waters, and include the ones which remain when a flora has been reduced by heavy plankton growth and pollutants, as an Irondequoit Bay. *Myriophyllum** is important and at times greatest in bulk, but it is never overwhelming. The density of eel grass, *Vallisneria americana* Michx.,

---

* The species is uncertain. Some investigators report *M. spicatum* L. a European species in the region while the Conesus population has been referred to as *M. exalbescens* Fern. The tentative choice was made because Fernald's study is the most substantial of those published. In the appendix to the 1957 edition of Fassett's Manual, the suggested name for *M. exalbescens* was *M. spicatum* v. *exalbescens* (Fern.) Jepson. The latest taxonomic review was that of Ogden (1974b). Here Ogden regarded *M. exalbescens* as a subspecies of *M. spicatum,* the older name having priority. He not only states that both are present in New York, but that both were "troublesome weeds . . . (in) . . . calcareous or brackish water of ponds and quiet streams." Consequently, there would be no ecological distinction. Investigators have reported informally that they can easily distinguish European from American *Myriophyllum,* and that the European plant is spreading rapidly. However, there has been no definitive taxonomic study, and the matter is unresolved.

frequently equals that of *Myriophyllum* at the height of season, although it may be more or less conspicuous to eye judgment. Also, coontail, *Ceratophyllum demersum* L., and water stargrass, *Heteranthera dubia* (Jacq.) MacM., are abundant, high in crop production, and sometimes dominant. The greatest quadrat densities have been recorded for *Ceratophyllum*. *Elodea canadensis* Michx. and *Potamogeton richardsonii* (Benn.) Rydb. are well represented locally. The naiad, *Najas flexilis* (Willd.) Rostk. & Schm., and species of pondweeds, *Potamogeton,* range from common to rare. One species not usually considered as a submerged aquatic, the star duckweed, *Lemna trisulca* L., actually grows in extensive mats on the bottom and covers other plants in the northern part of the lake instead of floating.

Because of its varied, stable flora, Conesus Lake engendered interest in some accepted generalities relating species presence to nutrient enrichment. There is an extensive literature on control of "water weeds." A recent survey (Baston and Ross, 1975) and experimental report (Peverly *et al.,* 1974) attest to this interest in New York State. For some time, there has been a presumption that the European species *Myriophyllum spicatum* L. and *Potamogeton crispus* L. were the "weeds" which grew in disturbed communities similar to terrestrial weeds, but the parallel is not suitable. Both species are present in the fertile and fertilized waters of Conesus Lake, but they do not monopolize the resources mobilized by rooted aquatics. Again, the characteristic homeostatic stability of Conesus Lake is demonstrated.

Both depth and standing crop have been investigated since 1967. Observations were made in at least 6 of the 9 years, the most recent in August 1975. Plants in over 150 quadrats were collected, weighed, and identified. The only data which approach this quantity is for Lake Opinicon, Ontario, with 60 computations from line intercepts (Crowder *et al.,* 1977).

Extreme depth is not limited by light and possibly temperature in most of the lake. On the lateral margins, a steep slope of about 45° limits depth to almost 2.4 m. At the ends, where gentle slopes extend deeply, the vegetation follows. Depth near the inlet is about 4.5 m, as Muenscher observed in 1926. At the northern end of the lake, Muenscher's tabular data indicate plants at 5.5 m (18 ft) and 6 m (20 ft). Since the information is contrary to Meunscher's general conclusion, it may be in error. After other lakes had been studied, it was realized that plants generally grow deeper and more luxuriantly at stream mouths, but the effect is less discernible in a fertile lake and more obvious in an infertile one. In the northern end of Conesus, by 1968 plants grew to 6.4 m, a depth equaled only in the crystal waters of Canadice Lake. During this period the major environmental change was the discharge of primary effluent from the Livonia Village treatment plant through Wilkins Creek, 1.3 km southeast of the outlet (1937–1972). It is logical that fertilization in clear waters resulted in denser growth at greater depth. The presumed area of influence is skewed northward in the typical pattern found for lateral streams in Canandaigua Lake, where the influence of streams on submerged plant communities is displayed as a museum. There may be also an eddy produced by outflow and southwest winds which distributes enriched waters further south on the west shore. Most important, the northern 10% of the lake is a broad shelf with shallow slope where the most luxuriant underwater garden could develop.

Even with the relatively large amount of data available, it is not easy to measure the crop in Conesus Lake with confidence. Fundamentally, the best analysis underwater is crude compared with competent terrestrial ecology. A single scale cannot be applied to all parts of lakes, greater depth is only generally related to higher crop, and patterns of seasonal increment vary from year to year. Aquatic vascular plants partly overwinter, but they behave neither as woody nor herbaceous terrestrial flora.

Experience in the field did provide some guidance. Differential over-wintering was apparent. *Vallisneria* is completely defoliated, although a considerable portion of its weight may be retained in corms which develop late or in rhizomes. *Myriophyllum* is conspicuous, but except for its red growing tips it is usually overcrusted and not necessarily viable. *Heteranthera* is persistent but appears lifeless. *Ceratophyllum* looks green and turgid, and *Elodea* certainly grows actively. The problem of determining increment for the growing season (or for the summer) is compounded by the expansion and contraction of the belt of vegetation. Vegetation persists chiefly between 2 and 4–5 m. It is scoured away from the shallower bottom by waves and ice and somewhat starved at lower depths to the extent of being unable to replace losses. Beginning in May or earlier, this refugeum expands shoreward and outward. Consequently, the seasonal increment above the bottom can range from 100 to perhaps 50%, but 50% of a dense community at 3 m is greater than 100% at 1 m. The available data are sufficient only to suggest a seasonal growth range. Almost nothing on the subject is published. Forsberg (1960) reported a 20% increase from July to August and a similar decrease to October during 2 years in Lake Osby, Sweden. It is conjectured that a greater increase, perhaps 30% in the overwintering zone, also occurs earlier in Conesus Lake in June and early July.*

All of the quantitative data reported here are expressed in the customary units of grams (or kilograms) of dry weight/m². Most were originally calculated as fresh weight in English units. Following the suggestion of Scul-

---

* Preliminary observations in 1976 indicate that the increase in crop may be above 90% except at the most extreme depth.

thorpe (1967) (and a number of experiments with local materials), the dry weight is arbitrarily set as 10% of the fresh weight. Dry weight varies widely, but the uncertainties balance reasonably well as long as relatively clean material is being considered. At this time, a slightly higher portion, 12–14%, might be justified, but this source of error is insignificant compared with other sources. It is little enough in the scale of standing crops which differ by two-, five-, or tenfold.

The maximum increment observed in shallow water is about 400 gm dry weight/m², but this is regarded as unusual, and about half that amount is probably more typical. In the middle depths of 2–4 m (1.8–3.7 m is more accurate), the increment can be twice as much. The addition depends greatly on the overwintering crop. To a degree, the overwintering crop provides a source of spread and growth in relation to its amount.

In order to provide a more controlled basis for crop comparisons, a "height of season" or maximum crop was designated. This is best chosen in late August, although it may remain stable or even increase slightly through a quiet, sunny September. In 1975, a year which had a very early fall (or rather, early cool and cloudy weather), the last week in August may have been a trifle past peak. Crowder et al. (1977) found the height of season during late July in Lake Opinicon, Ontario, and it probably does not come later than early August in northern Europe (England, Sweden), but the exact time is uncertain. Given this control, some comparison may be attempted between the maximum crop of Conesus and other lakes. Most of the scanty comparative data were assembled by Sculthorpe. Mendota, in Wisconsin, produced 202 gm dry weight/m² in the summer of 1921, but there are no modern records, and access by Scuba-diving gear was not possible then. Lake Osby, Sweden, reached a maximum of 240 and 680 gm dry weight/m² in August of 2 consecutive years. A better comparison is derived from Bristow's data. Although the average was 328 gm dry weight/m² for 2 years, the 1972 crop was double that of 1973, well over 400 gm. Since the line intercepts sampled many unproductive areas and different depths, the highest crops found are of particular interest: 1109.6, 1288, and 1454 gm dry weight/m². Thus, the high crops in Conesus Lake are similar in range to those found in Lake Opinicon. Table 21 includes a selection of crop weights for Conesus Lake. The maximum values range from 1400 to 1800 gm dry weight/m².

In the Finger Lakes region, the only data which may be compared are that of Peverly et al. (1974). The highest crop at a 1.3-m depth was 518 gm dry weight/m², but this is from a pond, and surprisingly, in May, while the August crop was only 83 gm. In Cayuga Lake, a high measurement of 878 gm dry weight/m² was recorded from 2.1 m on July 31, 1974.

A more general range for the productive belt of approximately 2–4 m is 6–900 gm. In shallow waters, the plants grow densely and attract attention

TABLE 21

Selected Crop Determinations of Aquatic Vascular Plants for Conesus Lake

Location, date, temp.	Depth (m)	gm dry weight/ m²
South end		
May 6, 1969;	2.1	24.9
Water temp. 10°C	3.7	32.6–38.3
Near Wilkins Creek		
Jun 16, 1970	1.5–2.4	56.7
Water temp. 16°C	3.0	284–965
Jul 25, 1970	2.0	885
(Water temp. in summer generally is over 20°C)	3.6	568
	4.5	261
Aug 28, 1970	1.0	431
	2.0	499
	3.0	1407
Sep 14, 1969	2.6–3.6	1816
South end		
Sep 2, 1969	1.9	777
	3.6	567
North end, south of Sand Point		
Sep 1968	3.7	1470

for aesthetic or utilitarian reasons, but the maximum crop is only one-quarter to one-half that of medium depth vegetation (2–4 m).

The final consideration here will be estimation of the total crop in the lake. It is not yet possible to begin serious assessment of the quality and quantity of its role in ecosystem dynamics. On the face of the matter, there must be a far greater role for the huge crop of Conesus Lake than for the limited plots of Hemlock and Canandaigua Lakes. Yet a small, single, luxuriant community such as that at the inlet of Canadice Lake may be critical in food chains. For Conesus, Oglesby *et al.* (1975) suggested that nutrients released by decomposing vascular plants may fertilize fall blooms of algae with soluble-reactive phosphate. For the present, a determination of the total crop, and knowledge of its variation, both seasonally and from year to year, would be desirable. Forest and Mills reported preliminary determinations based on average standing crop at the south end of the lake to a depth of 6.1 m. It was the equivalent of 270 gm dry weight/m², which would be low for the large north end area, but too high for the lateral margins. The preliminary computation was 8500 tons fresh weight, the equivalent of 26.23×10^6 gm or 2.63×10^4 kg dry weight. Subsequently, P. G. Savard (unpublished) used data from 40 quadrats along four lateral transects and longitudinal lines from each end and calculated the areas of the lake to depths of 3.05 and 6.1 m. His total was identical. If anything, the estimate is conservative. Three-quarters of the shallow water are at the north and south ends of the lake, and the lateral transects overrepresented the middle sections. Furthermore, 45% of the gently sloping area lies in the north where the highest macrophyte production has occurred during recent years. The north region of the lake was undersampled in relation to the south. An alternate determination can be made by allowing 400 gm dry weight/m² for the northern area alone, 1.34×10^6 m². The total crop would be twice the estimate from transect data, 5.35×10^4kg. This is impressive, but not unreasonable.

Hutchinson (1975) recently published "Limnological Botany" as the third volume of his monumental treatise on limnology. Included is a most comprehensive compilation of work on vascular aquatic plants, not generally available in limnological or botanical literature. This information was available too late for specific reference in this review. The information was considered, however and no compelling ground was found to make major alterations. There are substantial differences which should be resolved, but it is felt that the data on Conesus and neighboring waters will support the conclusions offered here.

DISCUSSION OF ECOSYSTEM

Analysis of Structure, Trophic State, and Stability

Conesus shares a common origin and some physical and chemical features with the other Finger Lakes of New York, but each Finger Lake constitutes a distinctly functioning ecosystem. An interplay of many environmental factors determines the degree of similarity in ecosystem structure and trophic state. A comprehensive comparison of Conesus Lake to other lakes in the region is desirable, but only a limited discussion is in order here. The only recently published comparative study of the basic physical and chemical features of the Finger Lakes is that of Oglesby et al. (1975). The graduate theses of Mills (1975) and Chamberlain (1975) included detailed analyses of plankton for Conesus, Hemlock, Owasco, and Skaneateles Lakes, but both contain comparisons with a wider group of lakes. Mills dealt specifically with the problem of comparative trophic state, particularly with the index of chlorophyll a level. A number of North American and European lakes are compared along with most of the Finger Lakes. These may be found in the report on Oneida Lake by Mills et al. in Volume II this treatise. Stewart and Markello (1974) similarly compared trophic states of Conesus, Canadice, and Honeoye Lakes with three smaller lakes in western New York, and a series of English lakes relating transparency to total phosphorus.

Chemically and probably biologically as well, Conesus most closely resembles Silver Lake, across the Genesee Valley to the southwest. The two lakes were surveyed concurrently by Muenscher (1927). K. M. Stewart (personal communication) concurs with this judgment on the basis of a considerable store of analysis. Silver Lake is smaller, shallower in extreme depth, but almost 150 m higher and its watershed is extraordinarily small, about five times the lake compared with 14:1 for Conesus. Silver is also unique among lakes in the Finger Lakes region in the close proximity of inlet and outlet streams, and it lacks both of the characteristic finger lake shape and bottom contour. While the flora of Silver Lake is not as well known as that of Conesus, evidence indicates similarity in algae and rooted aquatic plants, in composition, amount, and seasonal flux.

Hemlock Lake, which is alongside and parallel to Conesus Lake, is most similar in hydrographic characteristics. The size of Hemlock's watershed is almost identical in relation to the lake. Its surface area is about one-third smaller and mean depth about one-sixth greater. The altitude of the surface of Hemlock Lake is 25 m greater, extreme depth somewhat greater, and there is a marked difference in watershed land use and lake management. Conesus is a crowded suburban backyard and recreation area, and until recently it was a sink for municipal wastes and septage as well as agricultural contributions. Hemlock has received the maximum protection afforded by law throughout the twentieth century. As water supply for the City of Rochester, the level was slightly raised and stabilized, cottages removed for some distance back of the shores, and both boats and motor size limited. Furthermore, the watershed has been more heavily wooded than Conesus since early settlement. Nevertheless, the biota of the two lakes is startingly different although the generalized trophic state is similar. The algae of Conesus is larger in size range, and the blooms of blue–greens spectacular. Conesus contains one of the richest and most abundant communities of vascular plants known for the region, while Hemlock supports only a few plants in very shallow water. Although the number of species of zooplankton in Hemlock is greater, variety in terms of coexistence in samples is distinctly greater in Conesus. Conesus is clear, grazed to cleanliness by voracious microcrustacea, while Hemlock is murky from small phytoplankters. Resources for phytoplankton crops are differently mobilized, and the total would not average higher for Conesus Lake except for its very high bloom times.

Bounty of variety and amount and remarkable homeostatic stability of Conesus are revealed both in its terrestrial and aquatic community structures. In view of the urbanization of the lake periphery and the decades of gross insult with waste deposit, the aquatic ecosystem now appears to be much as it was in 1926, and probably in 1910. Thus, the example of Conesus compels attention to some difficult and general questions.

What Are the Limits of Lake Stability?

Six western Finger Lakes and two adjacent lakelike bays of Lake Ontario have been studied during the recent 8-year period (1967–1975) by over six investigators who exchanged information to a great extent. The comparison indicates that Irondequoit Bay has changed within 70 years, somewhat within 35 years, and Honeoye Lake changed measurably in less than 20 years. Sodus Bay seems nearest the edge of stability followed by Silver and Conesus Lakes. Clarity of water and depth of rooted vegetation are the critical indexes evoked for this judgment, but others would likely coincide. Judgment was reassessed following appraisal of changes following the Hurricane Agnes floods of 1972. A significant appreciation was gained on the role of catastrophy in maintaining, as well as altering, dynamically stable states. Both Irondequoit Bay and Honeoye Lake improved after the high water levels following Hurricane Agnes; Canadice Lake and Canandaigua Lake became more weedy. Evidently, the high nutrient level and phytoplankton crop is reduced in the more fertile waters, and the increased clarity increased the depth, abundance, and changed the relative species composition of the aquatic vascular plant community. A more infertile lake is temporarily fertilized by silt near stream mouths where submerged plants have already grown more luxuriantly than elsewhere. Consequently, the catastrophic flooding probably retards the succession of communities in later stages and accelerates change, at least temporarily, in earlier stages. In light of these findings, the standing crop must be viewed in terms of the time since the last major flood, or other highly unusual condition. Flood levels of 25-year probability are sufficient to produce changes of the magnitude observed. Floods of a higher level (50-, 100-, or 500-year interval) or of unusual duration might show still more drastic results. Also, the events are not evenly distributed but may occur in pulses of short and long intervals.

In Conesus, the crop of submerged aquatics was lower 3 years after the flooding, but meanwhile the perimeter sewer had been installed. Furthermore, natural perturbations in the crop are evident and their limits are not known. After the introduction of municipal effluent (1937), the crop attained slightly greater depth and probably a higher standing crop within former community limits. With sewage diversion (1972–1973), the changes may reverse to some extent. Total crop should decline quickly, but plants will not retreat from the depths for some time and might not do so within a predictable period. Algal blooms are reported as diminished, but caution must be applied since no substantial data are available after August 1973.* Changes in the high peaks would be more noticeable than a long range decline in mean annual crop. Nevertheless, blooms should not be expected to vanish. *Calothrix* (*Gloeotrichia*) was abundant in 1926, and it can be expected in 1976 or 1986.

Does the Example of Conesus Conform to Limnological Theory?

The modern concept of eutrophication is surprisingly recent in origin. It does not appear to have been stated clearly before Mortimer (1941–1942) and Lindeman (1942). Although it is now the central theme of limnology, visualization and the language employed to convey limnological ideas have evolved irregularly but continually both before and since 1942. Like Darwin's theory, limnology has been visualized differently in each generation.

A full discussion of the historical and current interpretations is not appropriate here, although a wide diversity of interpretation is recognized. Serious disparities have not been resolved, and indeed little attention given to the problem. For present purposes it is sufficient to presume that application of a general theory should enable ranking of lakes, prediction of rank by certain indexes, and possibly the alteration of rank by management.

Indeed, one of the most significant realizations of recent years is that lakes naturally slip both down and up the trophic scales which have been used to rank them. The elaborate work of Mills (1975) and Chamberlain (1975) were attempts not only to rank lakes but to quantify the ranking. Quantification of theory is quite another activity from the natural history of classification (including ranking) or the practical decisions of management; a computer program is rarely needed to justify elimination of sewage and septage from natural waters.

Moreover, a competent aquatic naturalist can integrate enough clues in a visit on a late summer day to rank lakes accurately, or such simply measured and calculated criteria as mean depth and total dissolved oxygen have been used by Mills and many others to rank lakes in correlation with pigment production. Furthermore, the ranking is regular enough to conform to mathematical derivations which should be of predictive value in unexamined lakes.

In gross judgments, lake investigators have had little difficulty in agreeing on ranking Conesus among the Finger Lakes regardless of differences in their specialty, regardless of whether their data are qualitative or massively

* Measurements of both plankton and macrophytes in the fall of 1976 confirmed the prediction that the crop of both would decline. However, natural perturbation cannot yet be eliminated as the real cause.

quantitative, and regardless of whether they use either simple or elaborate conceptual models.

Nevertheless, a tremendous effort is being exerted to find formulas which are realistic, general, and accurate. Mills and Chamberlain could easily agree on ranking Skaneateles Lake least fertile, Owasco Lake next, but Conesus Lake and Hemlock Lake refused to fall into line. For example, Mills found chlorophyll *a* higher in Hemlock during summer, but lower than Conesus in fall and winter, and the two lakes were equal in spring. Mills pursued the problem, comparing physical, chemical, and biological (phytoplankton) indexes. He was lucid and explicit on the anomaly. In retrospect, Birge and Juday's (1914) secchi disk reading should have indicated the puzzle, but Mills went further and explained the low transparency of Hemlock with its many small phytoplankters.

In practice, modern limnologists follow Rhode (1969) and others in accepting phytoplankton as the concrete evidence of the abstraction: eutrophication. The trophic state is the amount of photosynthetic pigment. Transparency is a consequence of the pigment level, and nutrients are a cause which must be related to the pigments resulting.

Both Stewart and Markello (1974) and Mills (1975) derived trophic rankings using phosphorus loading and concentration, and transparency, or chlorophyll *a* concentration in different formulations. Both compared Conesus not only with other Finger Lakes but with the better-known lakes of glaciated North America and Europe. In addition, the National Eutrophication Survey (USEPA, 1974) was directed specifically at trophic placement.

Although the efforts to bring lakes into a common theory were admirable, the devotion to abstract modeling has been accompanied by a lack of attention to the real ecosystem detail—complexity, variety, and the nuances of structure and behavior. In a specific example, concentration on mid-lake plankton has led to treating the system as though it were homogenized. Yet, biomass which includes hundreds of tons of rooted plants in Conesus Lake is not equivalent to the plankton of Hemlock Lake. Furthermore, within any lake there may be different trophic situations. The point is more amply demonstrated in a less fertile lake like Canandaigua where local influences produce sharp differences. Yet, even in Conesus fertilization by municipal effluent was restricted to a distinct part of the lake. Modeling formulations have given full mathematical credit for the contribution of two streams (Hanna's and Wilkins' Creeks) which could not possibly affect much of the lake. Consequently, added uncertainties are compounded with those already concealed in figures for volume, retention time, precipitation, and the pattern of mineral input. If Conesus does fall into a logical position in relation to other lakes, the achievement may only demonstrate that trophic placement is still basically a crude art, notwithstanding elegant models and masses of data.

Mills provided one example of the finer details which should be drawn when he distinguished between mean level of productivity and peak level. Mode might have been more descriptive than the mean, but the effort is in the right direction. With the discrimination of maximum and mean, the state of the lake is seen to be well within the mesotrophic range with respect to phytoplankton most of the year, with occasional incredibly high maxima. Other steps might include due attention to rooted aquatics and to other nonplanktonic organisms. Subdivision should be studied in those lake ecosystems with are in fact segmented.

Nutrient Budget

Elements related to the nutrient budget of Conesus Lake have been discussed individually in previous sections of this article: land use, population, lake morphometry, climatology, input of tributary streams, and seasonal flux of phosphorus, nitrogen, and silica. Specific attempts to derive budgets remain to be considered.

Phosphorus

Efforts to derive phosphorus budgets for Conesus Lake have been published by Stewart and Markello (1974) and the National Eutrophication Survey (USEPA, 1974). Both referred to the same basic quantification theory of Vollenweider (1968), but the two derivations are so different as to make detailed comparison impossible. In addition, Mills (1975) provided some data which are of summative nature, and phosphorus loading has been calculated. Oglesby and Schaffner (1975) provided the elements of the same calculation, with the results also included by Oglesby *et al.* (1975).

Unfortunately, none of the computations are entirely comparable to the others. They are summarized in Table 22. Only the summaries for total phosphorus provide an opportunity for discussion. The three sources differ in the ratio of approximately 4:2:1. Such differences cannot be explained by the diversions of sewage (1972) or septage (1973), as explained in the third footnote to the table. Indeed, both Mills and the USEPA collected data almost concurrently. The formulations of Stewart and Markello and of Oglesby and Schaffner contain two comparable categories (soil derived, population derived), but the latter is in "biologically available" phosphorus rather than total phosphorus. The only way in which comparison could be made is to accept the 67/46 ratio between total and available phosphorus reported by Oglesby *et al.* for the Conesus watershed. The 31% increase still leaves the two sources very far apart.

TABLE 22

Summary of Phosphate Loadings Calculated for Conesus Lake

Source	Data year	Amount (gm/m²)	
Stewart and Markello (1974)	1969	1.4	(total)
		0.838	(total projected after sewering)[b]
USEPA (1974)	1972–1973	0.36–0.38	(total)[c]
Mills (1975)[a]	1972–1973	0.053	(potential) gm/m³, not gm/m²
		0.67	(total, derived)
Oglesby et al. (1975)	1972–1973 (?)	0.67	(total)
		0.46	(soluble or available)

[a] Mills' data for soluble reactive phosphorus in the upper waters of the lake can be compared only for the period of summer stratification.

[b] Data were collected before diversion of sewage and septage from the lake. The projected figure was derived from Stewart and Markello's total by subtracting 75% of their population and recalculating by the same formula.

[c] Diversion of sewage and septage was in progress during 1972 and 1973. The USEPA noted that the municipal effluent from Livonia discharged through Wilkins Creek had been diverted during their sampling year (Oct 1972–Oct 1973), but no trend or abrupt reduction is discernible in their data for the creek.

Quite clearly, there are no present means of discriminating which , if any, of the calculations for phosphorus loading are representative. Of the three budgets, only that of the USEPA is based on some actual measurements of stream flow and chemical content. Nevertheless, in comparison with other lakes, all investigators have found their calculations satisfactory and meaningful. All agree on a trophic position for Conesus Lake in the range of upper mesotrophic or lower eutrophic. With respect to Conesus Lake itself, all of the derivations are subject to serious criticism. All sources are presumed to affect the lake as a whole. Yet Hanna's Creek is within a few meters of the outlet, and Wilkins' Creek, with its very high nutrient contribution for almost 40 years, is located only 1.3 km from the outlet. The USEPA study credited 15% of the phosphorus load to these two streams. It is difficult to imagine that either of these affects the southern basin of Conesus Lake as much as the lake is affected by its inlet stream. Yet most of the data on phytoplankton and water chemistry has been collected in the south basin. The budget of Oglesby and Schaffner may have employed an erroneous figure for the watershed size (including the lake), since this appears as 231 km² rather than the 180.5 km² which is accepted as approximately correct. This error would affect the calculation for soil contribution. Also, there is a 20% difference in the basin population between their figure and that which was derived by others. The two differences will balance each other to some extent.

Nitrogen

The seasonal flux and seasonal integrations of nitrogen content in the waters of Conesus Lake have been discussed appropriately under chemical limnology, and the chemical content of input streams considered as part of the basin hydrology. Annual nitrogen loading was calculated by the National Eutrophication Survey (USEPA, 1974) as 10.4 $gm/m^2/year$, of which 4.7 $gm/m^2/year$ was accumulated in the lake.

Limiting Nutrient

The National Eutrophication Survey reported that in 1972–1973, analysis of lake water indicated limitation by nitrogen because of a N/P ratio of less than 8/1 during May and October, but phosphorus limitation in July when the N/P ratio was 19/1. Oglesby et al. (1975) stated that phosphorus was critical in controlling the size of the summer phytoplankton crop in Skaneateles, Owasco, Cayuga, Hemlock, and Conesus Lakes. The unusual fall bloom of the blue–green alga *Aphanizomenon* together with the diatom *Melosira* received special attention in the report. There was a dramatic increase of soluble-reactive phosphorus at the time of the bloom, with the decay of macrophyte vegetation considered as its probable source. Nitrate nitrogen was low at the time. Nitrogen fixation by the *Aphanizomenon* was considered probable because of the high frequency of heterocysts. Heterocysts had been present only erratically and in small numbers during the spring and early summer. In addition, acetylene reduction tests indicated fixation in June, but not July and August. Thus, the two sources are clearly contradictory only with respect to the fall condition, but there is little substantial basis of comparison.

CONCLUSIONS AND RECOMMENDATIONS

Review and Outlook

Both terrestrial and aquatic aspects of the Conesus Lake ecosystem are notable for their variety and productivity, however, only the aquatic ecosystem will be considered here. The elements of the ecosystem, the species, food webs, physical settings, are not unique for the entire Finger Lakes region, but the presence of so many diverse elements within the watershed of a smaller lake is noteworthy.

The morphology of Conesus Lake is typical of Finger Lakes, but its depth (both mean and extreme) is less than any except Honeoye and Silver Lakes. The lake is dimictic and ice covers the water for about 3 months. In the late summer, oxygen is severely depleted at extreme depth (about 20 m), but only a very small part of the bottom is affected. In the anoxic area, anaerobic bacteria release a large amount of phosphorus from the sediments to the lower waters.

The water chemistry of Conesus Lake is typical with respect to ionic balance, but the total of dissolved solids is the highest among the Finger Lakes except for Seneca and Cayuga Lakes, which intercept salt-bearing strata. There is evidence that the nitrogen supply may limit phytoplankton growth in the spring and fall, but limitation by phosphorus is much more strongly indicated by available information.

The balance of agricultural and forest land uses in the watershed of Conesus Lake has been maintained for a century or more. Residential use, however, has increased enormously. As late as 1925, raw sewage entered the lake from privies near the shore. During the next half-century the population near the shore increased tenfold, but septic systems were installed. In addition, effluent from the Village of Livonia treatment plant was directed into the lake between 1937 and 1972. Both of these sources were directed outside the lake watershed in 1973. In recent years the chloride content of the water has doubled, possibly as a result of deicing salt spread on roads of the watershed.

There is a comparatively little evidence that the species of phytoplankton, aquatic vascular plants, or invertebrate animals have changed markedly within 50 or 65 years, or that oxygen depletion has increased in the deepest waters. It is principally on theoretical grounds and applications of experience gained elsewhere that changes are presumed to have been made which can be reversed. If the phytoplankton crop has increased in quantity with the added population of humans, it will decrease after sewage and septage have been diverted. There is somewhat better evidence that the crop of submerged macrophytes has markedly increased in the northern end of the lake, and the decline of the crop can be predicted more confidently. In addition, there has been a marked local growth of macrophytes and the green alga *Cladophora* at the mouth of Wilkins Creek, and this has declined since 1972. The amount of phytoplankton may be reduced directly because of decreased nutrient input, and indirectly through the reduction of the macrophyte crop. The most conspicuous reduction should be for the intense fall blooms, particularly if they are now enabled by soluble phosphorus from decay of the rooted plants. Nevertheless, it is unrealistic to expect that the diversion of nutrients will end blooms of phytoplankton or that Conesus

Lake will be free of weeds. On the contrary, dense growth of submerged vascular plants will remain common for an indefinite length of time.

There is no evidence to suggest a drastic change in the populations of invertebrates. *Daphnia* is characteristic of the zooplankton in deeper waters, while *Gammerus* dominates in shore areas where plants approach the surface. Both snails and clams are common on the bottom, and larvae of *Chironomus* are plentiful. The fish population has already been managed to provide sport fishery. Nevertheless, it has recently become evident that native species, reproducing themselves, constitute the overwhelming bulk of the community: perch, walleye, northern pike, and basses. The only exotic species stocked in this warm water ecosystem with even modest success has been the rainbow trout.

Virtually every segment of the aquatic ecosystem offers testimony to its overall stability, maintained during a period of tremendous increase in sewage and septage. High fluctuations, such as heavy fall blooms of phytoplankton, are followed by the restoration of a moderate operating range without destruction of important elements of the ecosystem. The limits of this stability are not known, but the decision to build a perimeter sewer for the lake rather than finding the limits of self-control by increasing the stress was indeed wise.

Thus, the most significant management decision of our time has been implemented. Conesus Lake is the more unique for this event. There remain two problems of lesser magnitude, both at the inlet area.

1. The 1- to 2-km² area adjacent to the lake has been mostly filled, bulldozed, and channelized with several hundred meters of canals which are now pondlike. The terrestrial surface is sparsely covered with grasses and herbs, but the owner of the property must envisage development. He has unsuccessfully proposed a heavy equipment repair area, although the design of the canals suggest residential areas such as those found in coastal Florida. Commercial or residential development could have a great impact on Conesus Lake.

2. The recently established state wildlife management area lies south of the inlet, bordering the inlet stream and its junction with McMillan Creek. This large area (about 5% of the watershed) is mostly covered by swamp forest over organic soil. Flooding to create a still shallow water marsh is contemplated, and trees wil die. The environmental impact of this plan on the lake and its biota has not been investigated. Although natural headland marshes are generally known to have a protective effect on lakes, such benefit should not be presumed in this case. There is some evidence that the present marsh area enriches the inlet stream markedly with phosphorus.

Flooding may increase the release of phosphorus by bacteria, and the added amount could reach the lake. Also, the environmental change may affect animal populations; its effect on game fish has not been assessed.

Research Recommendations

Conesus Lake and its watershed present unmatched opportunities for ecological study because of the combination of features included within a relatively small area. A parallel reinvestigation of all the phases of chemical and biological limnology discussed in this article would be most highly desirable. Data should be collected now, since the diversion of most human waste from the lake. A number of disparities in base data need to be resolved. These are minor problems, but they affect the correctness of other published data. In addition, there have been no investigations to establish a specific chain or web structure in the ecosystem.

Yet, perhaps the single greatest need for studies in the lake itself is for analysis of bottom sediments to derive a trophic history. It has been asserted here that Conesus Lake is remarkably fertile, varied, and stable, but the length of time it has remained in the modern state is unknown. The historical evidence for stability and change is indeed limited. Nevertheless, the present knowledge is sufficient to demand some accounting in terms of limnological generalities. Conesus has been listed among the lakes undergoing "rehabilitation," yet its quality could be considered good by a number of ecological indexes. A more intense comparative study with Hemlock Lake would be of high value both to concepts of limnology and as a guide to management. Here are two adjacent and morphologically similar lakes. One has been protected from man's waste to a great extent; the other is virtually a suburban backyard. Yet the aquatic ecosystems do not differ much in the summative measures of trophic rank. It is suggested that the structure or quality of the aquatic ecosystem has been neglected in the recent drive to quantify matter and energy flows. Qualitatively, Hemlock and Conesus are quite different as lakes, and their response to use, misuse, and management may be drastically dissimilar.

Finally, the terrestrial ecology of the Conesus Lake watershed is as interesting as the aquatic aspect, but only meager interest has been shown toward it. A comparative survey and mapping of the biotic communities and study of successional communities are the first research needs. Correlation with physical phenomena such as precipitation, soils, stream chemistry, and drainage would be logical as subsequent investigations. Ecological surveys are under way now, on a modest scale, and the results will be offered as a supplement to this review.

KNOWLEDGE USE, LAND USE, AND PUBLIC DECISIONS

Stout (1970) and Boyle (1974) are the only assessments of the Conesus watershed as a natural and social unit. These students made recommendations, which may have reflected the biases of their sponsors: college professors and, in the latter case, also the county planner. Yet students, professors, and planners do not make management decisions; they offer information and advice. Local governments and the Conesus Lake Association decided to build the perimeter sewer, led forward by the carrot of federal and state funding. Only when the critical bond issue for local costs approached a vote was there a call for information, or rather for support. The community conscience had, however, been raised sharply by the long, bitter controversy over destruction of the marsh at Sand Point, and the beginning of destruction in the inlet area. Shortly before the vote in December 1969, a vocal opposition developed. The need of a sewer was questioned to some extent, but almost everyone was firm in believing that the lake was polluted. By whom? A three-cornered denouncement emerged. Livonia residents blamed cottages and farms, cottagers blamed the village, and the farmer blamed the other parties. In this situation, a local journalist sought and well publicized the available objective evidence. Only the farmers were absolved, although subsequent studies of the nutrient budget would have implicated them as well. The scientific source made a disciplined effort neither to threaten a future without the sewer nor over-promise its benefits, and to remain disassociated from both promotion and opposition.

In the course of history, the bonds were voted and the new facilities built. The system is unique in New York State. Several million dollars were committed without a dollar to assess the state of the ecosystem or the impact of the project. Thus, the effects of the most important management decision likely to be made for 50 years may never be known in scientifically definable terms.

After this review was prepared a long dormant project unpredictably became active again. Strong community concern in 1977 has promoted dredging in the lake near its outlet and the outlet stream itself to the benefit of flood control. As in the case of the sewer project, there is no interest in predicting environmental impact before a decision is made or determining its consequences.

Other actions recommended for the watershed are water and land use decisions, not construction projects. Zoning came to the four peripheral towns after the perimeter of the lake was essentially saturated, and construction of multiple tiers of housing on the surrounding hills begun. Lot size is now controlled, but whether scattered housing or clusters come to be built, growth will not be controlled in amount. Only economics, salesmanship, and urban social pressures will set the level of growth.

Is growth in itself or unregulated change a threat to environmental quality? The immediate lake area is protected against human wastewater. The present pattern of farming and abandoned land is relatively stable, and erosion and sedimentation do not threaten land or water. Present motors apparently do not put damaging amounts of petroleum in the lake, but a marina or machinery repair area near the inlet might overthrow the remarkable filtering system which keeps the lake clear. Finally, even if the present condition is accepted, and no aesthetic improvement is considered, crowding can scarcely be avoided as an issue to be faced. Is the lake now crowded? Some residents say that they avoid the lake on weekends. If this is true, it may indicate a feedback mechanism already at work. As the number increases to a point, those who are annoyed sufficiently will leave. The level of use would rise, however, with users who tolerate more crowding. The only alternative to this "natural" regulatory system is some method of rationing use, but there is no suggestion of support for such a cause.

The possibility also remains that the resources are underused in some respects, as Boyle suggested. There is extremely little public access to the lake, if too much in the view of many residents. Only a very few boats may now be launched at the single-state site. Swimming is virtually restricted to residential lots, while the privately owned beach at Long Point is little used and other good beaches remained unused. Adjacent to the outlet, a shallow sandy beach is used for mobile homes. Adjacent to the inlet, the aborted development of lagoons has left an area which could be put to a multiplicity of public use if acquired. As much can be said of the watershed generally. Wisely conceived, there is environmental wealth available in the variety of terrain and vegetation, unique woodland, scenic home sites, productive farmlands, and recreational opportunities.

No rational land plan can be foreseen which would have a positive influence in the watershed. New York now has negative powers for increased prevention of damaging actions. Jurisdiction over stream beds and lake shore is now established, and precedent has been set for intervening in virtually any action which may affect water quality adversely—an extremely broad embrace if exercised. The requirements for impact analysis of "significant" projects and regulation of defined wetlands with a permit system were only added in 1975. The drainages and upland areas might be protected if a relationship to ecosystem dynamics can be recognized. Unfortunately, such recognition depends on broad interdisciplinary knowledge derived through professional study. Such knowledge does not exist, and this review is but a modest contribution to the need.

REFERENCES

Abraham, W. R. (1975). "Case History of Conesus Lake," Staff Paper. N.Y. State Dept. of Environmental Conservation, Reg. 8, Avon, New York. (Prepared for Warm Water Fisheries Workshop, 1976, mimeographed and appendices A-C.)

Anderson, H. T. (1976). "The Diamonds are Dancing. A History of Conesus Lake." Conesus Lake Association Inc., Livonia, New York.

Baston, L., Jr., and Ross, B. (1975). "The Distribution of Aquatic Weeds in the Finger Lakes of New York and Recommendations for Their Control," Public Serv. Legislative Study Program SS-506. N.Y. State Assembly, Albany. (Prepared under the direction of G. L. Miller, 1975.)

Berg, C. O. (1963). The middle Atlantic states. In "Limnology in North America" (D. G. Frey, ed.), pp. 191–237. Univ. of Wisconsin Press, Madison.

Birge, E. A., and Juday, C. (1914). A limnological study of the Finger Lakes of New York. Bull. U.S. Bur. Fish. 32, 524–609.

Birge, E. A. and Juday (1921). Further limnological observations on the Finger Lakes of New York. Bull. U.S. Bur. Fish. 37, 209–252.

Bond, M. S. (1947). "Census Data 1875–1945, Livingston County," Agric. Ext. Publ. No. 611. Dept. of Agric. Econ., Cornell University, Ithaca, New York.

Boyle, J. (1974). "Report on Conesus Lake. Present Trends and Directions," Report for Livingston County Internship Program. Dept. of Planning and Dept. of Geography, State University of New York College at Geneseo, (Contrib. No. 60-10/75. Environmental Resource Center at Geneseo.)

Bratton, C. A. (1954). "Census of Agriculture, Livingston County," Agric. Ext. Publ. No. 1080-23. Dept. of Agric. Econ., Cornell University, Ithaca, New York.

Bratton, C. A. (1964). "Census of Agriculture, Livingston County," Agric. Ext. Publ. No. 475-23. Dept. of Agric. Econ., Cornell Univeristy, Ithaca, New York.

Brown, E. S. (1970). An upstate lake gets a second chance. Health News, Albany, 47, 25. (Reprinted in Livonia Gazette, Livonia, New York, December 3, 1970.)

Carter, D. B. (1966). Climate. In "Geography of New York" (J. H. Thompson, ed.), pp. 54–78. Syracuse Univ. Press, Syracuse, New York.

Chamberlain, H. D. (1975). A comparative study of the zooplankton communities of Skaneateles, Owasco, Hemlock, and Conesus Lakes. Ph.D. Thesis, Dept. of Natural Resources, Cornell University, Ithaca, New York.

Child, D., Oglesby, R. T., and Raymond, L. S., Jr. (1971). "Land use data for the Finger Lakes Region of New York State." Cornell Univ. Water Resour. and Mar. Sci. Cent., Ithaca, New York.

Crowder, A. A., Bristow, J. M., King, M. R., and Vanderkloet, S. (1977). "Distribution, seasonality, and biomass of aquatic macrophytes in Lake Opincon (Eastern Ontario)." Naturaliste Can. 104 (No. 5) (in press).

DeLaubenfels, D. J. (1966a). Vegetation. In "Geography of New York" (J. H. Thompson, ed.), pp. 90–103. Syracuse Univ. Press, Syracuse, New York.

DeLaubenfels, D. J. (1966b). Soil. In "Geography of New York" (J. H. Thompson, ed.), pp. 104–110. Syracuse Univ. Press, Syracuse, New York.

Doty, L. (1925). "History of the Genesee Country," Vol. 1. S. J. Clark Publ. Co., Chicago, Illinois.

Drouet, F. (1968). "Revision of the Classification of the Oscillatoriaceae," Monogr. No. 15. Acad. Nat. Sci. Philadelphia, Pennsylvania.

Drouet, F. (1973). "Revision of the Nostocaceae with Cylindrical Trichomes," p. 292. Hafner Press, New York.

Drouet, F. and Daily, W. A. (1956). "Revision of the Coccoid Myxophyceae." Butler Univ., Bot. Stud. 12, 1–218.

Eschner, A. R. (1966). Water. In "Geography of New York" (J. H. Thompson, ed.), pp. 79–89. Syracuse Univ. Press, Syracuse, New York.

Fairchild, H. L. (1928). "Geologic Story of the Genesee Valley and Western New York," p. 215. Rochester, New York. The author.

Fassett, N. C. (1957). "A Manual of Aquatic Plants. (second edition with revision appendix by E. C. Ogden), p. 405. Univ. of Wisconsin Press, Madison.

Forest, H. S. (1976). Study of submerged aquatic vascular plants in northern glacial lakes. Folia Geobot. and Phytotax. (in press).

Forest, H. S., and Mills, E. L. (1971). Aquatic flora of Conesus Lake. Proc. Rochester Acad. Sci. 12, 110–138.

Forest, H. S., Kelly, J. W., Rugenstein, S. R. et al. (1977a). In preparation.

Forest, H. S., Jacikoff, T. M., and Nesbitt, C. B. (1977b). In preparation.

Forsberg, C. (1960). Subaquatic macrovegetation in Osbysjon, Djursholm. Oikos 11, 183–199.

French, J. H. (1860). "Gazetteer of the State of New York." I. J. Friedman Inc., Port Washington, New York (reissued in 1969).

Gilbert, B. K., and Kammerer, J. C. (1965). "Summary of Water Resource Records in the Genesee River Basin through 1963," Bull. No. 56. U.S. Dept. of Interior, Geol. Surv., and N.Y. State Dept. of Conservation, Water Resour. Comm., Albany.

Godfrey, P. J. (1971). "A Brief Summary of the Limnology of Four Finger Lakes, Owasco, Skaneateles, Conesus, and Hemlock," Mimeogr. Dept. of Natural Sciences, Cornell University, Ithaca, New York.

Greeley, J. R. (1927). Fishes of the Genesee region with annotated list. In "A Biological Survey of the Genesee River System," Suppl. to Annu. Rep., Vol. 16, pp. 47–67. N.Y. State Department of Conservation, Albany.

Greeson, P. E., and Robinson, F. L. (1970). "Characteristics of New York lakes. Part 1—Gazetteer of lakes, ponds, and reservoirs," Bull. 68, p 124. New York State Department of Environmental Conservation, Albany.

Greeson, F. L., and Williams, G. E. (1970a). "Characteristics of New York lakes. Part 1A—Gazetteer of lakes, ponds, and reservoirs by county," Bull. 68A, p 121. New York State Department of Environmental Conservation, Albany.

Greeson, F. L., and Williams, G. E. (1970b). "Characteristics of New York lakes, ponds, and reservoirs by watershed," Bull. 68B, p 122. New York State Department of Environmental Conservation, Albany.

Hall, D. J., and Waterman, G. C. (1967). Zooplankton of the Finger Lakes. Limnol. Oceanogr. 12, 542–544.

Harman, W. M., and Berg, C. O. (1971). The freshwater snails of central New York. Search: Agric., 1, 1–68.

Hutchinson, G. E. (1975). Aquatic Botany. In "A Treatise on Limnology," Vol. III. Wiley, New York.

Kelly, J. W. (1973). Introgressive hybridization between red and silver maples. M.A. Thesis, Dept. of Biology, State University of New York College at Geneseo.

Khan, K. R. (1974). Developmental morphology of Calothrix parietina (Nageli) Thuret.

[= *Gloeotrichia echinulata* (J. E. Smith) Richter]. Ph.D. Thesis, Dept. of Biology, State University of New York at Binghamton. (Contrib. No. 40-5/74. Environmental Resource Center at Geneseo, New York.)

Lindeman, R. L. (1942). The trophic-dynamic aspect of ecology. *Ecology* **23**, 399–418.

Lougeay, R. (1975). Possible inaccuracies of precipitation data for western New York State. *Assoc. Am. Geogr. Middle States Meet., 1975, Proc.* **9**, 53:58.

Lougeay, R. (1976). Adjustment of measured precipitation to account for gage under catch. *J. App. Meterology* **15**, 1097–1101.

McIntosh, W. H. (1877). "History of Monroe County, 1788 -." p. 320. Lippincott, Philadelphia, Pennsylvania.

Maxwell, T. F. (1972). Developmental morphology of *Gleotrichia echinulata* (J. E. Smith) Richter. M.A. Thesis, Dept. of Biology, State University of New York College at Geneseo. (Contrib. No. 58-12/72. Environmental Resource Center at Geneseo, New York.)

Mills, E. L. (1975). Phytoplankton composition and comparative limnology of four Finger Lakes, with emphasis on lake typology. Ph.D. Thesis, Dept. of Natural Resources, Cornell University, Ithaca, New York.

Mortimer, C. H. (1941). The exchange of dissolved substances between mud and water in lakes. *J. Ecol.* **29**, 280–329.

Mortimer, C. H. (1942). *J. Ecol.* **30**, 147–201.

Muenscher, W. C. (1927). Vegetation of Silver Lake and Conesus Lake. *In* "A Biological Survey of the Geneseo River Watershed," Suppl. to Annu. Rep. No. 16, pp. 66–71, and Appendix VII, p. 86. N.Y. State Department of Conservation, Albany.

Neil, J. H., and Owen, G. E. (1964). Distribution and environmental requirements and significance of Cladophora in the Great Lakes. *Pub. 11 Great Lakes Research Div.* Univ. of Michigan, Ann Arbor, Michigan. pp. 113–121.

New York State Department of Agriculture and Markets. (1974). "New York Agricultural Statistics." Albany.

New York State Department of Health. (1961). "Upper Genesee River Drainage Basin," Surv. Ser. Report No. 2. Water Pollution Control Board, Albany.

New York State Museum and Science Serivce. (1971). "Geological Map of New York. Finger Lakes Sheet," Map and Chart Series, No. 15. Albany.

Ogden, E. C. (1974a). Potamogeton in New York. *N.Y. State Mus., Bull.* **423**.

Ogden, E. C. (1974b). Anatomical patterns of some vascular aquatic plants of New York. *N.Y. State Mus., Bull.* **424**.

Oglesby, R. T., and Schaffner, W. R. (1975). Response of lakes to phosphorus. *In* "Nitrogen and Phosphorus, Food Production, Wastes, and the Environment" (K. S. Porter, ed.), p. 25–57. Ann Arbor Sci., Publ. Michigan.

Oglesby, R. T., Schaffner, W. R., and Mills, E. L. (1975). "Nitrogen, Phosphorus and Eutrophication in the Finger Lakes," Tech. Rep. No. 94, Cornell Univ. Water Resour. and Mar. Sci. Cent., Ithaca, New York.

Peverly, J. H., Miller, G. L., Brown, W. R., and Johnson, R. L. (1974). "Aquatic Weed Management in the Finger Lakes," Technical Rep. No .90. Cornell Univ. Water Resour. and Mar. Sci. Cent., Ithaca, New York.

Phillips, D. W., and McCullough, J. S. W. (1972). "The Climate of the Great Lakes Basin," Climatol. Stud. No. 20. Atmos. Environ. Serv., Toronto, Ontario.

Rhode, W. (1969). Crystallization of eutrophication concepts in northern Europe. *In* "Eutrophication: Causes, Consequences, Correctives," pp. 50 64. Nat. Acad. Sci., Washington, D.C.

Ritchie, W. A. (1965). "The Archaeology of New York State." American Museum of Natural History, Natural History Press, Garden City, New York.

Robertson, I., Blakeslee, C. A., and Blakeslee, C. L. (1948). The mollusca of the Niagara Frontier region. *Buffalo Soc. Nat. Sci., Bull.* **19**, 2–191.

Savard, P. G., and Bodine, D. N. (1971). Algae of Conesus Lake, Livingston County, New York. *Proc. Rochester Acad. Sci.* **12**, 146–159.

Sculthorpe, C. D. (1967). "The Biology of Aquatic Vascular Plants." St. Martin's, New York.

Stewart, K. M. (1972). Isotherms under ice. *Verh. Int. Ver. Theor. Angew. Limnol.* **18**, 303–311.

Stewart, K. M., and Markello, S. J. (1974). Seasonal variations in concentrations of nitrate and total phosphorus, and calculated nutrient loading for six lakes in western New York. *Hydrobiologia* **44**, 61–89.

Stout, G. J. (1970). Land use in the Conesus Lake watershed, Livingston County, New York. 1930–1970. M.S. Thesis, Dept. of Biology, State University of New York College at Geneseo. (Contrib. No. 29-6/70, Environmental Resource Center at Geneseo, New York.)

U.S. Department of Agriculture. (1956). "Soil Survey of Livingston County, New York, Ser. 1941, No. 15. Soil Conservation Service, in cooperation with Cornell Univ. Agric. Exp. Stn., Ithaca, New York.

U.S. Department of Commerce (1972a). "Characteristics of Population, New York," 1970.

U.S. Department of Commerce (1972b). "General Social and Economic Characteristics, New York," 1970.

U.S. Department of Commerce (1973). "Census of Agriculture, New York" 1969.

U.S. Environmental Protection Agency. (1974). "National Eutrophication Survey report on Conesus Lake, Livingston County, New York," EPA Reg. II, Working Pap. No. 156. Pacific Northwest Environmental Research Laboratory, Corvallis, Oregon, and National Environmental Research Center, Las Vegas, Nevada.

Vollenweider, R. A. (1968). "Scientific Fundamental of the Eutrophication of Lakes and Flowing Waters, with Particular Reference to Nitrogen and Phosphorus as Factors in Eutrophication," Tech. Rep. O. E. C. D., Paris (DAS/CSI/68. **27**, 1–159, 34 ill., 60 p. bibliography. Mimeogr.)

von Engeln, O. D. (1961). "The Finger Lakes Region: Its Origin and Nature," p. 156. Cornell Univ. Press, Ithaca, New York.

Wade, J. W., and Vasey, C. E. (1975). A preliminary study of the snails of Conesus Lake, Livingston County, New York. *Proc. Rochester Acad. Sci.* **12**, 414.

Walker, R. (1975). "A Survey of Some Aquatic Invertebrates of the Conesus Lake Basin." Livingston County, New York. (Contrib. No. 59-10/75. Environmental Resource Center at Geneseo, New York.)

Walker, S. (1975). "Fluctuations of Great Lakes Levels," Bull. No. 184. Rochester Committee for Scientific Information, Rochester, New York.

The Limnology of Canandaigua Lake

Stephen W. Eaton and Larry P. Kardos

INTRODUCTION

Canandaigua Lake is affectionately known to its devotees as the "Gem of the Finger Lakes." The Seneca Indian word for the lake, Kanandarque, is loosely translated, "the chosen spot." According to legend, either Bare Hill or South Hill, both just east of the lake, was the birth place of the Seneca nation.

The center of the lake lies 42° 46′ N latitude and 77° 18′ W longitude. It is the fourth largest of the Finger Lakes of central New York, exceeded in volume of water by Seneca Lake, Cayuga Lake, and Keuka Lake. It is a slightly larger western twin of Skaneateles Lake (Birge and Juday, 1921).

The lake basin is the least disturbed of the major Finger Lakes. At its inlet is a cattail (*Typha*) and elm–maple (*Ulmus americana* and *Acer rubrum*) wetland, now a New York State Game Management Area, called Hi Tor (542.3 hectares) (Anonymous, 1973). According to Child *et al.* (1971), the drainage basin is 49.7% forested, more forested than the basins of any of the five major Finger Lakes, and is 38.8% in active agriculture, the only basin with less being that of Keuka Lake with 33.5%.

The earliest humans to inhabit the basin were probably Paleo Indians, about 7000 BC. Richie (1965) recorded a fluted point from the Middlesex Valley, and axes of the Lamoka phase of the Archaic Indians (3500–1300 BC) have been found at three different places in the basin. The early Woodland Indians are represented by a type station at Vine Valley dated about 563 BC. These Indians were, according to Richie, northern offshoots of the Adena culture of Ohio. The Middlesex phase of this culture was known to make gill nets so these people probably influenced the lake. One of the largest Indian villages of the Late Woodland stage, Owasco culture, was located 2.4 km northwest of the north end of the lake. Richie gives the age of this village as about 1000 AD and suggested a maximum population of 300 to 350 persons at any one time. "Net fishing, probably with a variety of types which included the seine and set nets anchored in the stream or lake with flat notched pebble sinkers, was practiced throughout Owasco times, indeed, apparently over the entire span of Indian occupancy from

Lamoka to Iroquois" (Richie, 1965, p. 278). This Owasco culture was the earliest culture in New York State to cultivate corn and beans, and probably pumpkins and squashes. This cultivation certainly demanded clearing some of the forests. With this shift to an agrarian economy, the Iroquois tended to move off the Allegheny Plateau to the more fertile Ontario lowlands. Day (1953) has assembled from the seventeenth century literature the extent of some of these forest clearings. In 1669, Galinee visited a Seneca village near Victor, New York which stood in a clearing six miles in circumference. Sullivan's expedition in 1779 to subdue the Indians of central and western New York destroyed an orchard of 1500 trees in the Cayuga country. He estimated his forces destroyed corn at 40 villages which at a moderate computation must amount to 160,000 bushels. Day (1953) said that fire was used by the Iroquois of central New York for hunting and for clearing fields.

For nine years after the Senecas were scattered by General Sullivan's army, from 1779 to 1788, the Canandaigua Lake basin stood virtually unoccupied (Simpson, 1942).

The soldiers of Sullivan's army returned to New England and gave enthusiastic reports on the fertility and beauty of the Genesee country and before long a business venture unprecedented in American history was initiated, with the objective of opening this new land to settlement. Here the county–township–range system of land subdivision was used for the first time in America, and the first regular land office opened in Canandaigua. Oliver Phelps of Windsor, Connecticut and Nathaniel Gorham purchased from Massachusetts all the land in New York between the meridian near Seneca Lake and the Genesee River. For this 10,360 km² area they paid $100,000 to Massachusetts and $5000 to the resident Indians with an additional $500 annually "forever." The last of these payments was made about 1815. The Indians were to have hunting and camping privileges in the region for 20 years. But few remained after the settlers arrived.

Phelps arrived in Canandaigua in 1788, put up a storehouse, cut a sleigh road eastward to Geneva, and cleared a wagon road to the head of navigation on Canandaigua outlet about six miles to the northeast.

From November 21, 1788 to June 9, 1790 the entire area bordering on the lake was sold by Phelps and Gorham. It was not until 1794, however, that the Pickering Treaty was signed between the United States of America and the Six Nations and the fear of Indian raids was stilled. Development which had been retarded from 1788 to 1794 was renewed with vigor (Gelser, 1975).

The first to settle near the shores of the lake was Gamiel Wilder, who settled at Seneca Point and had a grist mill, saw mill, and distillery in operation by 1791 (Conover, 1893). Canandaigua was the center of activity and take-off point for development of the Phelps and Gorham purchase.

Development proceeded rapidly, particularly after 1825 with the completion of the Erie Canal which gave ready transportation to population centers such as Albany, New York, and Boston.

Transportation from the Naples area to markets was mainly by lake boats until 1892 when a railroad spur was completed from Geneva to Naples. With the advent of automobile travel and rural electrification, transportation shifted from lake boats and trains to trucks, and by the 1920's the lake began to develop rapidly into a summer recreational area. Robert G. Cook (1931, p. 14) described his early recollections of the lake. "In the early '70's when I first remember the trip on the lake, there were not many cottages or cabins, as they were then called, and most of them were above or south of Black Point."

The lake was used in the nineteenth and early twentieth centuries by lake boats for transportation. The last to run commercially for passengers and/ or freight was the Eastern Star, owned and operated by Wallace Reed, which was moved to Seneca Lake about 1937. The McKechnie Brewing Company stored 2000 tons of ice during the winter for their annual use in the 1870's (McIntosh, 1876). Farmers about the lake also cut ice to be stored in ice houses for use during summer. The Canandaigua Water Works Company, a private corporation, was organized in 1884. The pumping station was situated near the lake shore, at the foot of Main Street, Canandaigua. The water was taken from a crib 800 m distant from the mainland and pumped to a stand pipe at the head of Main Street. In 1893, there were 24 km of main pipe with 525 water takers and 90 fire hydrants (Conover, 1893). Today the lake water is used by Canandaigua City, Gorham, Rushville, Newark, and Palmyra whose combined maximum capacity in 1970 was 5.0×10^7 liter/day (Greeley and Hansen, 1970).

Today, beyond its use as a reservoir, the main use for the lake is recreation. Over 2000 cottages surround the lake, the great majority around its northern half.

GEOGRAPHY OF THE DRAINAGE BASIN

Physiography

The Canandaigua Lake basin is underlain by sedimentary rocks of the upper and middle Devonian age (Fisher et al., 1962; Clark and Luther, 1904). These have only a gentle dip to the south. The southern half of the basin is located in the glaciated part of the Allegheny Plateau, and the northern half on the upper till plain of the Great Lakes section of the Central Lowland physiographic provinces (Pearson and Cline, 1958). The

Portage Escarpment separates these two provinces, and its position may be seen in Fig. 1. As Birge and Juday (1914) pointed out, the hills around the southern part of the lake basin are higher and steeper, coming down close to the lake. No other Finger Lake has so many high hills adjacent to it.

Description of the Bedrock

Clark and Luther (1904) published their classic on the stratigraphy and paleontology of the Canandaigua and Naples Quadrangles as the result of extensive work conducted as early as 1836 by Professor James Hall. They stated that in 1836 Canandaigua and its lake were readily accessible and so were the numerous villages scattered through northern Ontario County, but about these northern villages rock exposures were hard to find because of the thick glacial debris. So the gulleys and rock exposures along the shores of Canandaigua Lake and the adjacent hills were ideal to work out the then unknown stratigraphy, particularly that of the Hamilton Beds. More recent studies have been assembled by Fisher *et al.* (1962) in a Geologic Map of New York (Fig. 1, 2; Table 1).

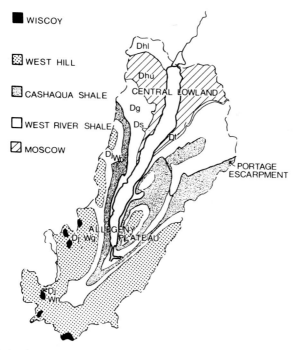

Fig. 1. Bedrock geology and physiographic regions of the Canandaigua Lake basin (redrawn from Fisher *et al.,* 1962).

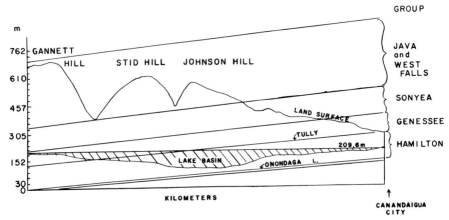

Fig. 2.   Section of the Canandaigua Lake basin and surrounding bedrock showing dip of rock to south, vertical scale exaggerated (after Clark and Luther, 1904).

The lake basin was formed by Pleistocene ice deepening several pre-viously existing north and south flowing stream valleys (von Engeln, 1962). Von Engeln stated that the rock structures of central New York consisted of a vast pile, 2450 m thick, of beds of limestone, sandstone, and shales gently inclined (dipping) southward. After an initial uplift of unknown height the top surface of the pile was worn down by surface runoff to a peneplain of low relief. These beds vary greatly in their resistance to weathering. This peneplanation was succeeded by another uplift about 30 million years ago which did not significantly disturb the arrangement or dip of the rock beds. The Pleistocene, which began about 1 million years ago, renewed the wearing down process. Von Engeln (1962) stated that there was probably a water parting at the southwestern base of the Adirondacks which stretched through central New York. From this divide, which was about opposite the central part of Canandaigua's present basin, the streams flowed north into Hudson Bay, and below the divide water flowed toward the Atlantic Ocean. The Susquehanna and Delaware River drainages are probably remnants of this southern drainage. The lake basin at the north is surrounded by low hills of the Ontario Plain, but to the south the embay-ment of the Ontario Lake Plain has resulted in steep slopes adjacent to the lake. This is due mainly to the weak friable shales, flags, and sandstones which were weathered by rain and streams and then gouged during the Pleistocene by ice up to 3050 m thick.

The oldest beds are at the north end and are made up of the bottom of the Middle Devonian shales and limestones of the Moscow and Ludlow formations (Fisher et al., 1962). Clark and Luther (1904) called these rocks

Canandaigua Shale and Limestone and they are exposed on Lower Gage and Deep Run and in the cliffs along the lake between Tichenor and Menteth Points. Next higher in the column are West River Shales and Tully Limestone of the Genesee group. These form the top of the Middle Devonian rocks of the basin and extend south above the lake level to about Black Point on the west shore and Long Point on the east shore (Fig. 2). These beds contain a 0.3 m thick crinoidal or encrinal limestone (Tichenor Limestone) exposed at the opening of Tichenor Ravine and along the shore cliffs to the south. Then follows a 22.5 m thick, soft, light bluish shale, Moscow Shale, with many crinoids. Within this shale bed is another limestone about 0.3 m thick, the Menteth Limestone. This is a layer with many fine trilobites, corals, and brachiopods and is exposed at the first falls in Tichenor and Menteth Ravines, in the Lake Cliffs between the two points, in Gage Creek, Deep Run, and Hope Point Ravine.

The Tully, a prominent limestone layer of the Genesee group to the east of the lake basin, lenses out in the hills just to the east. At Gorham, only 10 km east, it formed the north rim of the basin of Potter Swamp before being blasted to drain the area for muck farming. It is represented in the lake cliffs by thin shales and iron pyrites just south of Black Point.

Lying directly over the Tully are very black bituminous shales, exposed along the bluffs between Black and Seneca Points on the west shore and on the east shore along the lower slopes of Bare Hill. In these shales are

TABLE 1

Key to Fig. 1 Description of Bedrock[a]

Epoch	Group	Abbrev.	Formation and description
Upper Devonian	Java and West Falls	Djwn	Wiscoy formation: sandstone, shale
		Djwg	West Hill and Gordeau Formations: shale, siltstone, Rhinestreet shale, Grimes sandstone
		Djwr	Pre-Grimes shale, Lower Rhinestreet shale
	Sonyea	Ds	Cashaqua shale, Middlesex shale
Middle Devonian	Genesee	Dg	West River shale, Genundawa limestone, Penn Yan, and Geneseo shales
		Dt	Tully limestone
	Hamilton	Dhu	Moscow and Ludlowville formations: shale, thin limestone
		Dhl	Skaneateles and Marcellus formations: shale, thin limestone

[a] From Fisher *et al.* (1962).

located the Genundewa Limestone, here only about 25 cm in thickness. This thin limestone can be seen in cliffs between Black and Seneca Points and between Seneca and Hicks Points on the west shore, as well as in the cliffs opposite on the east shore. Cashaqua Shales of olive and gray color with occasional flags and sandstone layers are next in the series and mark the beginning of the Upper Devonian epoch. Concretions are typical in the upper part, and a thin limestone layer, the Parrish Limestone, occurs in rock outcrops above the Naples Valley. Above this layer are the Hatch shales and flags of blue and olive shales, thin black shales, calcareous concretions, and hard sandstone and flags. Clark and Luther (1904) said these are the "fundamental cause of the highlands" of the area. The remains of a Lepidodendron were found in these beds in Grimes Glen by D. Dana Luther.

Streams of the basin cut all the rocks so far described and also about 275 m of additional Devonian sandstones, flags, and shales. The Grimes sandstone, 18 m thick, is exposed in Grimes and Tannery Gully, and above this Westhill flags and shale (170 m thick) are exposed in Tannery and Grimes Ravine. A High Point (high cracking point) sandstone, 30 m thick, helped prevent erosion of some of the highest hills in the southern part of the basin, and on top of the highest hills are 61 or so meters of Prattsburg sandstones and shales. Flagstone from these were early quarried for sidewalks in Naples.

In summary, the rock structure of the basin and its surrounding hills is made up mostly of calcareous shales and a few thin limestone layers. On the higher hills to the south more sandstone occurs between these shales. At the south end of the Naples Valley a very prominent terminal moraine, Valley Heads Moraine of Wisconsin Time, blocked the flow of water to the south (Fairchild, 1926).

Soils

The north end of the lake basin is surrounded by the most productive soils. These are of low gradient, and surface runoff from them is not as severe as it is from soils in the southern half of the basin. These soils of the north are of the Odessa–Schoharie Association (OS) (Fig. 3) which Pearson and Cline (1958) said occur on areas of glacial clays. Slopes are mainly less than 10%. These are considered to be an association dominated by fair soils for crops.

From Canandaigua City south to about Menteth Point, and in areas near Roseland Park and from Cottage City south for about 8 km adjacent to the lake basin, are Honeoye–Lima (HL) soils judged good to excellent for crops. On most farms 60 to 80% of the acreage is good to excellent for

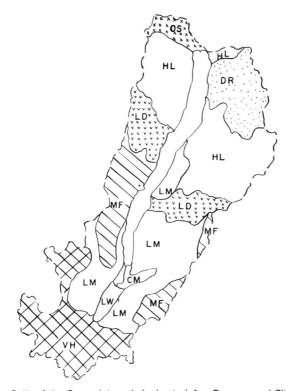

Fig. 3.   Soils of the Canandaigua Lake basin (after Pearson and Cline, 1958).

many kinds of crops, including hay, pasture, and vegetables. Most acreage is used for vegetables and other tilled crops. For most of the association, the supply of organic matter in the surface soil is about one-half to one-third of what it was originally. Most of these soils are nearly neutral in reaction and liming is important only in the southernmost parts of the association. Potassium deficiency may occur under intensive cropping but under the fertilization normally practiced in the region, phosphorus appears to be contained at a level satisfactory for efficient crop production.

An area to the east of the lake basin centered at Reeds Corner (Fig. 3) contains Darien–Romulus (DR) soils. These are moderately fine textured on glacial till derived primarily from shale on gentle slope. There is almost no good crop land in this association. The control of water is the chief management problem. Although the substratum is calcareous, liming is needed because the surface soils and subsoil are commonly acidic. Phosphorus is needed regularly. The best suited legumes are birdsfoot trefoil and the clovers.

Lansing–Darien (LD) soils are located on the west shore from along the northern edge of the Portage Escarpment in the vicinity of Cheshire down to the lake, and again on the east shore from Vine Valley to include Middlesex in the West River Valley. Fair cropland covers most of this association of soils. These soils must be limed at regular intervals. Both potassium and phosphorus are needed for best results.

Lordsdown–Manlius (LM) soils are located on much of Bare Hill and South Hill along the east shore of the lake and in the West River Valley downstream of Middlesex. These also occur along the west side of the lake near the south end above Woodville and extend south on the slope east and west of the Naples Valley. They are not suited for crops. Pearson and Cline (1958) said their best use is for forestry. Except where they are used for growing grapes, they are mainly in forest, though the tops of South Hill, Bare Hill, Stid Hill, and Johnson Hill are used for summer pasturage. Where they have been used for grape culture, very serious erosion has occurred.

Mardin–Fremont–Volusia (MF) soils are extremely variable in quality for cropping. They may be poor, fair, or good. They are located along the west slopes of the lake from Black Point south to near Bristol Springs. They are on the summit of the broad plateau remnants of the southern half of the basin. Maintaining fertility and drainage are the principal problems. The soils are strongly acidic and all need lime regularly. Phosphorus should be applied regularly and potassium is likely to become deficient under intensive cropping. Birdsfoot trefoil is the best legume.

Valois–Howard (VH) soils are on morainic topography near Naples, mainly on the terminal moraine of the Valley Heads Glacier. Fertility is moderate. Most require lime and all need phosphorus regularly. Alfalfa is probably best for stands intended to last three years or less, and for longer periods birdsfoot trefoil is best. Grapes are grown on these soils but the shift in grape culture is now mainly to the Lodell–Wayland (LW) soils of the Naples Valley. These are aluvial soils north of Naples. Drainage is the principal management problem of the Lodell–Wayland soils, and the water table cannot be lowered too much because they lie just above the level of Canandaigua Lake.

Carlisle–Muck (CM) occurs at the inlet of the lake and cannot be drained as the lake level is maintained at a datum of 209.6 m.

Ground and Surface Water

The bedrock and its mantle of glacial debris and soils do much to control the quality of the ground and surface flowing water. Data on this quality and flow in the western Oswego River basin have been prepared by Crain

(1975) for the Cayuga Lake basin and WA-ONT-YA basin Regional Water Resources Planning Boards. Crain stated, in summary, that dissolved solids concentration of precipitation in the area was about 10 mg/liter; sulfate was the predominant constituent. The dissolved solids concentration of small streams during a period with large quantities of overland flow generally ranged from 50 to 300 mg/liter. That of the ground water commonly tapped by wells in the Canandaigua Lake basin ranged from 150 to 500 mg/liter. The ground water in the Canandaigua Lake basin is of the bicarbonate type.

In the area about Canandaigua Lake basin, sulfate concentration is generally much less than 250 mg/liter, within suggested limits for domestic use. High chloride concentrations are a problem only in the deeper wells, especially in the valleys. Crain described an area of 33 km² where wells commonly tap ground water with more than 250 mg/liter in West River Valley upstream and downstream from the village of Middlesex (Fig. 4). The ground water of the basin is generally moderately hard (61 to 120 mg/ liter as $CaCO_3$) to very hard (greater than 180 mg/liter as $CaCO_3$) using the criteria of Durfor and Becker (1964). This is also true of the hardness of the lake water. Natural gas is common in the bedrock around Canandaigua Lake and was found in one well in the West River Valley where samples were collected in 1966 (Crain, 1975).

Demography

Simpson (1942) studied development of the Canandaigua basin area in terms of population growth and subdivided this into three periods. The first, from 1790 to 1840, was marked by rapid and continuous growth from 464 persons in 1790 to 17,200 in 1840 (townships of Canandaigua, Hopewell, Gorham, Middlesex, Italy, Naples, and South Bristol). All towns showed growth during this period but Canandaigua outstripped all the others from the beginning (Fig. 5). The next period, 1840 to 1880, was one of instability. During this period the population of the region oscillated downward then upward three times but resulted in a net increase of 14%. The period concluded with the largest population concentration in the basin up until 1880, namely, 19,705 or 26 persons per square kilometer.

The third period, 1880 to 1930, was one of decline in population except in the city of Canandaigua (Fig. 5). During this time the small general farms were being abandoned on the Allegheny Plateau and there was a shift from rural to urban living.

Simpson (1942) constructed maps of the distribution of individual housing units in the lake basin in 1860, 1900, and 1937 (Fig. 6). This shows quite nicely the rather regularly distributed general farms of 1860 with the same

Fig. 4. Locations of population centers, points about the lake, and the larger numbered tributaries often mentioned in the text. Numbers in lake indicate limnological stations, 1971–1975. M, marina.

concentrations of population which exist today to supply services to these farms: Cheshire, Bristol Springs, Middlesex, and Rushville. By 1900 the houses were more concentrated along roads and in the valleys, and also, there was a beginning of growth along the perimeter of the lake. These two trends continued with cottages and permanent summer houses increasing. A map of the basin dated 1900 (Clark and Luther, 1904) showed 244 houses or cottages around the perimeter of the lake. A Canandaigua Lake directory edited and published by the Canandaigua Yacht Club in 1933 (Anonymous, 1933) listed 776 cottages—539 along the east shore, and 237 on the west shore. By 1960–1961 there were 1371 "cottage residences" listed in a directory of the lake—743 along the east shore and 628 on the west shore (Anonymous, 1960–1961). Rochester residents had discovered the lake and, additionally, the quality of Lake Ontario had deteriorated. By 1968 a directory published by Ontario County Press listed about 1800 cottages around the lake (Anonymous, 1967–1968). In the last eight years considerable building, particularly higher on the hills, has continued. Most building adjacent to the lake has ceased, particularly where zoning ordinances are in effect, such as in the towns of Canandaigua and South Bristol. A condominium complex called Bristol Harbour is a recent development on the west shore just north of Seneca Point and has a population of about 300 persons at the present time.

According to estimates based on the 1970 census, about 7700 persons inhabited the basin (Oglesby, 1974). Approximately 2000 living units are

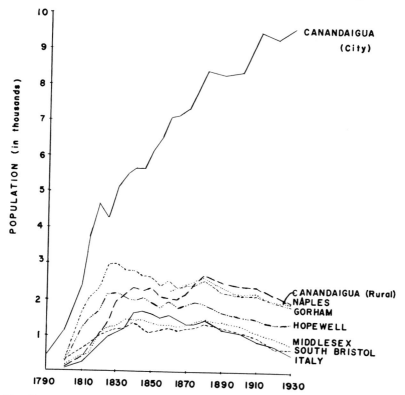

Fig. 5. Population growth in the Canandaigua Lake basin, 1790–1930 (after Simpson, 1942).

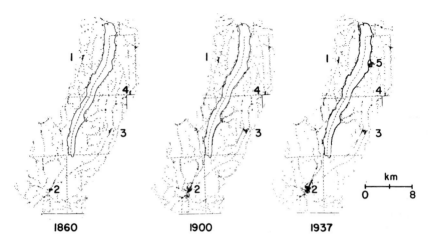

1860   1900   1937

Fig. 6. Population distribution in the Canandaigua Lake basin in 1860, 1900, and 1937. Each dot represents a single family dwelling (after Simpson, 1942). Numbers represent villages of (1) Cheshire, (2) Naples, (3) Middlesex, (4) Rushville, and (5) Cottage City.

located around the perimeter of the lake, which, as of this writing (although it is difficult to get an accurate count), empty their sewage into either septic tanks or, to a lesser extent, holding tanks. The city of Canandaigua, at the foot (north end) of the lake, has a population of a little over 10,000 persons, but the city's wastewater and storm water have secondary treatment and the effluent flows into Canandaigua Outlet, not into the lake.

Land Use

The original forest cover was hardwoods, pine, and hemlock, but this was removed by 1820 except for woodlots used by farmers. The first crops grown were wheat, corn, and barley. Sheep and cattle were raised, and milk was processed into cheese and butter (Pearson and Cline, 1958). After completion of the Erie Canal, grain and other field crops, dairy products, vegetables, and fruits became the important farm products and remain so today. In 1975 the top three income producers to farmers in Ontario County were diary, $13,200,000; vegetables, $10,170,000; and wheat and grain, $5,630,000 (Rodney S. Lightfoot, Ontario County Agricultural Extension Agent, personal communication). Grape culture still dominates the steep slopes from Seneca Point south along the west shore and adjacent to Vine Valley on the east shore. This culture began about 1850 and peaked about 1910 (Morris, 1955). By the end of Prohibition in 1933 nearly half of the 1214 hectares (3000 acres) of grapes present in 1910 was gone. This was built back up to 1012 hectares (2500 acres) by 1966 but with a new emphasis toward cultivation on the flats. It was estimated that 25% of the vineyards was on valley flats, 50% on moderate slopes (8–20% grade), and the remaining 25% on slopes in excess of 25% grade (Schultz, 1967). Following World War II and up until 1970, DDT was applied at least four times per year during the growing season and probably was responsible for many of the high levels found in the lake trout (Woldt and Gavagon, 1970).

Cornell University Water Resources and Marine Science Center has published a paper on land use data for the Finger Lakes Region of New York State (Child et al., 1971). These data were assembled from aerial photographs, interpreted by trained aerial photographers, and spot field checked for verification. Each land area was categorized into one of some 130 LUNR (Land Use and Natural Resources) classifications. The Canandaigua Lake basin was subdivided into four subunits—Naples Inlet, West River, an area at the southern half of the lake drained by small tributaries, and an area at the north drained by small tributaries (Fig. 7). One reason for such division was the variable retention times of water entering from different subunits. Table 2 shows nine principal land uses and their distribution in the four subunits.

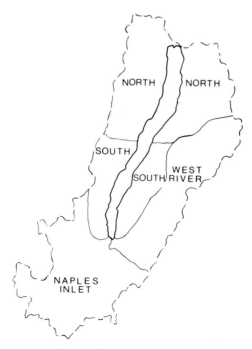

Fig. 7.   Land use subunits of the Canandaigua Lake basin (Child *et al.,* 1971).

TABLE 2

General Land Use in the Canandaigua Lake Basin[a]

Canandaigua Lake basin	Drainage basin (km²)	Active agriculture		Inactive agriculture		Forest		Residential		Commercial		Industrial		Outdoor recreation		Extractive		Public	
		km²	%	km²	%	km²	%	km²	%	km²	%	km²	%	km²	%	km²	%	km²	%
Naples Inlet	122.51	28.78	23.5	9.92	8.1	81.69	66.7	0.97	0.8	0.17	0.1	0.04	<0.1	0.41	0.3	0.26	0.2	0.21	0.2
West River	101.91	48.08	47.2	8.74	8.6	43.33	42.5	1.21	1.2	0.05	<0.1	0.11	0.1	0.07	<0.1	0.07	<0.1	0.25	0.2
South	63.04	13.20	20.9	4.90	7.8	41.60	66.0	2.01	3.2	0.04	<0.1	0.00	0.0	1.13	1.8	0.11	0.2	0.09	0.1
North	119.77	68.11	56.9	12.33	10.3	35.63	29.7	2.77	2.3	0.34	0.3	0.05	<0.1	0.13	0.1	0.16	0.1	0.12	0.1
	407.23	158.17	38.8	35.89	8.8	202.25	49.7	6.96	1.7	0.60	0.1	0.20	0.1	1.74	0.4	0.60	0.1	0.67	0.2

[a] From Child et al. (1971).

In treating the drainage basin as a whole, an area of 407.23 km²,* 38.8% was found to be in active agriculture, 8.8% in inactive agriculture, 49.7% in forest, 1.7% residential, less than 0.1% commercial, less than 0.1% industrial, 0.4% outdoor recreation, 0.1% extractive, and 0.2% public. Active agriculture is mainly confined to the north subunits while the most heavily forested areas are in the South and Naples Inlet subunits.

Vegetation Patterns

The forests of the basin are mostly Oak–Northern Hardwoods (Stout, 1969), which are made up mainly of oaks in pure stands or mixed with northern hardwoods.

The forested slopes leading down to the lake are mainly Oak-Hickory but in the more shaded situations of the gullies hemlock, birch, beech, and maple occur. Basswood (*Tilia americana*), black walnut (*Juglans nigra*), and white ash (*Fraxinus americana*) are also common. Red cedar (*Juniperus virginianus*) and white pine (*Pinus strobus*) are often present. In a few areas of the basin native red pine (*Pinus resinosa*) and white birch (*Betula papyrifera*) occur. Sycamore (*Platanus occidentalis*), willows (*Salix*) of various species, and cottonwood (*Populus deltoides*) are common trees about the perimeter of the lake. In Naples Swamp, *Acer rubrum* and black ash (*Fraxinus nigra*) dominate along with shrubs of *Cephalanthus occidentalis* and *Spiraea alba*. Dead American elm (*Ulmus americana*) trunks and branches still stand to show its previous dominance. In this swamp are good stands of cattail (*Typha latifolia*).

Recreation

About the lake are many public recreational areas. Most provide swimming, boating, and picnic facilities and some provide hunting and fishing.

The city of Canandaigua filled in a portion of the north end of the lake to form Kershaw Park about 1950. Here facilities for picnicking and swimming are provided. Ontario County Park is located on the east shore off East Lake Road about 4 km south of the foot of the lake. Another Ontario County Park is available on Gannett Hill at what is locally called the Jump-Off, 4 km west of Cook Point. Deep Run Park is a state-owned area along the east side of the lake approximately 6 km south of the foot. Swimming is the speciality here with lifeguards on duty. A total of five marinas are available around the lake: two at the north end, one 4 km south of the

---

* About 26 km² below the figure used by the United States Environmental Protection Agency (1974).

north end on the west side, one at the south end at Woodville, and one on
the West River. Gas may be obtained near Cottage City about 8 km south
of the north end on the east shore. Public launch sites are available at the
north and south ends of the lake, and High Tor is available at the south end
for hunting, fishing, and the general enjoyment of nature. A small public
beach is located near a general store at Vine Valley about 18 km south on
the east side (Piampiano, 1973). The rest of the shore is in private
ownership.

Point Source Pollutants

The main waste sources enter the lake at the south end by way of West
River. Rushville is a community straddling the Ontario and Yates County
line near the headwaters of West River. This village has a population of
about 570 persons and has storm sewers and private sewerage systems.
Comstock Foods, Inc. is a local beet and carrot processing and canning
industry which treats its effluent by passing wastes through a rotary screen
and then into lagoons. Farther down West River, Middlesex Valley Central
School serves a population of 640 persons with sand filters and septic tanks
(USEPA, 1974). Emerson Produce Company operates a chicken processing
plant near Middlesex Village and the effluent of this plant receives secondary treatment. Naples (Fig. 4), with a population of 1240, has a village
water supply taken from Reservoir and Grimes Creeks but no village waste
disposal system. Storm sewers empty into Naples Creek and adjacent tributaries, but only private subsurface waste disposal systems handle village
sewage. Widmer's Wine Cellars, Inc. has a primary treatment plant which
discharges into Naples Creek. In 1965, the Federal Water Pollution Control
Administration conducted a stream survey on Naples Creek. They found that
the water quality lowered considerably during the summer, with a noticeable
decline in dissolved oxygen and a corresponding increase in BOD. Tertiary
treatment will eventually be required because of the downstream proximity to
"AA" classified Canandaigua Lake (Camp, Dresser, and McKee, 1968).

There are, at present, no active town dumps leaching into the basin of
Canandaigua Lake. South Bristol Town Dump leaches into Mud Creek to
the west and Naples is hauling solid wastes to Livingston County. The
Town of Canandaigua did have a landfill which was covered and closed
down about 1973, because it leached into Menteth Creek (Tributary 43).
The landfill for Rushville and Middlesex areas is located to the east of the
basin and leaches into Flint Creek. The City of Canandaigua has a dump
whose leachate is received by Canandaigua Outlet (J. Carver,
personal communication).

Bristol Harbour, located just north of Seneca Point, has secondary treatment of its sewage which at present is serving about 300 persons (USEPA, 1974).

Most of the rest of the population living within the drainage basin has private septic tanks or holding tanks to handle wastes. There is presently under development, for which contracts are about to be finalized, a perimeter sewer which will handle sewage in the more populated areas around the northern half of the lake (J. Carver, personal communication).

The largest concentration of people in the basin is at the north end in the city of Canandaigua, but storm and sewer drainage is received by Canandaigua Outlet and affects the lake only during spills or exceptionally high water. The 2000 cottages surrounding the lake probably contribute greatly to the loading of nitrogen and phosphorus, as do farms surrounding the north and south ends.

The socioeconomic status of the basin is generally one of relative affluence, though the city of Canandaigua keeps its rural charm and has lost population since its heyday in the 1890's. Only light industry is present in Canandaigua and the outlying small population centers of the area. Recreation, primarily during the summer season, is second only to agriculture in bolstering the socioeconomic status of the basin.

Climatology

Approximately 25 km northeast of the center of Canandaigua Lake is a climatological bench mark station as designated by the United States Weather Service. Dethier and Pack (1965) have prepared a summary of the climate of Geneva, New York from data assembled at this station.

They said that the climate and weather of Geneva are governed predominately by atmospheric flow from continental areas. Air flowing from the Atlantic Ocean affects the area on infrequent occasions and is therefore of secondary importance. The prime source of atmospheric moisture and humidity is the Gulf of Mexico. The relative nearness of the Great Lakes, especially Lake Ontario, has a significant effect on Geneva's climate. The area's lakes have a moderating influence on temperatures, particularly during the months from early fall through March or April.

The climate is marked by warm summers and rather long, cold winters. Maximum precipitation occurs in the late spring and early summer months (Table 3). The freeze-free season ranges from 140 to 175 days in seven years out of ten. Annual precipitation ranges between 66 and 97 cm in seven years out of ten, and the 30-year annual mean is 84.1 cm. From early November through early April a total snowfall of 127 to 191 cm may be expected in

TABLE 3

Mean Monthly Temperatures and Precipitation at Geneva, New York 25 km NE of the Center of Canandaigua Lake for Period 1933–1962[a]

Months	Mean temperature (°C)	Mean precipitation (cm)	
		Rain	Snow
January	−3.3	5.6	30.5
February	−3.3	6.2	35.6
March	1.1	7.0	35.6
April	7.8	7.5	10.2
May	14.4	7.8	Trace
June	19.4	8.1	0
July	22.2	7.8	0
August	21.1	6.9	0
September	17.2	6.5	0
October	11.1	7.8	2.5
November	5.0	6.7	17.8
December	−1.7	6.0	25.4
		84.1	157.5

[a] After Dethier and Pack (1965).

the majority of winters, and the 30-year average was 157.5 cm. Persistent cloudiness during the late fall and winter months is a notable feature of the climate. On an annual basis there is an average of 170 cloudy days, of which about 20 occur in each of the months from November through February. Only 25 to 30% of maximum possible sunshine may be expected from November through February, but amounts increase to 65 to 70% in the summer months. The prevailing wind is westerly from November through April but it shifts to southwesterly the remainder of the year. The average velocity decreases from about 18 km/hr in the winter and early spring to about 13 km/hr in the summer and early autumn.

The weather of the growing season (April 1 to September 30) at Geneva, New York has been summarized for the years 1953–1967 (Peck *et al.*, 1968). Their station was on the Darrow or Fruit Breeding Farm 3.8 km due west of the Geneva Experiment Station or about 22.5 km from the center of Canandaigua Lake. Of significant interest to the limnologist are the pooled quadratic curves for solar radiation, wind, and evaporation. Solar radiation ranges from a low of about 300 gm cal/cm² per day on September 30 to a peak of 552 gm cal/cm² on June 28 about a week following the peak of daylength. The mean solar radiation was 469 gm cal/cm² (Fig. 8).

Winds during the growing season ranged from an average high of 164.2 km/day in 1964 to a low of 125.5 km/day in 1962 (Fig. 9). The highest cumulative wind for any day during the 15 years was 640.4 km on April 5, 1963, and the lowest was 0 km on August 20, 1961. Wind was the only weather factor recorded where the lowest numerical value occurred during the middle of the growing season. The highest evaporation of water from the Weather Bureau Class A evaporation pan for any day during the 15 years was 1.2 cm on July 4, 1966, and lowest was 0.003 cm occurring on 13 different days. The mean evaporation for the 15 years was 0.445 cm/day varying from a high mean of 0.503 cm/day in 1955 to the lowest of 0.366 cm/day in 1956 (Fig. 10). The growing season pooled quadratic curve for evaporation peaked on July 8.

Curves obtained from the pooled standard quadratic computer program are shown in Fig. 10. Units have been omitted from this chart so that the slopes and the peaks of the curves may be compared on a relative basis. Note that the solar radiation curve fairly closely follows daylength. The evaporation curve lags behind daylength and peaked on July 8. The mean air temperature lagged farthest behind daylength, peaking on July 24.

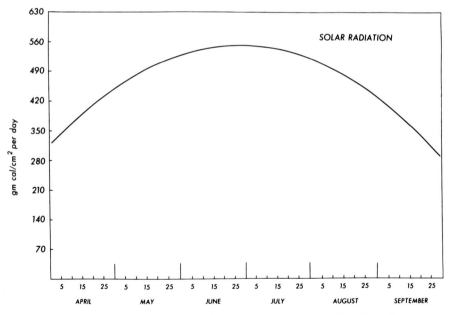

Fig. 8. Pooled quadratic curve for solar radiation near Geneva, New York (from Peck *et al.*, 1968).

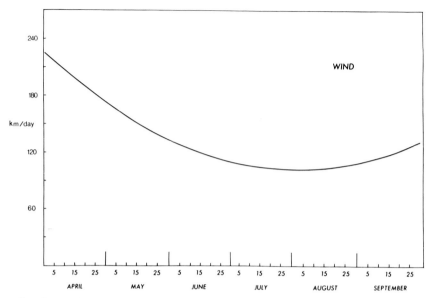

Fig. 9. Pooled quadratic curve for wind near Geneva. New York (from Peck *et al.*, 1968).

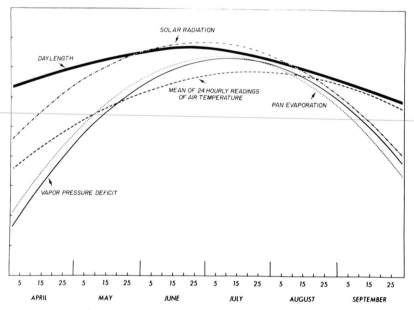

Fig. 10. Pooled quadratic curves for daylength, solar radiation, mean of 24-hr readings of air temperatures, vapor pressure deficit, and pan evaporation near Geneva, New York (from Peck *et al.*, 1968).

Ice cover on Canandaigua Lake is complete more often than on Seneca and Cayuga but less than on the smaller lakes to the west such as Honeoye, Conesus, and Hemlock (Table 4). The table indicates that the Lake froze over completely (or nearly so) in 8 out of the past 22 years. The winter with the lowest temperatures on record was that of 1933–1934, and the Naples Record of February 7, 1934 said "For several days Canandaigua Lake has been covered with ice over its entire length." On April 11 the same weekly reported that "the ice disappeared from Canandaigua Lake last Thursday" (April 5). Previous to that time the lake had frozen over completely in 1926 and before that in 1917 (Table 4). Both the north and south ends of the lake almost always freeze over for about a kilometer or two from their ends. In years when the lake freezes over completely this happens most often from late January to mid February, when the water temperatures are lowest, and ice-off usually occurs in early April.

Morphometry

The morphometry of the Canandaigua Lake basin was first described in detail by Birge and Juday (1914) (Table 5). Essentially, all hydrographic details are based on surveys made by successive classes of the College of Civil Engineering at Cornell University during the summers of 1888 to 1890. The bathymetric map (Fig. 11) is based on 395 soundings and its contours were determined by a careful trigonometric survey. The sounding line was of wire; an apparatus was provided for releasing the weight when the bottom was reached, and a registering apparatus recorded the depth. The soundings were well placed and the position of each was controlled by transit instruments on shore and a sextant in the boats.

With its maximum depth of 84.1 m (209.6 datum) Canandaigua is the sixth deepest lake on the continent east of the Rocky Mountains. It is long (24.9 km) and narrow (2.44 km) with a shore development index of 2.48. The basin is generally aligned in a north–northeast to south–southwest heading, tending to turn the southwest winds (typical of summer) approximately 12° to the north and the northwest winds (typical of winter, fall, and spring) into the south–southwest about 20°, reaching rather straight to the south end of the lake.

The mean slope of the shore is 7.0% (somewhat steeper than Cayuga at 5.2%), except at the extreme north and south ends where there are shallow areas and shoals. Along east and west shores there is a narrow shoal area totaling about 1.8 km². At the south end there is about 1.0 km² of water less than 6.1 m in depth, and at the north end a shoal area of about 4 km². This totals about 6.8 km² or 16% of the total lake surface area. Where the 50 tributaries enter the lake there are an equal number of deltas of varying sizes depending on the amount of flow in the tributary, its drainage area, vegeta-

TABLE 4

Data on Ice Cover on Canandaigua Lake[a]

Date	Comment[b]	Date	Comment
1796	Cayuga (1)	1955	50% lake open all winter (4)
1816	Cayuga (1)	1956	75% lake open all winter
1818	Cayuga (1)	1957	75% lake open all winter
1835–1836	Seneca, Cayuga (1)	1958	Open all winter but 90% frozen at times
1855–1856	Seneca, Cayuga (1)	1959	Frozen over except a large pond at Vine Valley
1868	Seneca	1960	Open all winter but 90% frozen at times
1872–1873	Isabella Grapes killed (2)	1961	Lake 100% frozen February 1–March 1
1875	Seneca, Cayuga (1)	1962	Lake 100% frozen February 12–March 27
1884–1885	Seneca, Cayuga (1)	1963	Lake 100% frozen February 25–March 25
1894	"Canandaigua remained covered with solid cap of ice until April 1"	1964	Practically no ice all winter
1899	"Many Canvasback Ducks killed about airholes that remained open" (3)	1965	100% frozen March 2 but opened up next day
1904	Seneca, Cayuga (1)	1966	Open all winter but within one still night of freezing twice
1912	Seneca, Cayuga, Skaneateles, "all lakes" (1)	1967	Practically no ice all winter
1914	Seneca	1968	Open all winter but 98% frozen twice
1917	"Cars drove on ice from Naples to Canandaigua"	1969	All frozen except big pond between Vine Valley and Seneca Point
1918	Seneca	1970	100% frozen January 20–March 20
1923	Canandaigua completely frozen over February 25	1971	Open all winter but 80% frozen February 4
1926	Canandaigua still covered with ice March 31—last ice left April 9	1972	Open all winter but 98% frozen March 4
1934	Ice cover complete late January to April 5	1973	Open all winter but 90% frozen February 27
1936	February 12—Canandaigua Lake was entirely closed by a smooth cap of ice, 18″ thick in some places. L. C. Lincoln drove his car from Deep Run across lake on ice	1974	Practically no ice all winter
		1975	Practically no ice all winter
1936–1946	No mention ice cover Naples Record	1976	100% frozen January 24 but opened up next day and all ice was gone February 23

[a] When Seneca and Cayuga Lakes froze over, it was assumed Canandaigua Lake also froze over. Years not shown probably indicate ice cover was not complete for any length of time.

[b] Numbers in parentheses indicate the following references: (1) Birge and Juday (1918); (2) Hedrick (1933); (3) Eaton (1909); (4) R. L. Case, data years 1955–1976, personal communication; no reference given before 1955 indicates information taken from Naples Record, Naples Library.

TABLE 5

Morphometric Details of Canandaigua Lake[a]

Parameter	Metric units	U.S. Customary units
Drainage area (2)	407.23 km²	157.2 mile²
Lake surface area (A) (2)	42.3 km²	16.3 mile²
Elevation above mean sea level	209.6 m	688.0 ft
Length	24.9 km	15.5 miles
Width	2.44 km	1.5 miles
Area	42.3 km²	16.3 mile²
Maximum depth (at datum 209.6 m)	84.1 m	276.0 ft
Mean depth (M) (at datum 209.6 m)	38.8 m	
Volume development = mean depth/max. depth	0.46 m	
Volume (V) (at datum 209 m)	1640.1 × 10⁶ m³	57,897.0 × 10⁶ ft³
Normalized outflow	122.4 × 10⁶ m³/year	
Retention time (normalized)	13.4 years	
Mean slope		7.0%
Shore development (or index)	2.48	
Volume development (or index)	1.39	
A/2 at M	44.0 m	(53% of max. depth)
V/2 at M	26.0 m	(31% of max. depth)
Littoral zone [less than 6.1 m (20 ft) of depth] (3)	16%	
Approx. areal extent of rooted aquatics (3)	777.8 hectares	1922 acres or 16% of surface area
Flood elevation	210 m	689 ft
Low level elevation	208.5 m	684 ft
Normal summer elevation	209.7 m	688 ft
Shoreline in public ownership	1.9 km	1.2 miles or 3%
Length of shoreline	57.8 km	35.9 miles
Housing units around perimeter (4)	2,000	
Private no. housing units/km or mile	17.3/km	35.9/mile
Frontage in meters or ft/housing unit	57.8 m	147.1 ft

[a] Key: A, area; M, mean; V, volume; (1), mostly after Birge and Juday (1914); (2), Child et al. (1971); (3), Gelser, personal communication (1975); (4), number units estimated. Most morphometric details from survey by Cornell University 1888 to 1890 when 395 soundings taken.

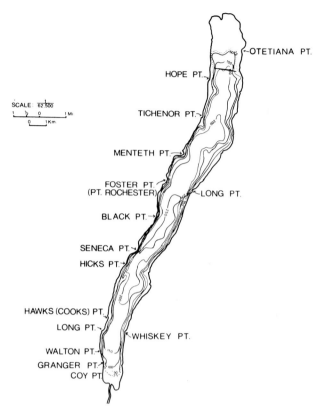

Fig. 11. Bathymetric map of Canandaigua Lake from survey by Cornell University 1888–1890 (from Birge and Juday, 1914).

tion, soil, and the rock structure in each individual basin. The largest delta exposed above lake level today is the swamp at the south end followed by Seneca Point, Menteth Point, Foster Point, Point Rochester, Cook Point, Long Point, and Granger Point, all on the west side of the lake. Deltas occur on the east shore at Whiskey Point, Vine Valley, Long Point, Gooding Landing, Green's Landing, and Otetiana Point, but these are all less extensive than those named on the west shore, perhaps due to the greater wave action from westerly winds.

As is true with all of the major Finger Lakes, little erosion and subsequent deposition, save the deltas mentioned above, seems to have taken place since the last glacial advance receded 8000 to 10,000 years ago. Most beaches are composed of coarse shale material eroded from cliffs located either along the west shore between Black Point and Granger Point or along the east shore between Long Point and a point a few kilometers north of Vine Valley (Fig.

4). The beaches at the north end are mainly of sand and gravel, reflecting the deeper mantle of drift occurring over the land at the north end.

The basin of the lake below the narrow shoal perimeter, where currents are less effective, contains great thicknesses of fine lake sediments of gray or sometimes brownish color (Eaton *et al.*, 1975).

Most beaches and shoal areas are covered by flattened dark grey to black shales but, in a few areas, gravel occurs. Such areas occur off Vine Valley and in areas opposite along the west shore. About Point Rochester, Foster Point, Menteth Point, and the north end in general there are considerable areas where glacial debris in the form of gravel occurs, sometimes making it difficult to use an Ekman dredge.

Hydrology

Canandaigua Lake is considered by the State Health Department and other government agencies as a multipurpose reservoir, being used for recreation, power (to a limited extent), water supply, and flow equalization (New York State Department of Health, 1955).

Stream Inflow

Tributary Number One in the numbering system established by the State Conservation Department for its Biological Survey in 1927 (Moore, 1928) is located about 0.6 km east of the natural outlet of the lake. This same numbering system has been used by the New York State Department of Health in stream and lake classification (New York State Department of Health, 1955). There are 17 small, numbered tributaries directly entering the lake on the east. The four with the largest drainage areas are tributary 1 (entering just south of Roseland Park), Deep Run (6), one unnamed just south of Crystal Beach (8), and Vine Valley stream (16). The largest tributary enters at the south end—West River (18). The largest single tributary of the West River joins as Tributary (2), Naples Creek. This, in turn, drains the entire southern part of the basin with its principal tributaries named Grimes Creek (8), Tannery Creek (9), Reservoir Creek (10), and Eelpot Creek (12a). Entering the west side of the lake are 32 small tributaries, the largest being Seneca Point Creek (33), Barnes Gully (39), Menteth Gully (43), Tichenor Gully (44), and Sucker Brook (50) entering from the north and west.

These tributaries carry precipitation into the lake basin from a drainage basin of 407.23 km². The lake itself has a surface area of 42.3 km² (Table 6). The Canandaigua Lake Pure Waters, Ltd. has organized a group of interested citizens (under the direction of Nelson Smalt) to record rainfall in the basin since 1969. These rainfall stations are located on Gannett Hill and at Vine Valley, Naples, and Granger Point. There is also a station at the

Canandaigua Water Works on the west shore of Canandaigua Lake. Smalt also records lake level and evaporation losses at Granger Point (near Woodville), and lake levels are also recorded at a United States Geological Survey guage at the end of the Canandaigua City Pier.

Precipitation records from the five stations over the past five years show a considerable spread for the basin as a whole. The north end gets considerably less precipitation than the hills of the Allegheny Plateau surrounding the southern end. This can best be illustrated by the greatest precipitation event to hit the area since records have been kept: Hurricane Agnes of June 21–23, 1972 (see tabulation below).

Canandaigua City Pumping Station	10.59 cm
Gannett Hill Ontario County Park	20.07 cm
Naples, Widmer's Wine Cellars	22.48 cm
Granger Point	21.57 cm
Vine Valley (Robert Ludlow)	21.03 cm

Another precipitation event of May 3, 4, and 6, 1975 shows the same trend (see tabulation below).

Canandaigua City Pumping Station	4.19 cm
Naples, Widmer's Wine Cellars	6.68 cm
Granger Point	5.46 cm

In interpreting precipitation data over the lake basin one must consider that about 62% of the watershed drains into the lake at the south end, and this part of the drainage area receives more precipitation (Gelser, 1974). The River Forecast Center (National Weather Service, 135 High Street, P.O. Box 688, Hartford, Connecticut 06101) prepares a Canandaigua Lake Runoff Index which predicts quite accurately inflow and how the inflow will affect lake level. A 2.5-cm rain over the entire lake basin, with soils well saturated, will increase the lake level by 30.0 cm (Anonymous, 1973). This index can be obtained from the Rochester Office of the United States Weather Service. Important to the limnologist, is the fact that particulates from soils of the southern part of the watershed will have the longest retention time in the basin before eventually entering the outflow at the north or being added to bottom sediments.

The New York District Office of the United States Geological Survey supplied the United States Environmental Protection Agency, National Eutrophication Survey with the normalized flows for several principal tribu-

TABLE 6

Drainage Areas of Subbasins and Lake, with Their Percent of Total Catchment Area and Mean Annual Discharge

Inflow	Drainage[a] area (km²)	% of total catchment	Mean annual discharge[b] (m³/sec)
West River (gauge below Middlesex)	75.9	16	0.49
Naples Creek	123.5	26	1.39
Seneca Point Creek	11.4	2	0.07
Barnes Creek	3.4	1	0.02
Menteth Creek	15.5	3	0.10
Vine Valley Creek	12.2	3	0.08
Minor tributaries	191.7	40	1.48
	433.6[c]	91	3.63
Area of lake	43.0	9	
Total catchment	476.6[c]	100	
Outflow			
Feeder Canal	476.6[c]	100	2.05
Muir Dam	——		1.83
			3.88

[a] Drainage areas after USEPA working paper 149. Large basins accurate to ±5%; small basins ±10%.

[b] Mean of normalized monthly flows provided USEPA by the New York District Office of United States Geological Survey. Normalized mean monthly flows accurate within ±15%.

[c] This somewhat higher than estimate in Child *et al.* (1971).

taries of the lake, particularly the tributaries used to estimate nutrient loading (USEPA, 1974) (Tables 6 and 7).

Outflow

Records on outflows from the lake basin have been kept by the United States Geological Survey by means of a stream gauging station at Chapin since November 1939, and normalized flows are shown in Table 7. This gauge was reconstructed in the fall of 1974. No structure of any size exists to store water during periods of high inflow before water enters the lake basin, but one has been proposed at some location in West River Valley.

From an examination of inflow into the lake basin and outflow from it, and considering that the amount lost through evaporation averages perhaps

TABLE 7

Monthly Normalized Inflows in m³ per Second from Major Tributaries and Normalized Outflows from the Feeder Canal and Muir Dam of Canandaigua Lake[a]

Inflow	Mean	Jan	Feb	Mar	Apr	May	Jun	Jul	Aug	Sep	Oct	Nov	Dec
West River below Middlesex	0.4852	0.3116	0.5948	1.756	1.529	0.7364	0.2408	0.0199	0.0114	0.0085	0.0510	0.1756	0.3965
Naples Creek	1.392	0.9346	1.699	4.815	4.248	2.096	0.7364	0.1558	0.1190	0.0369	0.2436	0.4248	1.218
Seneca Pt. Creek	0.0737	0.0482	0.0907	0.2549	0.2238	0.1105	0.0397	0.0085	0.0057	0.0020	0.0142	0.0227	0.0680
Barnes Creek	0.0196	0.0142	0.0227	0.0680	0.0595	0.0284	0.0114	0.0023	0.0017	0.0006	0.0029	0.0057	0.0170
Menteth Creek	0.1006	0.0680	0.1246	0.3399	0.3116	0.1530	0.0538	0.0114	0.0085	0.0026	0.0170	0.0312	0.0878
Vine Valley	0.0759	0.0510	0.0935	0.2634	0.2323	0.1133	0.0397	0.0085	0.0057	0.0020	0.0142	0.0027	0.0680
Unnamed[b]	1.480	0.9912	1.841	5.098	4.531	2.237	0.7647	0.1671	0.1275	0.0397	0.2606	0.4248	1.303

3.627 m³/sec

Outflow	Mean	Jan	Feb	Mar	Apr	May	Jun	Jul	Aug	Sep	Oct	Nov	Dec
Feeder Canal	2.047	1.529	1.841	3.965	5.381	3.682	1.954	0.9912	0.9912	0.9912	0.9912	0.9912	1.274
Muir Dam	1.832	1.529	1.841	3.965	5.381	3.682	1.954	1.048	0.2832	0.1416	0.3682	0.5664	1.246

3.879 m³/sec

[a] After USEPA Eutrophication Survey Working Paper 149, Appendix A. These are accurate within ±15%.
[b] All other tributaries not listed above.

0.455 cm/day from April 1 to September 30 (Peck *et al.*, 1968), there seems to be little loss through the rock structure of the basin. The fine sediments which cover the bottom of the basin in areas not exposed to currents help prevent leakage into bed rock, which might occur along aquifers dipping approximately 25 m/km to the south.

If one considers the evaporation from the surface of the basin and the outflow to approximately equal the inflow from tributaries during the year, then retention time for Canandaigua Lake can perhaps best and most simply be figured by dividing normalized outflows per year into the total volume of the lake. If the volume is 1640×10^6 m^3 (Birge and Juday, 1914) with datum at 209.1 m (686 ft) and the normalized outflow of the feeder canal and the natural outlet is figured to average 3.879 m^3/sec (122.4×10^6 m^3/year), then the volume divided by the outflow per year gives a retention time of 13.4 years (flushing rate if no inflow or evaporation is considered). Previous workers have determined this retention time to be 15 years (USEPA, 1974) and 7.4 years (Oglesby *et al.*, 1975).

Lake Levels

Canandaigua City operates the dams which control lake level under authority granted by the New York State Legislature on June 13, 1886. Dams at the north end, the Feeder Canal Dam and Muir Dam, can control lake levels from 208.5 to 210.3 m above MSL at above 28.32 m^3/sec (Hershey, Malone, and Associates, 1973). When this flow of 28.32 m^3/sec is exceeded then the 210.3-m level cannot be maintained, as happened with Hurricane Agnes when the lake crested at 210.95 m above MSL (Table 8).

Management of the lake level is under direct control of the Canandaigua City Manager and the City Engineering Department. By mutual agreement with the State Health Department, and because of existing inadequate sewage treatment in the greater Canandaigua area, the City maintains at least a flow of 1 m^3/sec in the feeder canal. The New York State Department of Environmental Conservation recommends maintaining at least a level of 209.4 m to maintain the best water level for wildlife in the wetland at the head of the lake; at this level (209.4 m) the freezing of some private water intakes begins. At a level of 210 m some damage to lakefront cottages on the flood plain begins so that, ideally, the levels, by mutual consent of most persons and governmental agencies, should be maintained between 209.4 m and 210 m MSL (Gelser, 1974). Canandaigua Lake Pure Waters, Ltd. has been instrumental in working out this mutually agreeable management program, and in 1974 the lake levels were maintained between a high of 209.93 (May 30) and a low of 209.37 (November 14), a spread of only 0.56 m. During the past 40 years the highest levels have occurred in March 1936 and 1956, June 1972 and March 1976. High levels in March occur

TABLE 8

Ten Highest Stages of Canandaigua Lake since 1911[a]

	Date	Elevation in meters above MSL[b]
1	June 1972	210.95
2	March 1936	210.50
3	March 1956	210.47
4	March 1976	210.46
5	April 1929	210.32
6	December 1927	210.32
7	April 1916	210.29
8	April 1950	210.29
9	March 1913	210.27
10	April 1961	210.26

[a] After Hershey, Malone, and Associates (1973) and B. M. Gelser (personal communication).
[b] MSL, mean sea level.

because precipitation in the form of rain and snow has been relatively high for three months and average temperatures in March are above freezing (Table 3) (B. M. Gelser, personal communication).

PHYSICAL LIMNOLOGY

Temperature Regime

The mean temperature of Canandaigua Lake August 20, 1910 as figured by Birge and Juday (Table 9) was 11.1°C. The epilimnion is usually about 10 m deep in July and gradually thickens to as much as 15 m in some years by mid September. The thermocline is normally about 8 m thick and the hypolimnion about 50 m deep depending on the location of the water column. Homothermy begins at about 8°C and generally mixing begins in late November to early December (Fig. 12). By January the lake is isothermal at 4°C and continues to mix (if not frozen over completely) until it reaches, in March, temperatures as low as 1.3°C from the surface to a depth of over 40 m. K. M. Stewart (unpublished data) took temperatures under ice cover on January 17, 1968 off Seneca Point. These data show that with ice cover and reduced mixing an inverse stratification may occur producing a dimixis with fall and spring turnover separated by a winter inverse stratification (Fig. 17).

TABLE 9

Thermal Data and Volumes in 10-m Strata of Canandaigua Lake[a]

III Strata (m)	IV Volume (m³ × 10⁶)	V Relative volume	VI Temp. (°C)	VII Heat Content IV × VI (°C × m³ × 10⁶)	VIII V × VI (°C)
0–10	362.2	0.221	21.3	7714.86	4.707
10–20	302.6	0.184	14.5	4387.70	2.668
20–30	271.7	0.165	7.4	2010.58	1.221
30–40	241.2	0.147	6.25	1507.50	0.907
40–50	206.3	0.126	5.85	1206.86	0.737
50–60	153.4	0.094	5.75	882.05	0.540
60–70	80.2	0.049	5.65	453.13	0.277
70–80	22.2	0.014	5.45	120.99	0.076
	1639.8	1.000	72.15	18,283.67 × 10⁶	11.13[b]

[a] After Birge and Juday (1914) and Cole (1975).
[b] Volume weighted mean temperature of lake near end of summer, August 20, 1910.

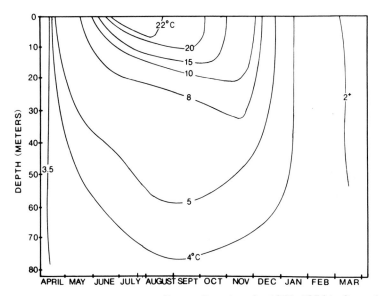

Fig. 12. Average temperature profiles at all stations for 1971–1974 in Canandaigua Lake.

Currents

During July and August 1975, special efforts were made to obtain data on the magnitude of currents using a Kahlsico current meter, of an improved Ekman type, giving speed and direction. On August 5, 1975, fairly typical wind patterns existed over the southern half of the lake and were monitored at our summer station by means of an anemometer. At 0805 hr the wind averaged about 13 km/hr from the south and gusted to 20 km/hr. During the next 2 hr the wind gradually decreased until it was calm at 1015 hr. Starting about 1230 hr the wind came from the north–northeast at 3–8 km/hr. Under these conditions currents were maximum off Seneca Point at 1418 to 1428 hr. Readings taken then showed the currents to be: 1 m below the surface, 15 cm/sec from the north; 5 m depth, 17 cm/sec from the north; 10 m depth, 3 cm/sec from the north. Off Black Point buoy the strongest current of the summer occurred at 1539 to 1549 hr. It was 21 cm/sec from the north at the surface. At 8 m, from 1553 to 1603 hr, it was about 3 cm/sec from the north.

At 1624 to 1634 hr off Long Point across the lake and to the north of Black Point (Fig. 4), the current at 1 m was 18 cm/sec to the south. Approximately opposite Seneca Point on the east side of the lake (Fig. 4) the current was 18 cm/sec to the south 100 m off shore at 1 m depth. These appeared to be typical currents of Canandaigua Lake during summer stratification and average wind conditions. Little is known about other currents of the lake.

Internal or temperature seiches certainly must occur; on a few occasions temperature profiles taken approximately 2 km north of the south end of the lake have shown the thermocline to be as little as 5 to 8 m below the surface, while 13 km north the thermocline was 10 to 15 m below the surface (S. W. Eaton, unpublished field notes, 1972). If the formula for the total periodicity of a seiche is considered to be

$$T = 2L/\sqrt{gh}$$

or twice the length divided by velocity, then the period for a seiche in Canandaigua Lake which is 25.6 km in length (L), with an average depth of 38.8 m (h), and considering acceleration due to gravity (g) to be 980 cm/sec^2, can be expected to be about 1 hr.

Surface waves have their greatest effect at the north end of the lake, and their amplitudes at various wind speeds have been determined by the United States Army Corps of Engineers (1967). These waves are usually most severe in causing beach erosion along the east shore near the north end of the lake during west winds (Table 10).

Langmuir circulation (Cole, 1975) has been observed at various times. Lines of floating debris moving out of the swamp at the south end have been

TABLE 10

**Maximum Wave Heights for the Northeast Portion of
Canandaigua Lake**[a]

Velocities (km/ hr)	Minimum wind durations required to produce waves (min)	Wave heights for minimum duration (cm)	Maximum wave heights (cm)
16	70	24	40
32	50	52	85
48	42	82	135
64	38	116	192
80	33	143	238
96	30	171	287

[a] The data given in this table for the northeast portion of the lake represent the highest wave heights that would occur in deep water for various overland wind velocities along Canandaigua Lake, and are presented as an example of the type of information that can be made available for the entire lake perimeter. (United States Army, Corps of Engineers 1977).

observed particularly after rapid rises in lake level. Following Hurricane Agnes, June 1972, a line of debris composed of *Ricciocarpus, Riccia fluitans, Lemna minor,* and other swamp floating aquatics came streaming through vertical gill nets in the vicinity of station 6a (Fig. 4). On a flight over the lake at about 305 m with A. M. Seymour, Jr., on August 1975, what must have been the surface slicks of Langmuir waves were seen several meters off both the east and west shores and trailing downwind along a north–south axis.

Color and Turbidity

Color and turbidity were recorded (New York State Department of Health, 1955) for 28 stations during July and August 1955. These were located in the lake, West River, Naples Creek, Sucker Brook, and the Feeder Canal.

The stations in the lake were taken in two series, one on July 5 and 7 and another on August 2, 5, 8, and 24. Five samples were taken along an east–west line passing through the gauging station at the north end of the lake, three samples were taken along an east–west line across the lake off Tichenor Point, three samples were taken off Grange Landing, and three were taken off Cooks Point (Hawk's Point) (Table 11, Fig. 11).

TABLE 11

Color and Turbidity of Lake and Tributaries in July, August, and September 1955[a]

Station	Date	Time	Color (ppm)	Odor[a]	Turbidity	Suspended matter (SM)	pH
Gauging Sta. W[b]	Jul 5	12:00	10	Mu-2	10	3	8.4
Gauging Sta. W	Aug 2	11:05	5	Vg-2	5	2	8.4
Gauging Sta. CW	Jul 5	11:50	10	E-3	5	2	8.4
Gauging Sta. CW	Aug 8	10:55	5	E-2	5	2	8.4
Gauging Sta. C	Jul 5	11:40	10	E-3	5	2	8.3
Gauging Sta. C	Aug 2	10:45	5	E-2	5	2	8.4
Gauging Sta. CE	Jul 5	11:30	5	E-2	5	2	8.4
Gauging Sta. CE	Aug 2	11:35	5	Vg-2	5	2	8.4
Gauging Sta. E	Jul 5	11:20	10	Mu-2	5	2	8.3
Gauging Sta. E	Aug 2	10:25	5	Vg-2	5	2	8.4
Tichenor	Jul 7 Aug 5	10:10	5	E-2	5	2	8.4
Grange Landing	Jul 7 Aug 28	10:30 11:50	5	Vg-1 Vg-2	Less than 5 to 5	2	8.4
Cook's Point	Jul 8 Aug 8	10:12 10:50	10 5	Vg-2 Ar-2	Less than 5 to 5	2	8.4 to 8.5
Naples Creek	Jul 26 Sep 2	9:55 11:30	6 18	E-2 Mu-2	Less than 5 to 23	2 3	7.8 8.3
West River	Jul 17	10:00	37	E-2	25	2	7.4
Sucker Brook	Jul 18	14:25	50	Sp-3	25	2	7.3

[a] Mu, musty; Vg, vegetable; E, earthy; Ar, aromatic; Sp, septic.
[b] W, CW, C, CE, and E: These mean, respectively, west, center-west, center, center-east, and east and represent sampling points along the line from west to east across the northern part of the lake. (Data from New York State Department of Health, 1955.)

Color ranged from 5 to 10 ppm in the lake and up to 37 ppm (Platinum-cobalt units) in West River at a pH range of 8.3 to 8.4; it registered 50 ppm in Sucker Brook at pH 7.3. In early July color was about 10 ppm, but at all stations in August it was 5 ppm. Turbidity in ppm was read as 5 in both July and August in the lake and up to 25 in West River and Sucker Brook.

Secchi Disk and Photometer Readings

Secchi Disk readings in December, January, and February have been taken by K. M. Stewart (unpublished data). They were 8.1 m on December 14, 1967, 7.8 m on January 17, 1968 (under ice cover off Seneca Point), and 8 m on February 2, 1968, when the lake had almost cleared of ice. One reading near station 7 on March 3, 1973 was 3.8 m, and a reading on April 2, 1972 was 5 m. From April to December there is a considerable fluctuation because of the changes in status of the plankton populations. Under average conditions transparencies range from 3 to 6 m from April until late November and then rise during December to 7.5 to 8 m (Table 12).

On June 27, 1972, following Hurricane Agnes, Secchi readings dropped to 0.5 m in the southern two-thirds of the lake, but they were back to near normal (3.5 m) by July 17. The clearest the lake has been reported from 1967 to 1975 was on September 2, 1974. At this time at station 7a the reading was 8.8 m. It was made at 1200 hr under clear skies, the winds were variable at 3–5 mph, and water temperatures registered 22.5°C at the sur-

TABLE 12

Secchi Disk Readings (m) in Canandaigua Lake in 1910, 1971–1975 during the Growing Season

Dates	Annual mean (m)	Range	Number of readings
1910[a]	3.7		1
Jun 22–Oct 30, 1971[b]	5.1	3.4–6.2	36
May 23–Oct 28, 1972[c]	3.0	0.5–8.0	61
May 15–Aug 16, 1973[d]	4.7	2.7–6.4	14
May 22–Nov 22, 1974[d]	4.6	2.0–8.8	18
Jul 10–Oct 4, 1975[e]	3.4	3.0–5.4	9
Average	4.2		

[a] Birge and Juday (1914); 10-cm disk.
[b] Kardos and Eaton (1972); 20-cm disk.
[c] Kardos (1975); 20-cm disk.
[d] Wilcox (1975); 20-cm disk.
[e] S. W. Eaton (unpublished field notes); 20-cm disk.

face and decreased to 21.7°C at 9 m. At that time the dissolved oxygen concentrations were only about 1 mg/liter below what they had been on three previous sampling periods (8.2 ml/liter), and the phytoplankton and zooplankton populations were also low (Fig. 13). Clearing of the lake also occurred on August 27, 1972 when the secchi reading was 7.2 m at station 7a. In 1971, the clearing occurred later, on October 2–3. At station 2 the reading was 8.5 m; at station 4, 7.8 m; at station 6, 5.2 m; at station 7, 4.8 m; and at station 9, 6.7 m. These readings remind one of the variability possible even in the middle of the lake away from turbidity produced by wave action on shore. The example above was not the usual situation. Normally from station 2 to station 8 there was only 1 m difference over 22 km of lake surface. In 1971 further clearing after October 3 did not occur until December 4, when we obtained readings of 6.7 m at station 6 and 7.5 m at station 7a. The cause of this late summer clearing has not yet been determined.

Submarine photometer readings were made by a Kahlsico photometer, No. 268WA300, on May 17, August 16, and October 3, 1973. The depths to which 2% of the surface light penetrated with a neutral filter are shown in Fig. 13. These depths were 11 m, 10.5 m, and 15 m, respectively, and represent approximately the depth of the euphotic zone.

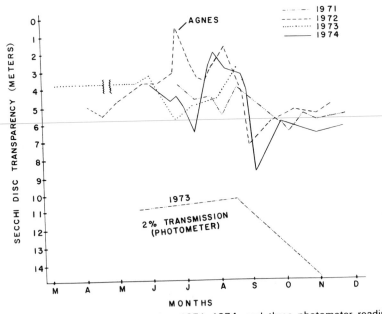

Fig. 13. Secchi disk transparencies, 1971–1974, and three photometer readings, 1973, in Canandaigua Lake. All stations available for single days averaged.

CHEMICAL LIMNOLOGY

Canandaigua Lake, with an average total hardness of 128 mg/liter as $CaCO_3$, can be considered a hard water lake (Durfor and Becker, 1964). Twenty-four tests for hardness made at the surface by the New York State Department of Health in July and August 1955 ranged from a low of 120 to a high of 156 mg/liter. Three tests made in April, July, and August 1973 by Cornell University ranged from 138 to 148 mg/liter as $CaCO_3$ (R. T. Oglesby, personal communication).

There is a well-developed calcium carbonate–bicarbonate buffering system like that of the rest of the large, deep Finger Lakes. Birge and Juday (1914) stated that fixed CO_2 in the waters of the Finger Lakes varied from a minimum of 6.8 cm³/liter in Canadice Lake to 24.0 cm³/l in Canandaigua Lake. As for degree of alkalinity, they stated it varied among the Finger Lakes from a minimum of about 0.5 cm³ in Canadice to a maximum of 3 cm³ in Canandaigua Lake. (These are amounts of CO_2 needed to bring sample to neutral.)

The methyl orange alkalinity measured in 1927 by the New York State Conservation Department Biological Survey was the highest of any of the Finger Lakes examined (including Cayuga, Seneca, Otisco, Skaneateles, and Owasco Lakes). On July 28, 1927, alkalinity ranged from 123 to 133 mg/liter as $CaCO_3$ from just below the surface down to 62 m. Total alkalinities taken at the surface during July and August in 1955 at 24 stations from the north end of the lake to a point about 3 km from the south end ranged from 91 to 104 mg/liter as $CaCO_3$ and averaged 96.5 mg/liter (New York State Department of Health, 1955) (Table 13). In April, July, and August 1973 Cornell University recorded 12 concentrations ranging from 94 to 117 mg/liter as $CaCO_3$ from depths between the surface and 60 m. There seems to have been little change from 1927 to 1973. The phenolphthalein alkalinity was negligible, as was to be expected. It was read as 4 mg/liter as $CaCO_3$ in April 1973; 9 mg/liter at the surface down to 4 mg/liter at 60 m in July; and in August it ranged from 7 mg/liter at the surface down to 0 mg/liter at 60 m (R. T. Oglesby, personal communication).

Of interest are the "sea biscuits" found near Squaw Island at the shallow north end. Along the shore of the island are pebbles of calcium carbonate averaging about 15 mm in diameter, the result of the high pH of the shallow north end. The calcium and magnesium ions in the local rocks and the carbon dioxide coming down Sucker Brook combine with just the proper wave action along the shallow shore of the island to produce a flattened pebble of marl (Fig. 14). There is much less NaCl than in Cayuga and Seneca Lakes whose basins lie deeper in the rock structure, with crypto-

TABLE 13

Alkalinities in Canandaigua Lake, 1910–1973

Year	Depth (m)	Total alkalinity as mg/liter of CaCO₃ (range)	No. of readings	Source
1910		"highest of Finger Lakes"		Birge and Juday, 1914
1927	1–62	123–133	5	Wagner, 1928 (Jul 28)
1955	surface	91–104	24	N.Y. State Dept. of Health (Jul and Aug), 1955
1967–1968	0–70	96–97	29	K. M. Stewart, unpublished data (Dec 14, Jan 17, Feb 2)
1972	0–67	83–120	18	USEPA, 1974 (May, Jul, Oct)
1973	0–60	94–117	12	R. T. Oglesby, personal communication (Apr, Jul, Aug)

Fig. 14. Photograph of sea biscuits from Squaw Island, Canandaigua Lake.

depressions, and which seem to be affected by Silurian salt deposits (Oglesby *et al.*, 1975).

pH

The hydrogen ion concentrations in the lake appear not to have changed from the New York State Biological Survey (Wagner, 1928) to the present. On July 28, 1927 pH ranged from 8.3 at 1 m down to 7.9 at 62 m. In 1955 (New York State Department of Health, 1955) pH values at the surface at 28 stations from the north end to a point south off Cooks Point ranged from 8.3 to 8.5 on July 5 and 7 and August 2, 5, 8, and 24.

Fairly complete pH data were collected in the lake from the surface to 60 m during 1972 and 1974 (Table 14). In 1972 following Hurricane Agnes, Gotham (1974) took a series of pH values near station 6a (Fig. 4) at almost weekly intervals from June 27 to November 25 and again on March 8, 1973. These are interesting in that they show the effect of rainfall on the chemistry of the lake. The data for April 25, July 18, and August 28, 1973 were from Cornell University (R. T. Oglesby, personal communication). The pH values in April are quite constant from 7.6 at the surface to 7.7 at 50 m and range in July from 8.6 at the surface to 8.2 at 60 m. On August 28 the values ranged from 8.4 at the surface down to 8.0 at 60 m. Wilcox (1975) took an excellent series of pH values from May to November 1974 in a study of the phytoplankton. His data are shown in Table 14.

With the increase in precipitation in October the pH values ranged from 7.2 to 7.6. Values were somewhat higher in December, January, and February. Stewart found the lake to be at a pH of 8.0 on December 14, 1967, 7.9 on January 17, 1968, and 8.0 on February 2, 1968. At about the same time that phytoplankton populations began to increase in late April and May the pH rose to 8.6 at the surface. The epilimnion showed vertical stratification with an average value of 8.3 at the 10 m level from May 27 to July 8, 1974, then dropped to 7.8 at all depths through the rest of July until August 12. On August 12, 1974 the pH returned to 8.0 at the surface and 5 m, and for the next three sampling dates it gradually increased at all depths down to 54 m. On September 2, 1974 the pH values were 8.5 at the surface, 8.7 at 5 m, and 8.5 at 20, 35, and 54 m, reflecting the effects of increased photosynthesis. On October 14, 1972 (USEPA, 1974) a series of pH readings at 3 stations in the center of the lake showed the epilimnion to be about 8.1 gradually decreasing to 7.6 at 67 m. All available pH readings per months were averaged and show the summer effect of photosynthesis in elevating the pH (Fig. 15). Table 14 shows that there has been little change in pH over almost a 50-year period. Variation can be explained by normal seasonal changes or by incidents such as Hurricane Agnes.

TABLE 14

Means and Series of pH Readings at Various Depths in Canandaigua Lake, 1927–1974

Year	Depths	pH
1927[a]	1	8.3
	10	8.3
	20	8.2
	30	7.9
	62	7.9
1955[b]	Surface	8.4

Year	Depths	pH	
		Jun 27	Mean Jul–Aug
1972[c]	Surface	6.8	7.4
	1	6.6	7.3
	3	6.8	7.4
	6	7.2	7.5
	9	7.3	7.5
	12	7.5	7.7
	15	7.8	7.6
	20	—	7.7
	25	—	7.6
	30	—	7.6

Year	Depths	pH		
		Apr 25	Jul 18	Aug 28
1973[d]	0	7.6	8.6	8.4
	5	7.7	8.6	8.5
	10	7.7	8.4	8.3
	50	7.7	—	—
	60	—	8.2	8.0

Year	Depths	pH				
		May 27	Jun[f]	Jul[f]	Aug[f]	Nov
1974[e]	0	—	8.6	8.0	8.1	7.6
	5	8.6	8.6	8.1	8.2	7.9
	10	8.5	8.4	8.0	8.0	8.1
	20	8.4	8.3	8.0	7.8	7.8
	35	8.4	8.2	7.8	7.8	7.6
	54	7.3	8.1	7.7	7.7	7.8

[a] From Wagner (1928). Taken July 28.
[b] From New York State Department of Health (1955). Twenty-eight stations averaged July 5, 7, Aug 2, 5, 8, 24.
[c] From Gotham (1974). Data based on 12 readings from center of lake except for depth at 20, where data are based on 6 readings from center of lake.
[d] From R. T. Oglesby (personal communication).
[e] From Wilcox (1975).
[f] Weekly means.

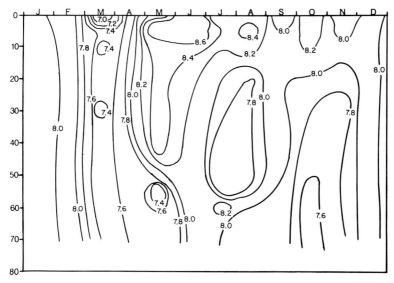

Fig. 15. Isopleths of average pH values for Canandaigua Lake for years 1967–1968, 1972–1974.

Dissolved Oxygen

The dissolved oxygen (D.O.) levels of Canandaigua Lake are almost always close to or above saturation in the epilimnion during summer stratification and have not changed appreciably in almost 50 years (Table 15). In the hypolimnion during stratification the D.O. levels drop to a low of about 8 mg/liter just before the fall turnover (Fig. 16). Mixing in late

TABLE 15

Percent of Saturation of Dissolved Oxygen in Canandaigua Lake in July and August

	Aug 20, 1910 (Birge and Juday, 1914)	Jul 28, 1927 (Wagner, 1928)	Jul 22, 1972 (Gotham, 1974)	Jul 22, 1974 (Wilcox, 1975)	Aug 19, 1975 (Eaton)[a]
0	106	95	107	101	103
5	105	—	115	101	104
10	108	98	89	104	97
15	104	—	90	87	94
20	94	95	—	99	96
30	93	85	—	96	90
40	93	69	75	82	90
50	—	—	—	—	—
60	90	—	—	—	73
80	72	—	—	—	—

[a] S. W. Eaton (unpublished data).

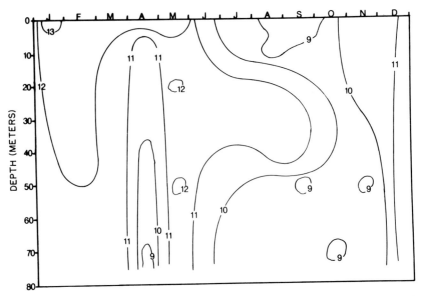

Fig. 16. Isopleths of monthly mean dissolved oxygen concentrations for Canandaigua Lake for years 1967–1968, 1972–1975. Values are in mg/liter.

November and through the winter rejuvenates D.O. levels and high concentrations are reached before the start of biological activity and summer stratification. Dissolved oxygen concentrations were measured in December 1967 and in January and February 1968 by K. M. Stewart (Fig. 17) and ranged from 13 mg/liter at the surface down to 11.8 mg/liter at 70 m.

The vertical distribution of D.O. near the end of stratification is shown in Fig. 17 and is typically of the orthograde type (Hutchinson, 1957).

Average mid to late August measurements of dissolved oxygen in the epilimnion over a three-year period (1972, 1974) indicated an average of 8.7 mg/liter. The only lower average readings of the epilimnion were recorded by Gotham (1974) on July 2, 1972 following the increase in water level (1.04 m) between June 20 and 24 (Fig. 18).

Nitrogen

Canandaigua Lake is a medium–high nitrate (NO_3–N) lake. Sixteen of 26 lakes assayed for nitrogen in New York State in 1972 had less mean inorganic nitrogen (USEPA, 1974). Berg (1963) indicated that, of the Finger Lakes, Canandaigua Lake was most like Skaneateles and Seneca Lakes in levels of inorganic nitrogen.

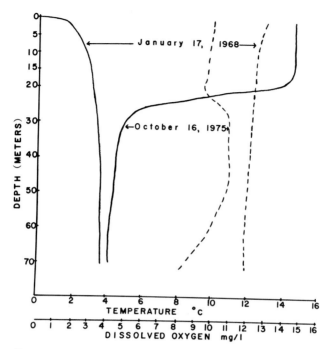

Fig. 17. Temperature and dissolved oxygen profiles in Canandaigua Lake, January 17, 1968 and October 6, 1975. Dotted lines are dissolved oxygen and solid lines temperature in degrees Celsius.

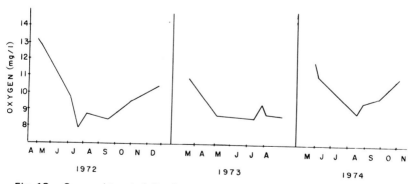

Fig. 18. Seasonal trend of dissolved oxygen in Canandaigua Lake, 1972–1974.

Near the end of stratification NO_3–N levels were low in the epilimnion (Fig. 19) but did not seem to be low enough to be limiting for biological activity at that time. The USEPA (1974) conducted a limiting nutrient study in 1972 and found that spiking with 10 mg/liter of NO_3–N alone did not increase the controlled yield of the assay alga *Selenastrum capricornatum*, but that adding phosphorus did.

There seems to be little change in nitrogen levels since 1955 (Table 16), though there are no data for the earlier years. Nitrogen seems to be recharged during the winter, reaching a high of 0.79 mg/liter in April. Then fluctuation with general depletion takes place during the summer in the epilimnion, but levels stay relatively high in the hypolimnion. Stewart (Fig. 19) measured a concentration of 144 μg/liter at 70 m on September 2, 1969.

Phosphorus

It has been shown that the nutrient exerting primary control over standing crops of phytoplankton in the Finger Lakes is phosphorus (Oglesby *et al.*, 1975). Although this may not have been true for 1972 with the enrichment of the lakes following Hurricane Agnes (Tables 17 and 18), it seems to

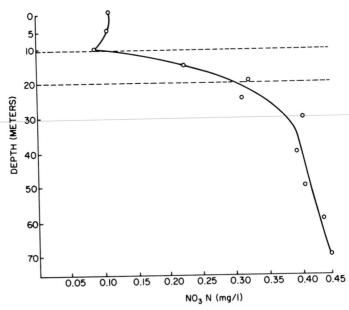

Fig. 19. Nitrate nitrogen concentrations for September 2, 1969 in Canandaigua Lake (K. M. Stewart, unpublished data). Circles represent actual measurements. The solid line represents approximate trend with depth. The dashed lines approximate the location of the thermocline.

TABLE 16

Concentrations in mg/liter of NO_3-N in Canandaigua Lake, 1955–1973

	NO_3-N		NH_3-N		
Year	mean	(range)	mean	(range)	Source and Comments
1955	0.39				Berg, 1963
1969	0.20	(0.11–0.44)			K. M. Stewart, unpublished data; Sep 2, 1969; 0–70 m; 12 readings averaged
1971	0.14	(0.06–0.30)			Grow, 1971; Aug 19; 7 stations; 0–50 m; averaged
1972 Apr	0.35	(0.31–0.41)	0.03	(0.01–0.05)	USEPA, 1974; 3 stations; 52
Jul	0.31	(0.27–0.45)	0.07	(0.04–1.1)	readings averaged; 0–66
Oct	0.32	(0.26–0.48)	0.03	(0.03–0.04)	m
1973 Apr	0.79				R. T. Oglesby, personal
Jul	0.25				communication surface; 3
Aug	0.45				readings averaged
1973 Jul	0.32	(0.24–0.69)			I. J. Gotham, III, personal communication; 28 readings averaged; 0–15 m

TABLE 17

Concentrations of Soluble Reactive Phosphorus (mg/liter) in Canandaigua Lake, 1972 and 1973

Year	Apr	May	Jul	Aug	Oct
1972[a]					
0	—	0.004	0.009	—	0.006
1–5	—	0.006	0.009	—	0.004
6–10	—	0.007	—	—	0.006
11–20	—	—	—	—	—
21–30	—	—	0.009	—	0.005
31–40	—	—	0.009	—	0.006
41–50	—	—	—	—	0.005
51–60	—	—	0.010	—	0.007
1973[b]					
0	0.0054	—	0.0014	0.0000	—
5	0.0047	—	0.0008	0.0000	—
10	0.0054	—	0.0008	0.0006	—
50	0.0047	—	—	—	—
65	—	—	0.0017	0.0012	—

[a] From USEPA, 1974. Three stations averaged for depth range in meters indicated.
[b] From R. T. Oglesby (personal communication). Three series, one station.

TABLE 18

Concentrations of Total Phosphorus in Canandaigua Lake in mg/l, 1972–1973

Year	Apr	May	Jul	Aug	Oct
1972[a]					
0	—	0.008	0.014	—	0.007
1–5m	—	0.009	0.032	—	0.008
6–10	—	0.011	—	—	0.007
11–20	—	—	—	—	0.007
21–30	—	—	0.015	—	0.006
31–40	—	—	0.031	—	0.007
41–50	—	—	—	—	0.009
51–60	—	—	0.029	—	0.029
1973[b]					
Surface	0.010	—	0.010	0.008	—

[a] From USEPA, 1974.
[b] From R. T. Oglesby (personal communication).

be true for more normal years such as 1973 (Table 17). In that year measurements of soluble reactive phosphorus at all depths sampled in July and August (0–65 m) in Canandaigua Lake were below 2 μg/liter (R. T. Oglesby, personal communication). During the cold months levels of soluble reactive phosphorus appear to increase so that by April levels are up to 5 μg/liter (Table 17).

Dissolved Solids and Specific Conductance

Specific conductance as reported by Berg (1963) was 271 μmhos/cm in 1955 (Table 19). Unpublished measurements made during winter 1967–1968

TABLE 19

Specific Conductivity in micromhos/cm in Canandaigua Lake

	Depths	Specific conductivity	Stations	Source[a]
1955	—	271	—	Berg, 1963
1967–1968	0–70	270–293	3	K. M. Stewart, unpublished information; Dec 14, Jan 17, Feb 2
1973	Surface	279–290	3	R. T. Oglesby, personal communication; Apr 25, Jul 18, Aug 28

[a] Date indicates when data were obtained.

by K. M. Stewart showed a slight increase as did those of R. T. Oglesby in 1973 (personal communication). This conductivity is slightly higher than Skaneateles Lake (231) and Owasco Lake (253) but much less than Seneca (646) and Cayuga (596); both the latter are lower in the rock structure and probably their basins transect more limestone and salt-bearing strata (Table 1). Various categories of solids in Canandaigua Lake are shown in Table 20.

Nutrient Loadings by Tributaries

The United States Environmental Protection Agency with the help of the New York State National Guard collected monthly (and in April and May semimonthly) near-surface grab samples from six important tributaries of Canandaigua Lake from November 1972 to October 1973 to determine nutrient loadings by nitrogen and phosphorus (Tables 21 and 22). Stream flow estimates for the year of sampling and a "normalized" or average year were provided by the New York District Office of the United States Geological Survey (Table 7). Nutrient loadings for unsampled minor tributaries and the immediate drainage were estimated by using the means of nutrient loads in Tributary 16, Tributary 39, and Tributary 43 (Fig. 4), and multiplying by the area of the ungauged streams. Loadings for Naples Creek were calculated in the same manner (USEPA, 1974).

The Village of Naples was the highest single contributor to the phosphorus load. It was followed by the minor tributaries and immediate drainage, direct precipitation, Middlesex, Naples Creek, and West River.

TABLE 20

Concentrations of Various Categories of Solids in Canandaigua Lake, 1973[a]

Parameter	Depth (m)	Apr 25	Jul 18	Aug 28
Total suspended solids (mg/liter)	0	2.05	1.08	0.40
	5	2.15	5.40	2.25
	10	2.00	3.15	0.96
	50–65	2.81	2.86	1.35
Total volatile solids (mg/liter)	0	0.30	1.46	0.40
	5	0.44	2.11	1.52
	10	0.44	2.10	0.84
	50–65	0.63	2.03	0.87
Nonvolatile suspended solids (mg/ liter)	0	1.74	0.38	0.00
	5	1.71	3.29	0.73
	10	1.56	0.83	0.48
	50–65	2.19	0.83	0.48
Total solids (mg/liter)	Surface	187	191	184

[a] R. T. Oglesby (personal communication).

The total estimate of loading for the lake was 6042 kg P/year. The output in the Feeder Canal and Outlet amounted to 2100.2 kg P/year, leaving a net accumulation of 3941.8 kg P/year in the lake. This was about 65% of the total input (Table 21).

Annual total nitrogen loading of Canandaigua Lake is shown in Table 22. The greatest single contributor was direct precipitation. It contributed an estimated load of 29.2% of the total or 46,426 kg N/year. This was followed in decreasing order by the minor tributaries and immediate drainage, Naples Creek, West River, and several contributors not exceeding individually more than 3% of the total.

It was determined that the outlet carried a load of 79,180.4 kg N/year, which left an annual net accumulation of 79,675.2 kg N/year in the lake. This accumulation amounted to 50% of the input of nitrogen. Oglesby *et al.*

TABLE 21

Average Annual Total Phosphorus Loading[a] of Canandaigua Lake and Net Accumulation in kg P/year[b]

Source	kg P/year	% of total
Inputs		
West River	476.3	7.9
Seneca Pt. Creek (33)[c]	31.8	0.5
Vine Valley Stream (16)[c]	145.2	2.4
Barnes Creek (39)[c]	9.1	0.2
Menteth Creek	27.2	0.4
Naples Creek	712.2	11.8
Minor tributaries and immediate drainage	1106.8	18.3
Known municipal sewage treatment plants		
Bristol Harbor Village	285.8	4.7
Middlesex	725.8	12.0
Naples	1406.5	23.3
Rushville	63.5	1.0
Septic tanks[d]	299.4	5.0
Direct precipitation	753.0	12.5
	6042.0	100.0
Outputs		
Feeder Canal and Outlet	2100.2	
Net accumulation P/year in lake	3941.8 Kg P	65% (of input)

[a] Loading of P in gm/m² of lake surface per year = 0.14.
[b] From USEPA, 1974.
[c] Numbers in parentheses are tributary numbers.
[d] From 1052 cottages about perimeter.

TABLE 22

Average Annual Total Nitrogen Loading[a] of Canandaigua Lake with Output and Net Accumulation in kg N/year

Source	kg N/year	% of total
Inputs		
West River (18)[b]	23,433.0	14.8
Seneca Point Creek (33)	1,878.0	1.2
Vine Valley Creek (16)	3,252.3	2.1
Barnes Creek (39)	535.3	0.3
Menteth Creek	2,767.0	1.7
Naples Creek	26,027.6	16.4
Minor tributaries and immediate drainage	40,379.5	25.4
Known sewage treatment plants		
Bristol Harbor Village	852.8	0.5
Middlesex	2,177.3	1.4
Naples	4,218.5	2.7
Rushville	2,422.2	1.5
Septic tanks[c]	4,486.1	2.8
Direct precipitation	46,426.0	29.2
	158,855.6	100.0
Outputs		
Feeder Canal and Outlet	79,180.4	
Net annual accumulation kg N/year in lake	79,675.2	50% (of input)

[a] Loading in total N per gm/m^2 of lake surface per year = 3.7.
[b] Numbers in parentheses are tributary numbers.
[c] Based on 1,052 lakeshore residences.

(1973) estimated that 45% of the nitrate nitrogen and 87% of the molybdate reactive phosphorus was retained in Owasco Lake, a slightly smaller though roughly comparable Finger Lake. This agrees rather well with these estimates for Canandaigua, though in different measurements. Canandaigua's figure represented total nitrogen and total phosphorus.

Major Ions of the Lake

The major ions of the Finger Lakes have been summarized by Oglesby *et al.* (1975). They point out that the chemistry of the Finger Lakes is strongly influenced by limestone escarpments and by limestone debris deposited in the cross valleys and in the lake itself. This is probably somewhat less true with Keuka and Canandaigua Lakes whose basins are higher in the rock structure.

TABLE 23

Major Ion Concentrations Reported in 1955 and 1973 in Canandaigua Lake as mg/liter and Below in Parentheses as mEq/liter

	Cations			Anions		
Year	Ca^{2+}	Mg^{2+}	$(Na^+ + K^+)$	HCO_3^-	SO_4^{2-}	Cl^-
1955[a]	35	9.7	7.5	126	28	5.1
	(1.75) +	(0.80) +	(0.33) = 2.88	(2.07) +	(0.58) +	(0.14) = 2.79
1973[b]	38.6	11.7	5.7	130	29.2	13.2
	(1.93) +	(0.96) +	(0.25) = 3.14	(2.13) +	(0.61) +	(0.37) = 3.11

[a] New York State Department of Health.
[b] R. T. Oglesby (personal communication).

The major cations are Ca^{2+}, Mg^+, and Na^+, and the major anions are HCO_3^-, SO_4^{2-}, and Cl^-. Their composition in the lake as reported in 1955 and 1973 may be seen in Table 23, with their milliequivalents in parentheses below. These indicate that some increase in major ions may be taking place. In April, July, and August, 1973 (R. T. Oglesby, personnal communication) soluble reactive silicon was 0.855 mg/liter, 1.153 mg/liter, and 1.119 mg/liter.

Trace Elements of the Lake

Iron was determined as being 0.02 mg/liter at the surface in April, 0.14 mg/liter in July, and 0.006 mg/liter in August, 1973 (R. T. Oglesby, personal communication).

Delta Laboratories of 34 Elton Street, Rochester, New York 14607 took core samples in Canandaigua Lake and in West River and Canandaigua Outlet to determine levels of the metals Cd, Cu, Pb, Zn, Co, and of PO_4–P in sediments (Pike, 1975). Samples were taken by a Phleger coring device which had a maximum length of 30.5 cm. Extraction of metals was done on core material and is shown in Table 24. These were extracted using a United States Environmental Protection Agency technique (USEPA, 1969).

Highest levels of mercury appeared in sediments of the West River just upstream of the beginning of Naples Swamp, in sediments off Tributary 16 (Willow Grove), and in sediments off the Canandaigua City pier. Zinc was highest off the Canandaigua City pier (103 μg/gm) and generally exceeded 50 μg/gm. Additional trace element data (United States Department of the Interior, 1974) are available for the water supplies which draw upon Canandaigua Lake. These data are presented in the Appendix.

TABLE 24

Heavy Metals Extracted by USEPA (1969) Method in Canandaigua Lake, 1972 by Delta Laboratories

Sites	Site Designations	Core[a] section	Cd (μg/gm)	Cu (μg/gm)	Pb (μg/gm)	Zn (μg/gm)	Co (μg/gm)	Hg (μg/gm)
4	West River at bridge	20.3	0.68	30.65	24.81	93.90	8.81	0.078
		30.5	0.79	24.94	20.05	86.07	9.54	0.093
5	Granger Point	10.2	1.04	65.20	14.71	75.11	9.46	0.063
6	Hicks Point	20.3	7.53	15.07	12.29	42.02	4.93	0.035
		30.5	2.80	13.06	13.21	35.56	5.22	0.033
7	Seneca Point	10.2	1.23	21.21	22.32	84.27	8.48	0.050
9	Willowgrove Point	10.2	0.88	22.89	20.26	71.52	8.06	0.084
11	Foster Point	10.2	1.81	40.13	18.84	89.68	10.17	0.060
		20.3	1.51	36.47	15.98	75.84	7.82	0.041
12	Long Point	10.2	1.04	12.76	14.65	53.39	9.24	0.052
		20.3	0.93	24.70	12.55	82.96	11.58	0.032
13	Menteth Point Beach	10.2	0.75	27.24	19.07	82.68	9.60	0.052
		20.3	0.33	7.92	6.54	34.36	4.78	0.033
14	Crystal Beach	10.2	1.02	21.90	15.89	74.28	10.93	0.045
15	Cottage City	10.2	1.02	11.83	102.27	25.80	7.05	0.019
16	Deep Run Park	10.2	0.70	9.50	12.72	25.13	4.64	0.045
17	Tichenor Point	10.2	0.90	15.84	15.84	45.63	8.84	0.032
		20.3	1.02	13.61	14.73	40.58	8.38	0.033
18	Otetiana Point	10.2	0.99	9.46	17.82	29.75	6.35	0.026
19	Hope Point	10.2	0.88	16.21	14.71	29.50	6.36	0.038
			0.68	9.28	11.95	22.89	5.09	0.070
20	City Pier Cove	10.2	2.01	28.15	90.55	103.02	11.48	0.0922
25	Naples Creek	10.2	0.95	23.04	16.16	76.81	7.90	0.039

[a] Core length (cm).

BIOLOGICAL LIMNOLOGY

Phytoplankton

Diatoms and chlorophyceans dominate the algal flora of Canandaigua Lake with rather weak myxophycean maxima occurring in the warm months, particularly from late August to mid October (Presutti, 1973; Wilcox, 1975). Birge and Juday (1914, 1921) evaluated the phytoplankton community on August 20, 1910 and found it dominated by blue–green algae, *Clathrocystis* and *Coelosphaerium* being the most abundant forms. On July 17 they found *Microcyctis* and *Staurastrum* dominating, and on July 28 *Aphanocapsa* was dominant. *Asterionella* and *Fragilaria* were the characteristic diatoms on August 20, 1910 and July 17, 1918. On July 28, 1918 another sample showed the blue–green genus *Aphanocapsa* ($9.8 \times 10^7/$ m³) and the diatoms *Navicula* ($10^7/$m³) and *Stephanodiscus* ($8.3 \times 10^7/$m³) to be present in relatively high numbers at 20 m.

The phytoplankton of the lake were studied in 1972 (Gotham, 1974; Presutti, 1973), in 1973 (Williams, 1974), and the most complete study was done from May 27 to November 22, 1974 (Wilcox, 1975). Most of Table 25 and the discussion here is from Wilcox. He sampled with a 3-liter PVC sampler at the surface, 5, 10, 20, 35, and 54 m at weekly intervals from May to September, and once a month during October and November. His samples were fixed in acid Lugol's solution but were not dyed with analine blue, and some of the deeper counts may include dead cells drifting toward the bottom.

All major taxa identified on each of the 18 collecting dates at 5 m are shown in Fig. 20 in organisms/liter. This shows a peak of green algae and diatoms occurring June 10 and 17 made up mainly of *Gleocystis ampla, Rhizosolenia gracilis*, and *Fragilaria crotonensis* (Fig. 21). Another peak on August 12 was made up of the same three species. Chrysophyta, including diatoms, showed two characteristic peaks, one on June 6 and the other on November 22. The Myxophyceae peaked on September 28 and this maximum was made up mostly of *Oscillatoria prolifica* and *Chroococcus limneticus*.

Following Hurricane Agnes (June 20–24, 1972) an unusual flora was present for a short period when the epilimnion was slightly acidic (pH 6.4 at 6 m) (Gotham, 1974). At this time *Tetraëdron, Chlorella, Staurastrum, Cosmarium, Dispora, Leuvenia, Synedra acus, Diatoma*, and *Botryococcus* were present. They were not encountered after July 12.

The dominant phytoplankton following this early 1972 acidic phase were the diatoms *Asterionella formosa, Nitzschia palea*, and *Navicula*. Other organisms which appeared during late July and August 1972 were *Dinob-*

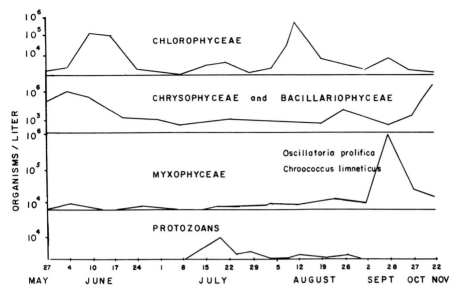

Fig. 20. Seasonal occurrence of the phytoplankton divisions and protozoans in Canandaigua Lake, 1974.

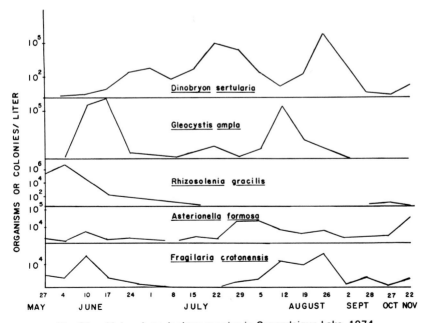

Fig. 21. Major phytoplankton species in Canandaigua Lake, 1974.

TABLE 25

Species of Phytoplankton Collected in Canandaigua Lake, 1972, 1973, and 1974 and Their Seasonal Occurrence

Organism	Seasonal occurrence
Chlorophyceae	
Ankistrodesmus convolutus Corda	common May–Nov
Ankistrodesmus fractus (Corda) Ralfs.	rare May
Ankistrodesmus falcatus (Corda) Ralfs.	rare Jul
Characiopsis pyriformis (A.Br.) Borzi	Aug–early Sep
Characium gracilipes Lambert	Jul–Sep
Chlorella elipsoidea Gerneck	Jul, Sep, Oct
Closteriopsis longissima Lammermann	uncommon Jun–Nov
Closterium acerosum (Schrank) Ehrenberg	uncommon Jun–Aug
Closterium dianae Ehrenberg	rare Jun
Coelastrum microporum Maegeli	rare Jun, Jul
Cosmarium botrytis Meneghini	uncommon Jun–Sep
Crucigenia rectangularis (A. Br.) Gay	rare Sep
Crucigenia tetrapedia (Kirch) W&GS West	rare
Dictyosphaerium ehrenbergianum Naegeli	[a]
Gloeocystis ampla (Kuetz) Lagerheim	common Jun, Jul, Aug
Gloeocystis gigas (Kuetz) Lagerheim	rare Aug 12 (14,436 indiv.)
Gloeocystis major Germechex. Lammermann	rare Sep 28 (2279)
Golenkinia radiata (Chod.) Wille	uncommon Jul–Nov
Oedogonium sp. Link	uncommon Jun–Jul
Oocystis lacustris Chodat.	uncommon Sep
Quadrigula lacustris (Chodat) GM Smith	uncommon May, Jun, Jul, Nov
Scenedesmus bijuga (Turp.) Lagerheim	uncommon Jul, Aug
Sphaerocystis schroeteri Chodat.	uncommon Jul, Aug
Tetraëdron minimum (A.Br.) Hansgirg	common Jun, Jul, Nov
Ulothrix sp. Kuetzing	common May–Aug, Oct, Nov
Chrysophyceae	
Dinobryon bavaricum Imhof	common late Jun, Jul–Nov
Dinobryon sertularia Ehrenberg	common May, Jun, Jul
Dinobryon divergens Imhof	[a]
Monocilia flavescens Gerneck	common late Jul–Sep
Mallomonas alpina Pascher & Ruttner	common May, Jul
Mallomonas caudata Iwanoff	rare Jul
Uroglenopsis americana (Calkins) Lemmermann	rare Aug
Synura uvella var. *Danica* (Kutz) GM Smith	uncommon Jun, Jul
Chromulina ovalis Klebs	[a]
Rhizochrysis limnetica GM Smith	rare Jul
Bacillariophyceae	
(phylogenetic order)	
Centrales	
Melosira granulata (E.) Ralfs	uncommon Jun, Jul
Cyclotella sp. (13μ)	common Jul–Nov
Cyclotella sp. (26μ)	common May–Jul, Sep, Nov

TABLE 25 *(Continued)*

Organism	Seasonal occurrence
Stephanodiscus astraea (F.) Grun.	uncommon Jun, Aug, Sep, Nov
Rhizosolenia eriensis H.L. Smith	uncommon Sep, Oct, Nov
Rhizosolenia gracilis H.L. Smith	common May–Nov
Pennales	
Tabellaria fenestrata Kutz	common May, Jun
Meridion circulare (Grev.) Ag.	common Jun, Sep, Oct, Nov
Diatoma tenuevar elongatum Lyngb.	uncommon Oct, Nov
Diatoma vulgare Bory	uncommon Jun, Jul
Fragilaria virescens Ralfs.	common Jun, Jul, Sep, Oct
Fragilaria crotonensis Kitton	common May–Jun, Sep, Oct
Fragilaria brevistriata Grun.	common Jul–Nov
Synedra radians Kutz.	common May–Aug, Oct, Nov
Synedra rumpens Kutz.	common May–Nov
Synedra ulna (Nitzsch) Ehb.	common May–Nov
Synedra acus Kutz.	common May, Aug, Sep, Oct
Asterionella formosa Hassall	common May–mid Aug
Rhoicosphenia curvata (Kutz) Grun.	uncommon Oct, Nov
Cocconeis placentula Ehr.	common Jul–Nov
Navicula capitata Ehr.	common May, Jun, Jul, Aug, Oct, Nov
Navicula cryptocephala Kutz	common May, Jun, Nov
Navicula viridula var. rostratella (Kutz)	uncommon May, Jun, Jul
Navicula minima Grun.	rare Jun
Stauroneis sp. Ehr.	rare Sep
Amphipleura pellucida Kutz	uncommon Jun, Jul, Oct
Gyrosigma spenceri (Queckett) Cleve	rare Aug
Pleurosigma delicatulum W. Smith	rare Jun
Amphiprora ornata Bailey	uncommon Aug
Gomphonema olivaceum (Lyngb.) Kutz	uncommon Sep, Oct
Cymbella ventricosa Kuetz	common May, Jun, Jul, Sep, Nov
Nitzschia acicularis W. Smith	uncommon Jun, Jul
Nitzschia palea (KQ) W. Smith	uncommon Jun, Jul, Aug, Oct
Surirella linearis W. Smith	uncommon May, Nov
Dinophyceae	
Gymnodinium caudatum Prescott	common Jul, Aug, Nov
Glenodinium gymnodinium Penard	rare
Glenodinium kulczynskii (Wolosz) Schitler	common Jul, Aug, Nov
Ceratium hirundinella (OFMuell) Dufardin	common May–Nov
Peridinium cinctum (Muell) Ehrenberg	[a]
Flagellates—*Incertae sedis*	
Cryptomonas spp. Ehrenberg	common May–Nov
Monads	common May–Nov
Myxophyceae (phylogenetic order)	
Chroococcus dispersus (V. Keissler) Lemmermann	rare Aug
Chroococcus limneticus Lemmermann	common Jun–Nov

(Continued)

TABLE 25   *(Continued)*

Organism	Seasonal occurrence
Chroococcus turgidus (Kuetz) Naegali	uncommon Sep–Oct
Microcystis aeruginosa (Kuetz)	uncommon Sep
Merismopedia elegans A. Braun	uncommon Jul, Sep
Merismopedia glauca (Ehr.) Naegeli	uncommon Sep, Nov
Gomphosphaeria aponina Kuetzing	rare Aug
Gomphosphaeria lacustris	uncommon Jul
Oscillatoria prolifica (Grev.) Gomant	common Jun–Nov
Sticosiphon regularis Geither	rare Oct
Lyngbya contorta Lemmermann	[a]
Lyngbya limnetica Lemmermann	[a]
Anabaena spiroides Klebahn	(present Jul 13 and 23, 1973)
Anabaena planktonica Brunnthaler	(present Oct 8, 1972)
Anabaena flos-aquae (Lyngb.) DeBrebisson	(present Aug, 26, 1972 North end of lake station 2)

[a] R. T. Oglesby (personal communication).

ryon sertularia, Pandorina morum, Stephanodiscus, Glenodinium, Scenedesmus, Eudorina elegans, and *Ceratium hirundinella.*

Samples taken September 17, October 8, October 28, and November 11, 1972 (Presutti, 1973) indicate a myxophycean bloom began in September 1972 and peaked on October 28, 1972. This bloom was made up of *Chroococcus limneticus, Oscillatoria splendida,* and *Anabaena limnetica.*

Studies in 1973 (Williams, 1974) showed the lake to be populated by species similar to those reported in 1972. Diatom, chlorophycean, and chrysophycean blooms dominated in July and August; in September and October blue–greens appeared more frequently but never dominated the phytoplankton. Table 25 is a list of phytoplankters collected in the lake during 1974 with a few added which were known to be present in the lake in 1972 and 1973 (R. T. Oglesby, personal communication; Gotham, 1974; Presutti, 1973).

The list shows 25 species of planktonic green algae, only four species of which appeared in six or more samples from May to November at 5 m. These were *Ankistrodesmus convolutus, Gleocystis ampla, Tetraëdron minimum,* and *Ulothrix* sp.

Ten species of Chrysophyceae were identified. The common species were *Dinobryon bavaricum, Dinobryon sertularia, Monocilia flavescens,* and *Mallomonas alpina* (Fig. 21).

Diatoms in the order Centrales were represented by six species of *Cyclotella* and *Rhizosolenia gracilis.* In the order Pennales 28 species were identified, and the common species were *Cocconeis placentula, Navicula*

TABLE 26

Concentrations of Various Forms of Plant Pigments and Fluorescence in Canandaigua Lake, 1973, April 25, July 18, and August 28[a]

	Depths (m)	Apr 25	Jul 18	Aug 28
Chlorophyll *a*	0	2.2	2.1	0.7
(mg/m^3)	5	2.5	2.7	1.5
	10	2.5	3.5	1.3
	50–65	1.6	1.1	0.7
Phaeophytin	0	1.2	1.0	2.1
(mg/m^3)	5	1.5	0.1	1.8
	10	1.1	1.1	1.0
	50–65	1.7	0.3	0.7
Carotenoids	0	5.0	1.2	0.9
(mg/m^3)	5	4.9	1.4	1.1
	10	3.4	1.5	0.9
	50–65	2.9	0.3	0.3
Chlorophyll *b*	0	0.4	0.5	0.7
(mg/m^3)	5	0.5	0.1	0.3
	10	0.4	0.2	0.7
	50–65	0.6	0.3	0.0
Chlorophyll *c*	0	1.7	2.7	2.2
(mg/m^3)	5	2.0	1.7	1.0
	10	2.3	3.0	1.9
	50–65	2.8	1.4	0.5
Fluorescence	0	20	17	17
(fluorescence	5	21	17	18
units × 30)	10	21	24	19
	50–65	19	15	16

[a] R. T. Oglesby (personal communication).

radians, N. rumpens, N. ulna, N. acus, N. capitata, N. crytocephala, and *Cymbella ventricosa.* This assemblage exhibits a Centrales–Pennales (C/P) ratio of 1:4.7.

The Dinophyceae were represented by five species but only *Gymnodinium caudatum, Glenodinum Kulczynskii,* and *Ceratium hirundinella* were common. *Cryptomonas* was common in samples from May to November.

The Myxophyceae were represented by about 15 species, but in 1974 only *Chroococcus limneticus* and *Oscillatoria prolifica* were common, although in a study by Wilcox (1975) net plankton was not examined.

Plant Pigments

The concentrations of chlorophylls *a, b,* and *c,* phaeophytin, and carotenoids were determined at depths ranging from the surface to 65 m during April, July, and August 1973 (Table 26). These show similar trends to those which occurred in Cayuga Lake in 1968 and 1969 (Wright, 1969) except that they were considerably lower, reflecting lower phytoplankton standing crops in Canandaigua Lake than in Cayuga Lake. There was a slight increase in phaeophytins in August in the epilimnion (Table 26), and low values occurred in April when chlorophyll *a* values were high.

Rooted Aquatics

Forest (1971, 1976) has studied the rooted aquatics of several lakes in western New York, including Canandaigua, and he has taken 35 quadrat samples in Canandaigua to determine standing crops of maximum growth in late summer. The principal species of rooted aquatics in 1971 were *Ceratophyllum demersum* L., *Anacharis canadensis* Michx., *Heteranthera dubia* (Jacq.) MacM., *Myriophyllum exalbescens* Fern., *Najas flexilis* (Willd.) Rost and Schmidt, *Ranunculus longirostris* Gods, *Potamogeton amplifolius* Tucherm., *P. crispus* L., *P. gramineus* L., *P. pectinatus* L., *P. pusillus* L., and *P. Richardsonii* (Benn.) Rydb. The attached alga, *Chara,* was also present.

The largest areas for rooted aquatics were much of the 1 km² shoal area at the south end, the 4 km² shoal at the north end, and an embayment near the mouth of Vine Valley Stream (Tributary 16). Other areas where rooted aquatics occurred were in the coves located to the north of the deltas formed by tributaries (Fig. 22).

The greatest species diversity was at the south end adjacent to the mouth of the West River. Rooted aquatics there extended down to a depth of 4.2 m (14 ft). Along the narrow shoal areas of the lake not protected from south winds the typical *Najas* and *Chara* community occurred, particularly from Black Point to the north end on the west shore and on the east shore near Otetiana Point (Baston and Ross, 1975; Fig. 22). This community was quite extensive just south of Black Point and in the 4-km² shoal area at the north end of the lake. *Chara* grew at a maximum depth of 5.7 m (19 ft) off Cottage City and *Myriophyllum* reached a maximum depth of 4.8 m (16 ft) at the same location (Forest, 1971).

In most areas where the rooted aquatics form "weed beds," *Myriophyllum* sp. and *Potamogeton Richardsonii* L. were the most abundant species in late summer, as was *P. crispus* in early summer during 1974 and 1975. Table 27 shows the average of three areas representing the most

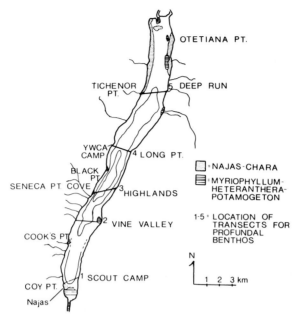

Fig. 22. Location of major communities of rooted aquatics and five transects for profundal benthos in Canandaigua Lake.

productive, the next most productive, and the least productive weed bed areas to average 659 gm/m² dry weight. The small area of shoals (16% of the total surface area) and the vigorous wave action brought by south and west winds keep these weed bed areas within reasonable bounds in most areas.

During the summers of 1929, 1930, and 1931, E. H. Eaton and F. A. Young collected *Potamogeton* spp. in several of the Finger Lakes including

TABLE 27

Rooted Aquatics Biomass in 19 Quadrats Sampled in Canandaigua Lake, August 20, 24, and 29, 1971[a]

Location	Number of quadrats averaged	Depth (m)	Gross dry weight gm/m²	(range)
Near inlet	9	1.8	719	(454–908)
Off Cottage City	4	3.3	499	(377–549)
North of Seneca Pt.	6	1.5–1.8	100	
Average			659	

[a] From Forest (1971).

Canandaigua. In their unpublished old notes are tables listing the occurrence of certain species in the areas where these rooted aquatics were best represented: the head and foot of the lake, off Vine Valley (Tributary 16), and just north of Seneca Point and Black Point (Table 28). These specimens were deposited in the herbarium at Hobart College. The nomenclature is that of "The Cayuga Flora" (Wiegand and Eames, 1926). There appears to have been a greater variety of species in 1929–1931 than in 1971, with the lack of *P. crispus* in the earlier study being most interesting (Table 28).

Bacteria

During the summer of 1973, Timothy R. Winship with two assistants took samples in tributaries to locate point sources of pollution on Menteth Creek (Tributary 45), Seneca Point Creek (Tributary 33), Hick's Point (Tributary 31), Long Point Creek (Tributary 25), Granger Point Creek (Tributary 21 or 22), Vine Valley Creek (Tributary 16), Crystal Beach Creek (Tributary 7, 8, or 9), and two other unidentified tributaries. They concluded (Winship, 1973, p. 31) that every stream listed above showed levels of bacterial contamination high enough to be considered an indication of a potential health hazard.

Delta Laboratories, 18 West Main Street, Webster, New York, 14580 (Anonymous, 1972) took a series of coliform counts at 28 test sites in summer 1972 around the lake which included four sites on West River above and below Middlesex. There were 7500 coliforms/100 ml above Middlesex, and 16,000/100 ml below. Counts dropped to 0 at the mouth of the river.

The State of New York Health Department takes samples of lake water at three Canandaigua Lake beaches before and during the swimming season, which is essentially June, July, and August. Coliform counts are seldom high enough to close beaches for any length of time. Following Hurricane Agnes in 1972 the beaches were closed for a considerable period but reopened by mid July. Counts are taken at Bristol Harbour (Town of South Bristol), Kershaw Park (Canandaigua City), and Ontario County Park (Town of Gorham). Except for one count of greater than 20,000 coliform colonies/100 ml on July 10, 1972 at Bristol Harbour, most counts were less than 2400/100 ml, the concentration that may not be exceeded in swimming areas under New York Law. The highest fecal coliform count was 310/100 ml on May 5, 1975 at Kershaw Park, but usually these counts in July and August were less than 5/100 ml (Table 29).

In a series of 28 coliform samples taken in the lake during July and August 1955, ten samples from the shallow area at the north end averaged 19 colonies/100 ml, (MPN). Twelve taken in a series from east to west across the lake off Tichenor Point and Granger Landing averaged less than

TABLE 28

Species of *Potamogeton* Found by E. H. Eaton and F. A. Young in Canandaigua Lake during the Summers of 1929, 1930, and 1931, and by Forest in 1971

Species of Potamogeton	Woodville	Seneca Pt. Cove	Black Pt. Cove	Vine Valley	Canandaigua City pier area	Forest
natans	infrequent				abundant	
epihydrus	frequent				common	
americanus	frequent		b		fairly common	
amplifolius	frequent			b	common/local	a
gramineus	a	b	b	b	fairly common	a
angustifolius	scarce	c	c			
richardsonii	frequent	b	b	abundant	b	a
bupleuroides	frequent	b	b	common	b	a
crispus						a
zosterifolius	abundant	b	b	b	common	
friesii	infrequent	b	b	b	b	
pusillus	common	b	b	b	b	
foliosus					b	a
filiformis		(shale cliff) b	b	b	b	
pectinatus	common	b	b	b	common	

[a] Found by Forest (1971) in the lake during 1971.
[b] Present.
[c] Questionably present.

TABLE 29

Raw Water Bacterial Water Quality as Determined in 1972, 1974, and 1975 by the State of New York Department of Health, District Office, Geneva at 9 and 121 m Off Shore (Canandaigua City) and at 15 and 335 m Off Shore (Gorham Town)[a]

Canandaigua City intake	1972 SPC[c]	COL[d]	Aug 20–Dec 1974 SPC	Jan–Jul 1975 COL
Jan–Apr (ave. 4 counts[b])	293	27	530	12
May–Sep (ave. 5 counts)	1350	161	2000	3
Oct–Dec (ave. 3 counts)	840	67	2100	14
Gorham Town water district intake	1972 SPC[c]	COL[d]	1974 SPC	1975 COL
Jan–Apr (ave. 4 counts)	49	3	51	7
May–Sep (ave. 4 counts)	1356	24	593	3
Oct–Dec (ave. 3 counts)	182	7	190	16

[a] Personal communication from Roger A. France, August 15, 1975.
[b] None of the individual determinations were greater than 2000.
[c] Standard plate count of total bacteria/ml (SPC).
[d] Coliform colonies/100 ml (COL).

3.6 colonies/100 ml (MPN). In a series off Cook's Point, 4 km north of the swamp at the south end, six counts averaged 20 colonies/100 ml (MPN) (New York State Department of Health, 1955).

In summary, bacterial pollution of the lake, insofar as this is indicated by coliform tests, has changed little since 1955 and seldom poses a problem except occasionally after high water such as that left by Hurricane Agnes in June 1972. Bacterial and fungal studies related to the decomposers of the ecosystem have yet to be done; some indication that a bacterial plate existed near the thermocline in 1972 was shown by Gotham (1974).

Benthos

The littoral benthos has not been systematically studied, but collections of hatches of aquatic insects during the 1960's and 1970's have been made, so some qualitative information is available. A large dark *Ephemera* usually emerged in May by crawling up on the beach. This hatch is often avidly collected by common grackles (*Quiscalus quiscula*) to feed their young. A large *Hexagenia* and a species of *Heptagenia* emerged in June and July, and a very small mayfly hatched near the end of July and early August. One species of stonefly hatched in June along with the neuropteran *Sialis*, often common in alewife stomachs. Several species of *Odonata* emerged from the lake. Many species of caddisflies (*Tricoptera*) and flies (*Diptera*) also were present.

Harman and Berg (1971) gave a good general account of the snails of the Finger Lakes region (and beyond), but little collecting was done in Canandaigua Lake. They mention only *Helisoma anceps* as being present in Canandaigua Lake. Littoral collections in 1975 by Ekman dredge were made at the shallow ends of the five transects shown in Fig. 22. In addition to *H.|Physa sayii, Gyranlus parvus, anceps, Goniobasis livescens, Amnicola limnosa, Valvata tricarinata,* and *Valvata lewisi* were found.

The profundal benthos were studied by E. H. Eaton (unpublished notes, 1927), and his study was repeated in 1974 (Eaton *et al.,* 1975) (Fig. 22, Table 30). Forty-five Ekman dredge (6 × 6 × 6 in.) grabs were taken along five transects of the lake from Deep Run to Tichenor Point at the north to the transect from Old Scout Camp to Coy Point at the south. These grabs were on bottom 30 to 83 m in depth, and they contained three dominant groups: Oligochaete worms of at least two species, *Pontoporeia affinis,* and dipterid larvae (Tendipedinae) of at least three species. In lesser numbers were fingernail clams of the genus *Sphaerium* and a small pale ostracod. Table 30 indicates some possible enrichment since 1927, particularly with regard to the oligochaetes.

The number of *Pontoporeia affinis* and dipterid larvae, taken as separate entities, never exceeded an average transect value of 2000/ml². If all taxa were added to obtain total organisms/m² per transect and the totals averaged to give an average count for that year, we find that in July 1927 there were 1773 organisms/m² and in July 1974 there were 2124 organisms/ m². A study of the profundal benthos of Cayuga Lake in 1952 (Henson, 1954) showed a similar community of five species with an average count on July 1 of 4730.4 organisms/m². These organisms are associated with the profundal sediments, where they feed on detritus falling from above, and show promise of being good indicators of the rate of eutrophication of the Finger Lakes.

Zooplankton

The zooplankton of Canandaigua Lake have been studied by Birge and Juday (1914, 1921), Hall and Waterman (1967), Eaton and Moffett (1971), Kardos and Eaton (1972), Eaton and Kardos (1973), and Wilcox (1975). They have all reported a typical plankton community for oligotrophic lakes of this size, depth, and latitude.

During the winter and spring the crustacean community was composed mainly of *Mysis relicta, Diaptomus sicilis, D. minutus,* cyclopoids [Hall and

TABLE 30

Mean Number of Organisms per m² in Sediments of Canandaigua Lake at Depths Ranging from 30–83 m in Brown to Dark Gray Muck[a]

Organism	Transect 1		Transect 2		Transect 3		Transect 4		Transect 5	
	1927	1974	1927	1974	1927	1974	1927	1974	1927	1974
Oligochaeta (Annelida)	17	576	0	143	75	296	43	375	29	416
Pontoporeia affinis (Crustaceae)	937	1952	725	638	1021	1140	1401	1179	1247	1054
Ostracoda (Crustacea)	17	0	7	22	11	11	49	154	14	201
Tendipedinae (Diptera)	335	722	538	534	564	312	514	326	1118	48
Sphaerium sp. (Pelecypoda)	34	9	43	79	11	43	80	74	36	315
	1340	3258	1306	1416	1682	1802	2087	2108	2444	2034
Grabs[b]	5	5	7	7	8	8	7	7	6	6

[a] Average total number of organisms per m² in 1927, 1773; in 1974, 2124.
[b] Number of grabs along transect that were averaged.

Waterman (1967) identified *Cyclops vernalis* and *Mesocyclops edax* from Canandaigua Lake, but their samples were not collected in winter], and the copepodite stages of *Limnocalanus macrurus* (Kardos and Eaton, 1972; Eaton and Kardos, 1973; Kardos, 1975). *Pontoporeia affinis* was also present in abundance on the bottom of the lake (Eaton *et al.*, 1975), but its winter status remains to be investigated. The others were distributed more or less homogeneously in the water column at this season.

As a temperature gradient began to develop in the spring, gradual concentrations of cyclopoids and the two *Diaptomus* species were noted in the upper part of the water column. *Limnocalanus macrurus* and *Mysis relicta,* on the other hand, began their seasonal diurnal vertical migrations at this time (Eaton and Kardos, 1973; Kardos, 1975). The most common spring rotifers were *Keratella cochlearis, Kellicottia longispina, Keratella quadrata,* and *Asplanchna priodonta* (Wilcox, 1975). These forms were uniformly distributed in the water column.

During the summer the cladoceran species appeared and increased to their greatest numbers. Most of these species peaked in August or September (Kardos and Eaton, 1972; Wilcox, 1975; Kardos, 1975). *Bosmina* was the most common, attaining numbers of $10^5/m^3$ in the epilimnion. *Ceriodaphnia pulchella, Diaphanosoma leuchtenbergianum,* and *Daphnia retrocurva* reached more modest peaks. *Leptodora kindtii* was recorded in the $100/m^3$ range, and *Polyphemus pediculus* was recorded only sporadically and in very low numbers. The summer assemblage of rotifers was characterized by *Keratella cochlearis, Conochilus unicornis, Pleosoma truncatum, Polyartha vulgaris,* and *Pompholyx sulcata.* As a group they peaked in early August, the most common rotifer being *K. cochlearis* ($4.0 \times 10^5/m^3$). Rotifers were much more common above the hypolimnion than in it (Wilcox, 1975).

Cladocerans were concentrated in the epilimnion and thermocline during the summer and early fall along with the cyclopoids and *Diaptomus minutus.* *Diaptomus sicilis* had a somewhat lower distribution, and *Limnocalanus macrurus* was found almost exclusively in the hypolimnion during summer days and no higher than the top of the metalimnion at night. It also reached population peaks in August. *Diaptomus sicilis* peaked during the winter. Cyclopoid numbers and those of *Diaptomus minutus* increased during the summer and were maintained at those levels well into the fall. *Pontoporeia affinis* also displayed a strong population during the summer, when the only samples were taken (Table 30). They remained much closer to the lake bottom over the 24-hr cycle than did *Mysis* (Eaton and Kardos, 1973). *Mysis relicta* underwent a vertical migration similar to that of *Limnocalanus macrurus,* but was more restricted to the hypolimnion than that species.

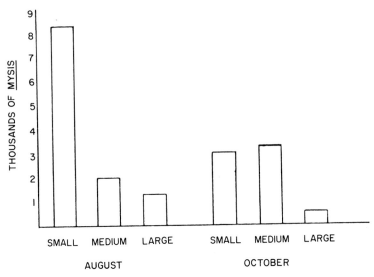

Fig. 23. Population structure of Canandaigua Lake *Mysis* in August and October 1972.

During the fall season the cladoceran species decreased to near zero in most cases. *Bosmina* and *Daphnia retrocurva* sometimes had late fall increases, but by winter their populations were at very low levels. Adult *Limnocalanus macrurus* were also very reduced in numbers or absent. *Mysis relicta* of all age groups maintained good numbers throughout the year (Eaton and Kardos, 1973; Kardos, 1975). Both they and *L. macrurus* were thought to be univoltine (Fig. 23) in Canandaigua Lake. As fall progressed and general mixing began, these species became more homogeneously distributed in the water column.

Fish Populations

The fish populations of the lake have been surveyed by Eaton (1928), Heacox and Stone (1945), Hartman (1958, 1959), Eaton and Moffett (1971), Eaton and Kardos (1972), and Kardos (1975). The Avon office of the New York State Department of Environmental Conservation takes annual inventories of the trout streams tributary to the lake and has planted marked trout in the lake on a regular basis since 1970. This is being done to determine, in the future, the percentage of the fisherman's catch of planted as opposed to naturally spawned trout, particularly lake trout.

One of the first species which appeared to be extirpated from the lake in recent times was the American eel (*Anguilla rostrata*). Prior to 1880,

Canandaigua was a "lake trout–cisco lake" (*Salvelinus namaycush–Coregonus artedii greelei*) (Eaton, 1928; Heacox and Stone, 1945). The whitefish (*Coregonus clupeaformis*) may have been native to Canandaigua Lake, but there is evidence of heavy stocking as early as 1897 and possibly prior to 1870 (when they were stocked in Hemlock Lake; New York State Forest, Fish, and Game Commission, 1897). By 1894, 3600 kg of whitefish were commercially harvested from Canandaigua Lake (Youngs and Oglesby, 1972). Both perch (*Perca flavescens*) and pickerel (*Esox niger*) were present in the lake prior to 1880, and they were large and numerous (Eaton, 1928). During the period 1898 to 1915 artificial plantings had made the pike perch or walleye (*Stizostedion vitreum*) one of the two most numerous game fishes in the lake (Table 31); the other was the pickerel. The walleye held this position until 1930 (Heacox and Stone, 1945). Burbot and whitefish were also plentiful. Bass were not so common as they later were in the 1940's and trout were relatively scarce. The rainbow smelt (*Osmerus mordax*) was first stocked in 1925 in an attempt to provide a stable forage base, since the cisco had declined (Heacox and Stone, 1945).

Eaton (1928) reported that in 1927 the whitefish, burbot (*Lota lota*), and walleye were still abundant, but lake trout remained relatively scarce. He felt that the introduction of alewives from either Seneca or Keuka Lake would supplement the forage base provided by smelt and cisco and aid in increasing the lake trout population. Eaton (1928) also reported that the smallmouth bass (*Micropterus dolomieui*) was common on near-shore rocky bottoms, but perch and pickerel were less common and smaller than in the 1880's. He noted that rainbow trout were present but scarce in Canandaigua Lake, but by 1934 the Naples Record stated that many large rainbow trout were running up Naples Creek to spawn.

From 1935 through 1944, 43 million walleye fry and about 475,000 lake trout fingerlings and yearlings were stocked (Table 31). In 1942, lake trout were exceedingly abundant throughout the entire lake, and walleyes were relatively scarce (Heacox and Stone, 1945). Smallmouth bass were plentiful and perch and pickerel had increased since 1927. Both the whitefish and the burbot had declined to very low levels, and this appeared to be correlated with the introduction of the smelt (Heacox and Stone, 1945).

By 1935 smelt had increased to the point where they were thought to be too abundant, and special permits were issued for the first time to allow their netting. State netting for the species followed in 1936. However, the smelt continued to increase, and record spawning runs occurred in 1940 and 1941. These were followed by a crash decline in 1943. The smelt population had recovered slightly at the time of the 1945 survey (Heacox and Stone, 1945). Even at their low numbers smelt were preferred by lake trout and walleyes over the far more numerous minnows (*Hybognathus nuchalis,*

TABLE 31

Fish Planted in Canandaigua Lake by the New York State Department of Environmental Conservation, 1895–1974

	1895–1899	1917–1926	1935–1944	1945–1961	1962–1970	1971–1974
Lake trout						
Fry	50,000	400,000	0	0	0	0
Fingerlings	73,400	541,250	390,382	1,277,750	346,560	140,688
Yearlings	7,413	0	84,490	127,980	48,595	30,460
Rainbow trout						
Fingerlings	0	0	0	0	23,280	0
Yearlings	0	0	0	6,000	8,053	0
Brown trout						
Fry	15,000	0	0	0	0	0
Fingerlings	1,000	0	0	0	0	37,000
Yearlings	0	0	0	0	0	20,837
Atlantic salmon						
Fingerlings	0	0	0	0	1,000	0
Yearlings	0	0	0	0	1,090	0
Whitefish						
Eyed eggs	0	0	1,020,000	0	0	0
Fry	10,000,000	5,750,000	1,080,000	0	0	0

Smelt	0	9,500	0	0	0
Muskellunge	100,000	0	0	0	0
Buckeye Shiner (Emerald Shiner)	0	9,500	0	0	0
Walleye[a]					
Fry	2,250,000	2,925,000	43,800,000	0	0
Yearlings	212	0	0	0	0
Adults	10	0	0	0	0
Yellow Perch[a]					
Fry	1,800	0	0	0	0
Fingerlings	0	43,500	366,000	0	0
Smallmouth bass[a]					
Fry	275	6,500	60,000	0	0
Fingerlings	250	6,000	11,588	0	0
Largemouth bass[a]					
Fry	2,700	0	0	0	0
White bass					
Fry	225	0	0	0	0

[a] Most were planted in the outlet but some were planted in the lake.

Notropis analostanus, Pimephales notatus, and *Notemigonus crysoleucas*), which were present in the lake at that time. The silvery minnow, *Hybognathus nuchalis,* was extremely abundant and netted commercially as bait. Heacox and Stone noted in 1945 that the trout condition appeared to have worsened since 1942, when the smelt were at their peak.

Walleyes were not stocked by New York State after 1941 (Heacox and Stone, 1945). Since then lake trout, a few rainbow trout, and (in the 1970's) brown trout (*Salmo trutta*) have been stocked (Table 31). Alewives (*Alosa pseudoharengus*) were stocked in 1953 without authorization from New York State (Eaton and Moffett, 1971).

The 1971 fish survey revealed that the alewife was the most important forage fish (Eaton and Kardos, 1972) at that time. The smelt was still at a low level, and only two specimens of the cisco were taken, one each in 1970 and 1971 (both found floating dead at the surface). Whitefish were not collected at the time of the survey, but three were reported caught in the previous five-year period (Eaton and Kardos, 1972). Among the minnows only the bluntnose (*Pimephales notatus*) could be called abundant. The silvery minnow was rare (one specimen). The Ninespine stickleback (*Pungitius pungitius*) was still present.

The lake trout was the best-represented predator. Rainbow trout were common as were yellow perch, largemouth bass (*Micropterus salmoides*), smallmouth bass, and the chain pickerel. The walleye was scarce and represented only by very large (650 to 800 mm) specimens, and only one burbot (an adult) was examined during the course of the study. A variety of panfish were common or abundant, e.g., rock bass (*Ambloplites rupestris*). Tomaka (1973) showed that yellow perch in Canandaigua Lake have a good growth rate (reaching 203 mm by their third year) and were in good condition.

Kardos (1975) studied the distribution of alewives and their population structure over a three-year period. He found them to be concentrated into large, slow moving, or stationary school(s) near the center and bottom of the lake during winter (Fig. 24). As spring approached and the water column became isothermal the alewives quickly dispersed upward. As the surface waters became warmer the alewives tended to concentrate there and may have run up tributaries flowing into the lake during April and May, but no good data on this are available. When a metalimnion formed, alewives remained in or above it, rarely being found in the hypolimnion. Alewives 140 mm or less in total length (yearlings or younger) displayed a shallower distribution in the water column than the larger ones. A reverse movement occurred in the fall as the water cooled and stratification disintegrated. Alewives dispersed downward in the water column and were homogenously distributed by December or January. From May to mid August there also occurred typical nightly spawning movements from mid lake to in shore.

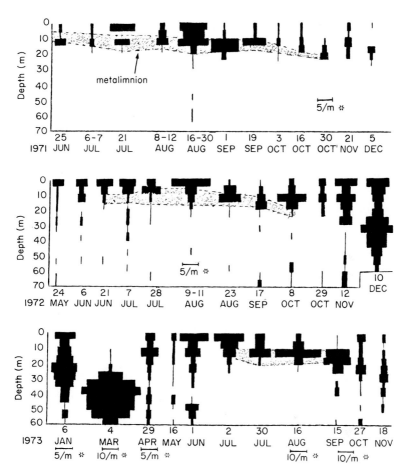

Fig. 24. Vertical distribution of alewives in Canandaigua Lake, 1971–1973. The asterisk denotes that the scale is for either five or ten alewives per vertical meter.

The spawning schools were predominantly composed of males, and it was hypothesized that males spawned much more frequently than females.

These yearly distribution patterns of alewives affected their food habits, excluding certain food species from them during the summer and making others more available (Fig. 25). *Mysis relicta* appeared to be the preferred food during the winter, spring, and late fall, but during the stratified period this species was restricted in most instances to the hypolimnion, while alewives were nearly all located above that zone. *Pontoporeia affinis,* another inhabitant of the depths of Canandaigua Lake, was not important in the alewife diet. During the stratified period the main food of limnetic alewives became zooplankton. The larger zooplankters such as *Limnocalanus* and

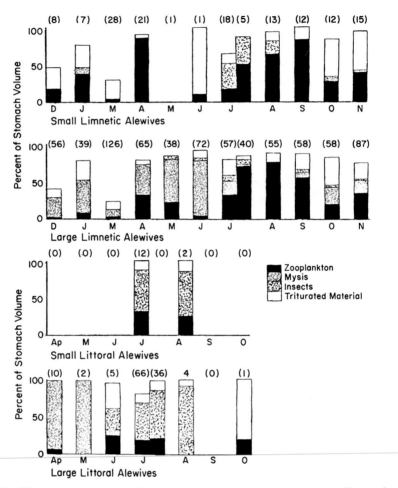

Fig. 25. Average percent of cardiac stomach volume occupied by different food organisms for small and large alewives during 1973 in Canandaigua Lake.

Diaptomus sicilis were preferred over smaller ones such as *Bosmina* and *Ceriodaphnia,* but other factors also seemed to influence the selection of a zooplankter such as method of locomotion, transparency, and abundance. Rotifers were ignored except for the colonial *Conochilus,* which became more positively selected as the colonies became larger. Alewives also consumed their own fry. These planktonic young were readily available, being found mainly in the epilimnion. This phenomenon was probably responsible for differences which were noted in the strengths of year classes, and has been noted for alewives in the Great Lakes (Brown, 1971; Webb,

1973). Alewives caught in the littoral zone had eaten insects, and these formed a substantial part of their diet.

Alewives in Canandaigua Lake now reach a maximum average length of approximately 170 mm (Fig. 26). Most of the energy for this growth is obtained during the summer months when zooplankton is the main food. *Mysis* appears to contribute mainly to the formation of reproductive products, since little growth is attained at the time of year when *Mysis* is important in the diet.

During the course of the alewife study (1971–1975), the 1969 and 1970 year classes were the dominant year classes in the population (age determined by reading scales) (Fig. 27). The 1971 class was very weak, the 1972 class was relatively strong, and the 1973 class was not found (Kardos, 1975). It was hypothesized that the 1972 year class profited from the Hurricane Agnes-induced turbidity (increased phytoplankton blooms and sediments); the 1971 and 1973 year classes not having this cover were more heavily preyed upon. A sample of alewives taken subsequently in 1975 by S. W. Eaton (Fig. 27) indicated that the 1969 and 1970 year classes had died out, the 1971 year class was still very weak (5.5%), the 1972 year class relatively strong (31.5%), the 1973 year class apparently weak (14.0%), and the 1974 year class apparently strong (49.0%). We would also expect that the

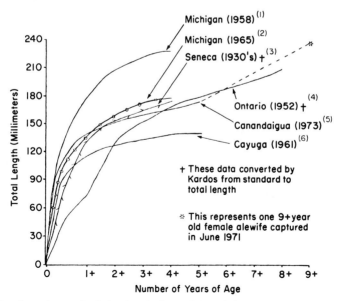

Fig. 26. Annual growth of alewives in Canandaigua Lake and other Finger Lakes and Laurentian Great Lakes. Numbers in parentheses refer to (1) Brown, 1971; (2) Norden, 1967; (3) Odell, 1934; (4) Graham, 1956; (5) Kardos, 1975; (6) Rothschild, 1962. The years in parentheses following lake name are actual years when data were collected.

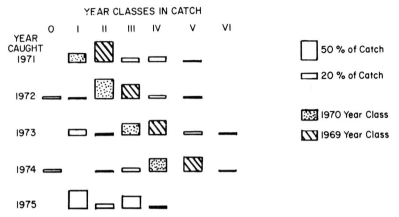

Fig. 27.   Age composition of the alewife population in Canandaigua Lake, 1971–1975.

total number of alewives in the lake in 1975 was less than when the 1969 and 1970 year classes existed (1970–1973). Data from two samples taken in 1974 have been presented in Fig. 27 for the purpose of continuity, but the gear used lacked 1.6-cm bar mesh netting and was set only over the littoral areas, which makes the data questionable. There was some indication from these 1974 samples that the 1969 year class was diminished after the summer spawnings.

Zooplankton composition has changed little qualitatively since the studies of Birge and Juday (1914, 1921); a few cladoceran species have been added (Kardos, 1975). However, *Bosmina* has increased 1000-fold since their studies. This is probably related to generally increased productivity, but also to selective predation by alewives on the larger zooplankters (Kardos, 1975). *Daphnia, Diaptomus,* cyclopoids, and *Limnocalanus* have remained steady in numbers.

Alewives composed 97% of the recognizable food of lake trout captured in Canandaigua Lake (Eaton and Kardos, 1972), and the trout were in much better condition than in previous surveys, although about 95% were infested by a tapeworm (Eaton and Kardos, 1972). Rainbow trout also exhibit good growth after migrating into the lake following their first two years in Naples Creek, and in 1976 a brown trout weighing 5.7 kg was caught in the lake. These trout are also utilizing the alewife as a major source of food after reaching sufficient size. Pickerel, bass, and perch populations are strong as of this writing.

The previous discussion suggests that the zooplankton and alewife populations in Canandaigua Lake can be expected to oscillate periodically, approximately every five to six years. A strong year class of alewives should remain dominant in the population for four or five years until it dies off. This dominant year class will reduce the size of succeeding year classes through cannibalization of fry. When the dominant year class dies out the next year's class should be strong, but there will be a period of about a year when the total number of alewives in the lake will be decreased. During times of strong alewife numbers there will be more of the smaller zooplankters. When the alewife population is at an ebb there will be more of the larger zooplankters. Also, when alewives are scarce trout should be more susceptible to angling.

DISCUSSION OF ECOSYSTEM

The lake still maintains its oligotrophic nature but only in the 1970's have nutrient loadings of the system been studied and algal assays undertaken to obtain base line data on the lake's trophic state (USEPA, 1974; Oglesby et al., 1975).

Dillon (1974) and Mills et al. (unpublished manuscript) have attempted to assess trophic levels of Canadian and New York lakes by relating phosphorus loading of the lakes to the average depth of the lake basins divided by the hydraulic retention time. If Canandaigua is considered to have a loading of phosphorus equal to 0.14 g P/m^2 per year (Table 21), a mean depth of 38.8 m (Table 5), and a retention time of 13.4 years (Table 5), then the lake may be considered oligotrophic, based on the criteria in both papers.

Secchi disk transparency, a practical but perhaps less precise measure of productivity as reflected by phytoplankton standing crop, has been used by Rawson (1955) and Oglesby (1974) to assay the trophic status of lakes by relating average annual secchi disk summer readings to average annual chlorophyll a levels in the top 10 m of water. If these readings are plotted for Canandaigua Lake along with other north temperate lakes the plot falls closest to Conesus Lake, Skaneateles Lake, and Owasco Lake in decreasing order.

If one selects morphometry from which to judge productivity (Rawson, 1955), Canandaigua Lake again falls into an oligotrophic category. Again, of the Finger Lakes it is most similar to Owasco and Skaneateles Lakes.

The algal populations have apparently changed from Birge and Juday's time (1910–1918) from dominance of blue–green algae to a dominance today of green algae and diatoms. The diversity of rooted aquatics has been reduced to a dominance of *Myriophyllum, Potamogeton crispus,* and *P.*

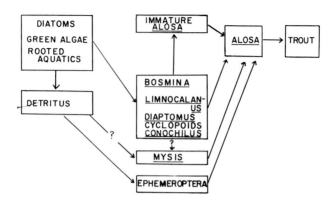

Fig. 28.   The food web of Canandaigua Lake from 1970 to 1975.

Richardsonii, though in many areas the *Najas–Chara* plant community is still extensive. This may be a transient trend following siltation caused by Hurricane Agnes in June 1972 (H. S. Forest, personal communication).

The total number of zooplankters has increased due to the 1000-fold increase in *Bosmina,* but other cladocerans and copepods have maintained comparable populations to those found by Birge and Juday (1914, 1921). The profundal benthos (sediments below 30 m) have shown some quantitative changes of a minor degree in 50 years. The same five taxa are present and these are typical of holarctic, oligotrophic bottom faunas. Oligochaetes have increased in these sediments since 1927 (Table 30) by factors of 3 and 4 but are still usually less than 500/m² in early July. Henson (1954) found a mean of 2504 oligochaetes/m² in Cayuga Lake on July 1, 1952. *Mysis* was also important at the second and possibly at the third trophic levels (Fig. 28).

The silvery minnow, the cisco, and the introduced rainbow smelt have all declined and the alewife has become dominant at the third trophic level. A major influence in maintaining stable alewife populations appeared to be cannibalism by adult alewives on their own immature stages. The alewife, in turn, has become the main forage of the fourth trophic level, constituting over 95% of the food of the lake trout. The alewife, through its great reproductive potential and ecological amplitude, seems to be reducing the diversity of the third trophic level, which was characteristic of the lake as recently as 1945 (Fig. 28).

If one follows Vollenweider (1973) in estimating "dangerous" or "permissible" loading rates for phosphorus for a lake of Canandaigua's mean depth and hydraulic retention time, the figures would be 0.32 gm P/m² per year for "dangerous rate" and 0.16 gm P/m² per year for "permissible rate." With the rate of 0.14 gm P/m² per year for Canandaigua Lake

(Table 21), the loading appears to be permissible to maintain its oligotrophic state. The loadings were based on 1972 data when severe flooding during June must have exceeded typical annual figures. With zoning laws restricting building around the lake's perimeter, and with completion of a sewage disposal system around the northern part of the lake, the lake's oligotrophic state should be maintained for many years.

TABLE 32

Summary of Knowledge Concerning the Descriptive Limnology of Canandaigua Lake in 1975

	Good	Moderate	Poor
Basin morphometry	X		
Geology and soils	X		
Climate	X		
Vegetation	X		
Population (1970)	X		
Land use	X		
Water use	X		
Waste discharge	X		
Hydrology			
Lake volume and area	X		
Sediments			X
Stream flow and retention time	X		
Precipitation	X		
Currents			X
Temperature	X		
Specific conductance		X	
Light		X	
Color		X	
pH	X		
Dissolved oxygen	X		
PO_4			X
NO_3			X
Alkalinity		X	
Major ions		X	
Trace metals			X
Phytoplankton		X	
Pigments			X
Primary production		X	
Algal assays		X	
Population description		X	
Zooplankton		X	
Benthos		X	
Fish		X	
Bacteria			X
Submerged aquatics		X	

PRESENT STATE OF KNOWLEDGE

Much is known about the general physical limnology of the lake and information related to land use and hydrology but little is known about lake currents. The bottom sediments, the seasonal flux of nutrient salts in the epilimnion, primary production, algal assays, and studies on the lake's decomposers are areas needing further attention. The second, third, and fourth trophic levels are fairly well understood, but much still needs to be done on the benthos of shoal areas and the periphyton (Table 32).

From work done by Kardos on the alewife we have come to realize the great potential of Canandaigua Lake as a model to help in the further understanding of the population dynamics of oligotrophic lakes. The lake is small enough to be circumnavigated readily, lacks the often equipment damaging currents of the larger Finger Lakes and Great Lakes, and yet it has maintained the chemical, physical, and biotic characteristics needed to be studied. Further work on the alewife, the first and second trophic levels of the ecosystem, and the decomposers would lead to a more perfect understanding of the lake and result in more sophisticated management of Canandaigua Lake, other Finger Lakes, and Laurentian Great Lakes.

SUMMARY

Canandaigua Lake lies in a narrow basin of Devonian rocks at 42° 46′ N latitude and 77° 18′ W longitude. It is the fourth largest in volume (1640.1 × $10^6/m^3$) and surface area (42.3 km²) of the Finger Lakes of central New York.

Approximately 7700 people inhabit the lake basin and approximately 2000 living units are located around the perimeter of the lake. The lake is utilized primarily as a water source for five municipalities in the area and the residents around the lake perimeter. Secondarily, it is used for recreation, flow equalization, and, to a limited extent, power.

Records on outflows from the lake basin have been kept since 1939 and the normalized outflows of the Feeder Canal and Muir Dam total 3.879 $m^3/$ sec. The volume of the lake at a datum of 209 m above mean sea level is 1640.1 × $10^6/m^3$. Assuming inflow during the year to about equal evaporation and outflow, the retention time (or flushing rate) is 13.4 years. Most water enters the basin in March and April.

The climate and weather are mainly governed by atmospheric flow from continental areas, but the prime source of moisture is from the Gulf of Mexico. The Great Lakes have a moderating influence on temperatures from early fall though March or April. The Secchi Disk readings from 1971 through 1975 averaged 4.15 from April to October and were typically from 7 to 8 m from December to March.

The lake becomes well stratified by late June and homothermy begins at about 8°C in mid November. When ice cover is complete an inverse temperature gradient develops and homothermy again occurs as the ice breaks up and gradually reaches stratification by June.

The lake is a hard water lake averaging 128 mg/liter as $CaCO_3$ or higher. There is a well-developed carbonate buffer system and the pH during the summer in the epilimnion averages about 8.3. In December, January, and February, the pH may drop to 7.6 at most depths but returns to higher levels as algal populations increase in spring. Dissolved oxygen is near saturation at most depths. It becomes lowest in the hypolimnion (70 m) in October at about 8.0 mg/liter. The dissolved oxygen profile at this time is typically of the orthograde type. The major cations are Ca^{2+}, Mg^{2+}, and Na^+, and the major anions are HCO_3^-, SO_4^{2-}, and Cl^-. The conductivity ranges from 279 micromhos/cm at the surface in April to 290 in August, much less than Cayuga and Seneca.

It is considered a medium–high nitrate lake, probably never being nitrate limited. Two studies have shown the lake to be phosphate limited, at least in late summer and perhaps throughout the growing season.

Diatoms and Chlorophyceae dominate the algal maxima of summer with relatively weak myxophycean maxima occurring from late August to mid October. Rooted aquatics are present in the shoal areas of the south and north ends. The coves on the north sides of the points also harbor small patches. A *Chara–Najas* community is characteristic of the narrow shoal areas of the east and west shores and the shallow north end. The profundal benthos shows only slight enrichment since 1927 and the same species groups are present.

Little change in composition of the zooplankton has occurred qualitatively since 1910–1918, but *Bosmina* has increased 1000-fold. This is believed to be an effect of the alewife, introduced in 1953, selecting larger cladocerans in the epilimnion during the summer. The alewife is now the main planktivore, having replaced the cisco and the introduced rainbow smelt. The lake trout, and, more recently, brown trout are being stocked, but the rainbow trout shows natural spawning adequate to maintain a good sport fishery. The lake is by most critera considered an oligotrophic lake showing some early signs of cultural eutrophication.

APPENDIX

Chemical Analysis of Distribution Water Taken from Canandaigua Lake at 9 m near Canandaigua City and along the East Shore, Town of Gorham, at 15 m, 1972 and 1973[a]

Chemical parameter	Canandaigua City 1972 (mg/liter)		Rushville, 1973 (mg/liter)
	Jan 17	Jan 11	Feb 28
Silica	2.9	3.2	2.5
Calcium	40.0	41.0	41.0
Magnesium	10.0	10.0	9.6
Sodium	6.0	6.2	7.4
Potassium	2.0	2.0	2.0
Bicarbonate	134.0	127.0	133.0
Carbonate	0.0	0.0	0.0
Sulfate	32.0	38.0	33.0
Chloride	8.7	11.0	12.0
Fluoride	0.1	1.0	0.1
Total Kjeldahl N	0.26	0.14	0.22
Nitrates as N	0.39	0.36	0.40
Phosphorus as PO_4	0.10	0.02	0.02
Dissolved solids sum	169.0	178.0	175.0
Hardness total	141.0	144.0	142.0
Hardness N C[b]	31.0	39.0	33.0
Specific conductivity	296.0	301.0	311.0
pH	8.1	7.6	6.2
Cyanide	0.0	0.0	.01

Chemical parameter	Canandaigua City, 1972 (μg/liter)		Rushville, 1973 (μg/liter)
	Jan 17	Jan 11	Feb 28
Phenols	0.0	0.0	0.0
Arsenic	1.0	2.0	0.0
Cadmium	0.0	1.0	0.0
Mercury total	<0.8		0.5
Selenium	0.0	3.0	1.0
Aluminum		88.0	38.0
Barium		22.0	19.0
Beryllium			2.0
Bismuth		4.0	8.0
Boron		18.0	16.0
Chromium		4.0	10.0
Cobalt		4.0	5.0
Copper		22.0	30.0
Gallium		0.8	3.0
Germanium		4.0	10.0

APPENDIX *(Continued)*

Chemical parameter	Canandaigua City, 1972 (µg/liter)		Rushville, 1973 (µg/liter)
	Jan 17	Jan 11	Feb 28
Iron		160.0	590.0
Lead		4.0	23.0
Lithium		10.0	0.0
Manganese		8.0	5.0
Molybdenum		0.9	3.0
Nickel		2.0	7.0
Silver		0.8	1.0
Strontium		90.0	310.0
Tin		4.0	10.0
Titanium		3.0	10.0
Vanadium		2.0	7.0
Zinc		360.0	0.0
Zirconium		0.8	15.0

[a] United States Department of the Interior (1974).
[b] Noncarbonate hardness.

REFERENCES

Anonymous. (1933). "Canandaigua Lake Directory." Canandaigua Yacht Club, Canandaigua, New York.

Anonymous. (1960–1961). "Canandaigua Lake Directory." Ontario County Press, New York.

Anonymous. (1967–1968). "Directory of the Canandaigua Lake and Town of Canandaigua," pp. 1–82. Ontario County Press, New York.

Anonymous. (1972). "Canandaigua Lake Study. Part I: Coliform, Phosphate, Nitrate." Bull. 2 Delta Laboratories, Inc. Webster, New York.

Anonymous. (1973). "Interboard Plan for the Greater Finger Lakes—Oswego River Basin," WA-ONT-YA Basin Regional Waters Resources Planning Board. Ithaca, New York.

Baston, L., Jr., and Ross, B. (1975). "The Distribution of Aquatic Weeds in the Finger Lakes of New York State and Recommendations for Their Control," Public Service Legislative Studies Program. N.Y. State Assembly.

Berg, C. O. (1963). Middle Atlantic States. *In* "Limnology in North America" (D. G. Frey, ed.), pp. 191–237. Univ. of Wisconsin Press, Madison, Wisconsin.

Birge, E. A., and Juday, C. (1914). A limnological study of the Finger Lakes of New York. *Bull. U.S. Bur. Fish.* **32,** 525–609.

Birge, E. A., and Juday, C. (1921). Further limnological observations on the Finger Lakes of New York. *Bull. U.S. Bur. Fish.* **37,** 211–232.

Brown, E. H., Jr. (1971). Population biology of alewives, *Alosa pseudoharengus,* in Lake Michigan 1949–1970. *J. Fish. Res. Board Can.* **29,** 477–500.

Camp, Dresser, and McKee. Consulting Engineers (1968). "Comprehensive Sewerage Study for Ontario County New York," Proj. WPC-CS-167. Boston, Massachusetts.

Child, D., Oglesby, R. T., and Raymond, L. S. (1971). "Land Use Data for the Finger Lakes Region of New York State," Publ. No. 33, pp. 1–29. Cornell Univ. Water Resour. and Mar. Sci. Cent., Ithaca, New York.

Clark, J. M., and Luther, D. D. (1904). Stratigraphic and paleontologic map of Canandaigua and Naples Quadrangles. *N.Y. State Mus. Bull.* **63**, 3–76.

Cole, G. A. (1975). "Textbook of Limnology." Mosby, St. Louis, Missouri.

Conover, G. S. (1893). "A History of Ontario County, New York." D. Mason and Co., Syracuse, New York.

Cook, R. G. (1931). "A History of Canandaigua Lake." Privately printed.

Crain, L. J. (1975). "Chemical Quality of Ground Water in the Western Oswego River Basin, New York." Prepared for Cayuga Lake Basin and WA-ONT-YA planning boards by U.S. Dept. of the Interior, Geol. Surv. in cooperation with New York State Department of Environmental Conservation.

Day, G. M. (1953). The Indian as an ecological factor in the northeastern forest. *Ecology* **34**, 329–344.

Dethier, B. E., and Pack, A. B. (1965). "The Climate of Geneva, New York." New York State College for Agriculture, Ithaca.

Dillon, D. J. (1974). "A Manual for Calculating the Capacity of a Lake for Development." Water Resource Branch, Ontario Ministry of the Environment. Toronto.

Durfor, C. M., and Becker, E. (1964). Public Water supplies of the 100 largest cities in the United States, 1962. *U.S. Geol. Surv., Water-Supply Pap.* **1812**, 1–34.

Eaton, E. H. (1909). The birds of New York. Vol. 1. *N.Y. State Mus. Mem.* **12**, 1–502.

Eaton, E. H. (1928). The Finger Lakes fish problem. *In* "A Biological Survey of the Oswego River Systems," Supp. 17th Annu. Rep., 1927., pp. 40–66. New York State Conservation Department.

Eaton, S. W., and Kardos, L. P. (1972). The fishes of Canandaigua Lake, 1971. *Sci. Stud.* **28**, 23–39.

Eaton, S. W., and Kardos, L. P. (1973). *Mysis* and some other large invertebrates of Canandaigua Lake, 1972. *Sci. Stud.* **29**, 3–15.

Eaton, S. W., and Moffett, L. J. (1971). A preliminary study of Canandaigua Lake in 1970. *Sci. Stud.* **27**, 69–85.

Eaton, S. W., Eaton, E. H., and Weglinski, P. (1975). Benthos of the deep basin of Canandaigua Lake. *Sci. Stud.* **31**, 27–33.

Fairchild, H. L. (1926). The Dansville Valley and Drainage History western New York. *Proc. Rochester Acad. Sci.* **6**, 217–242.

Fisher, D. W., Isachsen, Y. W., Richard, L. V., Broughton, J. G., Offield, T. W. (1962). "Geologic Map of New York, Finger Lakes Sheet," New York State Mus. and Sci. Ser. Geol. Surv. Map and Chart Ser. No. 5. Albany.

Forest, H. S., Grow, W. C. and Maxwell, T. F. (1971). "Some Sources of Input to Canandaigua Lake and Their Contribution to the Quality of the Environment—1971," Contrib. No. 14. Environmental Research Center at, Geneseo, New York.

Forest, H. S. (1976). Study of Submerged aquatic vascular plants in northern glacial lakes. *Folia Geobot. and Phytotax. Ptaha* (in press).

Gelser, B. M. (1974). Comments on Hersey-Malone water level study report. In letter to Gary L. Fritz dated Feb. 21, 1974.

Gelser, B. M. (1975). From Redskins to WASPS in 15 years. Unpublished manuscript.

Gotham, I. J., III. (1974). Effects of Hurricane Agnes on some physical, chemical, and phytoplanktonic characteristics of Canandaigua Lake, New York. *Sci. Stud.* **30**, 41–63.

Graham, J. J. (1956). Observations on the alewife *Pomolobus pseudoharengus* (Wilson) in freshwater. *Univ. Toronto Stud., Biol. Ser.* **62**, *Publ. Ont. Fish Res. Lab.* **74**, 1–43.

Greeley and Hansen, Engineers. (1970). "Wayne, Ontario and Yates (Counties of N.Y.) Report on Comprehensive Public Water Supply," Appendix Table C-31. New York and Chicago.

Grow, W. C. (1971). "A Water Quality Survey of West River, Naples Creek and the Inlet Area of Canandaigua Lake, Summer and Fall 1970," M. Ed. manuscript mimeo. State Univ. College, Geneseo, New York.

Hall, D. J., and Waterman, G. G. (1967). Zooplankton of the Finger Lakes. *Limnol. Oceanogr.* **12**, 542–544.

Harman, W. N., and Berg, C. O. (1971). The freshwater snails of central New York. *Search* **1**, 2–68.

Hartman, W. L. (1958). Estimation of catch and related statistics of the stream rainbow trout fishery of the Finger Lakes Region. *N.Y. Fish Game J.* **5**, 205–212.

Hartman, W. L. (1959). Biology and vital statistics of rainbow trout in the Finger Lakes Region, New York. *N.Y. Fish Game J.* **6**, 121–178.

Heacox, C., and Stone, U. B. (1945). "Canandaigua Lake 1945 Investigation," Mimeo. rep. for New York State Conservation Department. Albany.

Hedrick, U. P. (1933). "A History of Agriculture in the State of New York." New York State Agricultural Society. Albany.

Henson, E. B. (1954). The profundal bottom fauna of Cayuga Lake. Ph.D. Thesis, Cornell University, Ithaca, New York (Univ. Microfilms, Ann Arbor, Michigan).

Hershey, Malone, and Associates. (1973). "Report on Canandaigua Lake Water Level Study for Ontario County Planning Board." Rochester, New York.

Hutchinson, G. E. (1957). "A Treatise on Limnology," Vol. I. Wiley, New York.

Hutchinson, G. E. (1967). "A Treatise on Limnology," Vol. II. Wiley, New York.

Kardos, L. P. (1975). The effect of vertical distribution and other factors on the alewife *Alosa pseudoharengus* (Wilson) and its prey species in Canandaigua Lake, New York. Ph.D. Thesis, St. Bonaventure University, St. Bonaventure, New York.

Kardos, L. P., and Eaton, S. W. (1972). Zooplankton and benthic fauna of Canandaigua Lake. *Sci. Stud.* **28**, 45–61.

McIntosh, W. H. (1876). "History of Ontario County." Events, Ensign and Events. Philadelphia, Pennsylvania.

Mills, E. L., Forney, J. L., Clady, M. D., and Schaffner, W. R. (1976). "Oneida Lake—A Case History." *In* "Lakes of New York," Vol. 2. Academic Press, New York (in press).

Moore, E. (1928). Introduction *In* "A Biological Survey of the Oswego River System," Suppl. 17th Annu. Rep. 1927, pp. 1–16. New York State Conservation Department. Albany.

Morris, G. H. (1955). Rise of the grape and wine industry in the Naples Valley during the 19th century. M.S. Thesis in American History, Syracuse University, Syracuse, New York (unpublished).

New York State Department of Health. (1955). "Oswego River Drainage Basin Survey Series," Rep. No. 1. Finger Lakes Drainage Basin. Albany.

New York State Forest, Fish, and Game Commission. (1897). 3rd Annual Report. Albany.

Norden, C. R. (1967). Age, growth, and fecundity of the alewife, *Alosa pseudoharengus* (Wilson), in Lake Michigan. *Trans. Am. Fish. Soc.* **96**, 387–393.

Odell, T. T. (1934). The life history and ecological relationships of the alewife (*Pomolobus pseudoharengus* (Wilson) in Seneca Lake, New York. Ph.D. Thesis, Cornell University, Ithaca, New York.

Oglesby, R. T. (1974). "Limnological Guidance for Finger Lakes Management," Tech. Rep. No. 89. Cornell Univ. Water Resour. and Mar. Sci. Cent., Ithaca, New York.

Oglesby, R. T., Hamilton, L. S., Mills, E., and Willing, P. (1973). "Owasco Lake and its Watershed," Report to Cayuga County Planning Board and the Cayuga County Environmental Management Council.

Oglesby, R. T., Schaffner, W. R., and Mills, E. L. (1975). "Nitrogen, Phosphorus and Eutrophication in the Finger Lakes," Tech. Rep. No. 94. Cornell Univ. Water Resour. and Mar. Sci. Cent., Ithaca, New York.

Pearson, C. S., and Cline, M. G. (1958). "Soil Survey of Ontario and Yates Counties," Soil Survey. U.S. Department of Agriculture and Cornell Univ. Exp. Stn. US Govt. Printing Office, Washington, D.C.

Peck, N. H., Vittum, M. T., and Gibbs, G. H. (1968). Growing season weather at Geneva, New York. 1953–1967. N.Y. State Agric. Exp. Stn., Cornell University. Bull. 822, 3–26.

Piampiano, R. J. (1973). "Compendium of Laws Affecting Use of Canandaigua Lake and its Watershed." Canandaigua Lake Pure Waters, Ltd., New York.

Pike, E. G. (1975). "Canandaigua Lake Study," Part IV, Vol. 11, pp. 1–8. Delta Laboratories, Inc.

Presutti, L. (1973). "A Preliminary Study of the Phytoplankton of Canandaigua Lake," Senior Res. Proj. St. Bonaventure University, St. Bonaventure, New York.

Rawson, D. S. (1955). Morphometry as a dominant factor in the productivity of large lakes. Verh. Inst. Ver. Theor. Angew. Limnol. 12, 164–175.

Ritchie, W. A. (1965). "The Archeology of New York State." Natural History Press, Garden City, New York.

Rothschild, B. J. (1962). The life history of the alewife, Alosa pseudoharengus (Wilson) in Cayuga Lake, New York. Ph.D. Thesis, Cornell University, Ithaca, New York.

Schultz, P. E. (1967). The grape and wine industry of the Naples Valley of New York: A geographic interpretation. M.A. Thesis, University of Oklahoma, Norman.

Simpson, R. B. (1942). Studies in the geography of population change. Canandaigua Lake Region, New York. Proc. Rochester Acad. Sci. 8, 49–121.

Stout, N. J. (1959). "Atlas of Forestry in New York." State University College of Forestry, Syracuse, New York.

Tomaka, D. (1973). Growth rate of Perca flavescens in Canandaigua Lake. Unpublished. Res. St. Bonaventure University, St. Bonaventure, N.Y.

United States Corps of Engineers. (1967). "Flood Plain Information Report." Canandaigua Lake, Buffalo District, New York. (Prepared for Oswego River Basin Regional Planning Board.)

United States Department of the Interior. (1974). "Quality of Public Water Supplies of New York." US Govt. Printing Office, Washington, D.C.

von Engeln, O. D. (1962). "The Finger Lakes Region: Its Origin and Nature." Cornell Univ. Press, Ithaca, New York.

Wagner, F. E. (1928). Chemical Investigation of the Oswego watershed. In "A Biological Survey of the Oswego River System," Suppl. 17th Annu. Rep. 1927, pp. 108–139. New York State Conservation Department. Albany.

Webb, D. A. (1973). Daily and seasonal movements and food habits of the alewife in Indiana waters of Lake Michigan near Michigan City, Indiana in 1971 and 1972. M.S. Thesis, Ball State University, Muncie, Indiana.

Wiegand, K. M., and Eames, A. J. (1926). The flora of the Cayuga Lake basin, New York. N.Y. Agric. Exp. Stn., Ithaca, Mem. 92, 1–491.

Wilcox, G. (1975). Seasonal variation of phytoplankton in Canandaigua Lake from May to November 1974. M.S. Thesis, St. Bonaventure University, St. Bonaventure, New York.

Williams, A. (1974). "A Study of Phytoplankton of Canandaigua Lake from July to October 1973," Senior Res. St. Bonaventure University, St. Bonaventure, New York.

Winship, T. R. (1973). "Report of the Assistant Watershed Inspector (Canandaigua Lake) Summer 1973." Canandaigua.

Woldt, A., and Gavagon, J. E. (1970). DDT testing of lake fisheries continues. *N.Y. State Conserv.* **25,** 28–29.

Wright, T. D. (1969). Plant Pigments (Chlorophyll *a* and Phaeophytin). *In* "Ecology of Cayuga Lake and the Proposed Bell Station" (Oglesby, R. T. and Allee, D. J., eds.) Publ. 27. Cornell Univ. Water Resour. and Mar. Sci. Cent., Ithaca, New York.

Youngs, W. D., and Oglesby, R. T. (1972). Cayuga Lake: Effects of exploitation and introductions on the salmonid community. *J. Fish. Res. Board Can.* **29,** 787–794.

Limnology of Eight Finger Lakes:
Hemlock, Canadice, Honeoye,
Keuka, Seneca, Owasco,
Skaneateles, and Otisco

W. R. Schaffner and R. T. Oglesby

INTRODUCTION

The 11 Finger Lakes of New York occupy a series of long narrow basins trending in a north–south direction. The valleys are preglacial in origin, but have been profoundly altered by the invasion of ice (Birge and Juday, 1914).

The ridges that divide the valleys rise to a maximum elevation of about 600 m above mean sea level (a.m.s.l.) and 300–400 m above the lake surfaces. In general, the ridges are lower at the northern ends and the valleys are wider and less steep sided (Birge and Juday, 1914).

The lakes themselves vary considerably in altitude. Canadice, with an elevation of about 340 m (a.m.s.l.), is the highest, and Cayuga the lowest (120 m). The bottoms of two of the lakes (Cayuga and Seneca) extend below sea level (Fig. 1, Table 1).

The shores of the lakes are smooth and regular, and, in general, have a steep slope. Shoreline irregularities that do occur are small and are the result of flat deltas built by tributary streams and wave action (Birge and Juday, 1914). Most of the large tributaries enter at the southern ends of the lakes.

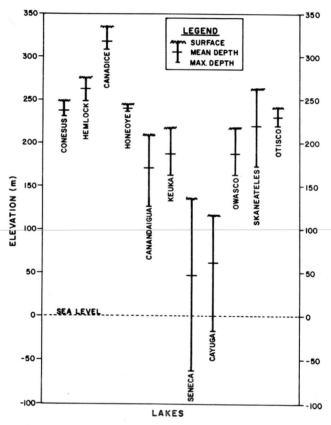

Fig. 1.   Surface elevation, mean depth, and maximum depth of the Finger Lakes.

TABLE 1

Morphometry of the Finger Lakes[a]

Lake	Location[b] Lat.	Long.	Elev. (m)	Surface area (km²)	Length (km)	Width Max. (km)	Width Mean (km)	Depth Max. (m)	Depth Mean (m)	Vol. (× 10⁶ m³)	Mean Slope %	Mean Slope deg min	Development Shore	Development Volume
Conesus	42°50′04″N	77°42′18″W	249.0	13.67	12.6	1.34	1.06	18.0[c]	11.5	156.83[c]	2.32	1 20	0.71	1.68
Hemlock	42 46 39	77 36 59	275.8	7.2[d]	10.8	0.80	0.70	27.5	13.6	105.89	7.8	4 28	—	—
Canadice	42 44 27	77 34 20	334.0	2.6	5.1	0.62	0.51	25.4	16.4	42.6	6.2	3 33	2.05	1.04
Honeoye	42 47 00	77 30 42	245.0	7.05	6.6	1.42	1.06	9.2	4.9	34.81	1.08	0 37	1.45	1.62
Canandaigua	42 52 30	77 16 20	209.7	42.3	24.9	2.44	1.70	83.5	38.8	1640.1	7.0	4 00	2.48	1.39
Keuka	42 39 22	77 03 40	217.9	47.0	31.6	3.32	1.15	55.8	30.5	1433.7	7.8	4 28	4.58	1.64
Seneca	42 52 06	76 56 26	135.6	175.4	56.6	5.20	3.10	198.4	88.6	15539.5	9.0	5 08	2.74	1.41
Cayuga	42 56 51	76 44 09	116.4	172.1	61.4	5.60	2.80	132.6	54.5	9379.4	5.2	2 58	3.35	1.23
Owasco	42 54 12	76 32 34	216.7	26.7	17.9	2.10	1.49	54.0	29.3	780.7	4.4	2 31	2.27	1.63
Skaneateles	42 56 42	76 25 47	263.0	35.9	24.2	3.25	1.48	90.5	43.5	1562.8	8.4	4 48	2.45	1.44
Otisco	42 54 16	76 18 47	240.2	7.6	8.7	1.22	0.89	20.1	10.2	77.8	2.3	1 21	2.04	1.52

[a] All information, unless noted otherwise, from Birge and Juday (1914).

[b] Greeson and Williams (1970).

[c] Maximum depth estimated to be 20.2 m as compared to 18.0 m given in Birge and Juday (1914). This may account for differences in volumes calculated, as Birge and Juday give a value of 13.4 km².

[d] Birge and Juday (1914) give this value which is correct for the lake after it was converted to a reservoir, but depth soundings were only available from the Canal Board Map of 1849.

Although similar in form and topography, the lakes differ widely in area and depth (Table 1), and as a result have been divided into two categories consisting of six major (Canandaigua, Keuka, Seneca, Cayuga, Owasco, and Skaneateles) and five minor (Conesus, Hemlock, Canadice, Honeoye, and Otisco) lakes, eight of which will be considered in this paper.

The larger lakes take in and give out enormous amounts of heat during the course of the year. Heat is absorbed by the water during the spring, and liberated in autumn, a factor that produces a considerable effect on the local climate, and which is intensified by the narrow steep valleys. The result is that frosts are delayed in autumn, and in the spring the cold water chills the valley air so that vegetation does not start to grow until the danger of a killing frost has passed. These conditions make the slopes of the valleys particularly well adapted for fruit raising, as is evidenced by the many vineyards and orchards found in the basins.

Historically, the region has been occupied by native Americans since the retreat of the Pleistocene glaciers some 9000 years ago. The first group to leave a lasting effect in the area were the Iroquois, who are thought to have arrived sometime during the thirteenth or fourteenth century. They held dominion over a large portion of upstate New York until the European conquest in the 1700's. Three member tribes of the Confederation of the Five Nations (the Iroquois) lived in the Finger Lakes region. They were the Senecas, Cayugas, and Onondagas (Rayback, 1966).

Permanent settlement by the white man did not occur until after the American Revolution. All of the area had at least the beginnings of colonization by 1809 (Rayback, 1966). As transportation improved, more people arrived, and by the 1880's almost all of the tillable land was in agriculture of one form or another. Since that time, there has been a steady decrease in agricultural land use due to the abandonment of farms.

NATURE OF THE BASINS

Geological Description

The most authoritative description of the morphogenic processes that shaped the Finger Lakes and their basins is that of von Engeln (1961). This has been summarized by Oglesby (1976), and that synopsis is recapitulated here for the Finger Lakes as a whole.

For a period of 325 million years during the Paleozoic age the Finger Lakes region was part of a vast inland sea. During this time evaporative processes, the precipitation of dissolved material, and sedimentation of particles resulted in the accumulation of bed after bed of sand, mud, lime, and salt. Eventually these materials were compressed into rocks with a total depth of some 2400 m (8000 ft). As a result, Upper Devonian sandstones and shales are to be found in the southern portion of the basins, changing to Middle Devonian limestones northward (Fig. 2a–c).

About 200 million years ago the land was uplifted. At this time drainage was to the south through the Susquehanna River system. Over the next 100 million years the unlifted land was gradually eroded into a peneplain which was then disrupted by additional uplifting. Post peneplain headwater erosion carved deep north–south gaps in the landscape.

About one million years ago this pattern of fluvial cutting was interrupted by the advent of the Ice Age. Great masses of ice ground and gouged their way into a large funnel created by the low relief of land to the north of the lakes, giving way to steep-sided valleys toward the south. Twice glaciers advanced in this fashion with a long interglacial period between. The first glacial invasion caused major modifications in the rock topography, while the second brought about great changes in the aspect of the countryside.

Postglacial time in the Finger Lakes region extends over a period of about the last 9000–10,000 years. During this time, sculpturing by running water has brought about notable landscape changes with the cutting of the many scenically famous rock gorges that are a characteristic feature of the region.

During the last period of uplift, rock formations in the Finger Lakes region were generally given a moderate inclination. The resultant cuestas, with their escarpment profiles accentuated by the downcutting of streams through the less resistant rock strata, are common landscape features of the area. Three major rock formations were of special importance in determining postglacial topography. From north to south these were the Onondaga limestone, 30–45 m (100–150 ft) thick; the shallow Tully limestone, 5–6 m (15–20 ft) in thickness (Fig. 3); and the Portage escarpment composed of sedimentary shales and sandstones. The latter, though not delineated by a sharp crest, is a dominant relief element in the topography of the region, and its front characteristically consists of a steep slope 275–300 m (900–1000 ft) high.

The most dramatic landform features in the Finger Lakes basins are the result of glacial action, and this is particularly true of the depressions in which the lakes themselves lie. Von Engeln describes these as true rock-scoured basins, citing their shapes and the fact that the bottoms of Cayuga and Seneca (Fig. 1) are below sea level as evidence. The east–west tributaries, lying athwart the direction of major ice flows, were excavated to lesser depths. As a result, they were left "hanging" far above the main valleys and now cascade into them through series of scenically spectacular, although individually not very high, waterfalls.

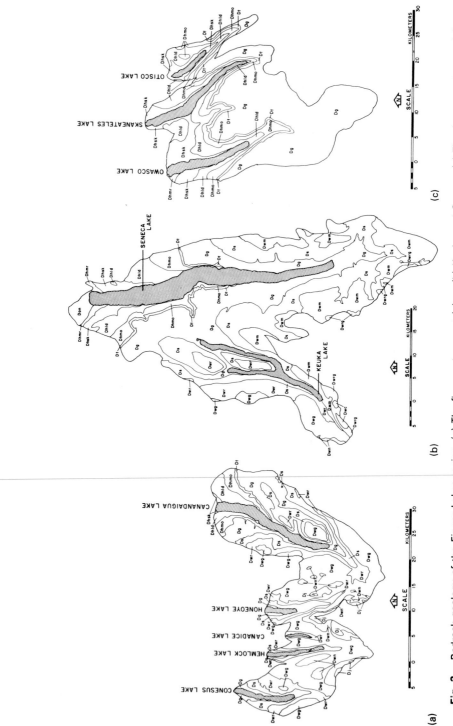

Fig. 2. Bedrock geology of the Finger Lakes region. (a) The five westernmost lakes. (b) Keuka and Seneca. (c) The three easternmost lakes. Redrawn from a map by Rickard and Fisher (1970). Geological formation symbols are identified in the key to map symbols tabulated on p. 319.

Fig. 2.   Map Symbols

Map symbol	Geological formations
West Falls group	
Dwn	Nunda Formation—siltstones, mudstones, and shales, calcareous siltstone concretions
Dwg	West Hill Formation—siltstones and shales; Garden Formation-shales, siltstones, mudstones; calcareous siltstones
Dwr	Lower Beers Hill—shales, mudstones, calcareous siltstone
Dwrg	Gardeau Formation (as above); Roricks Glen Shale— shales, calcareous siltstone
Dwm	Beers Hill Shale, Grimes Siltstone, Dunn Hill, Millport, and Moreland shales
Sonyea group	
Ds	Cashaqua Shale replaced eastwardly by Rye Point Shale, Rock Stream ("Enfield") Siltstone, Middlesex Shale replaced eastwardly by Sawmill Creek Shale, Johns Creek Shale, Montour Shale, the last three replaced eastwardly by Kattel Formation—shales and siltstones
Genesee Group and Tully Limestone	
Dg	West River Shale, Genundewa Limestone, Penn Yan Shale, Geneseo Shale, North Evans Limestone; all of above except Geneseo replaced eastwardly by Ithaca Formation—sandstones, siltstones, shales, mudstones, and Sherburne siltstone
Hamilton group	
Dhmo	Moscow Formation—in west shales and then interbedded limestone, crinoidal limestone; these replaced eastwardly by Cooperstown shales and siltstones and Portland Point Limestone; in east, "Manorhill" shales and sandstones
Dhld	Ludlowville Formation—shales and limestones, Wanakah and Ledyard shales and thin limestones, Centerfield limestone; replaced eastwardly by King Ferry and other shales and Stone Mill calcareous sandstone
Dhsk	Skaneateles Formation—Levanna shales, limestones; replaced eastwardly by Butternut shales, Pompey shale and mudstone, Delphic Station shale and sandstones, Mottville limestone, and calcareous sandstones
Dhmr	Marcellus Formation—Oatka Creek calcareous shales; replaced eastwardly by Cardiff shales, Chittennango shales, Cherry Valley limestone, Union Springs shales

(Continued)

Fig. 2. *(Continued)*

Map symbol	Geological formations
Onondaga Limestone and Ulster group	
Do	Oriskany sandstone—calcareous sandstone, siliceous limestone (replaced eastwardly by Glenerie Formation)
Java group	
Dj	Hanover shale—calcareous shales and mudstones, calcareous nodules and concretions; replaced eastwardly by Wiscoy Formation—siltstones, mudstones, and fine sandstones; at base, shales

The first of the two ice advances was responsible for most of the great glacial erosion phenomena of the region. While glacial erosion has played a part in the formation of the major gorges, the principal sculpturing agent has been downcutting by glacial meltwater and streams during the interglacial and postglacial periods. The second glaciation filled the interglacial gorges with debris, and this material is still being downcut in places.

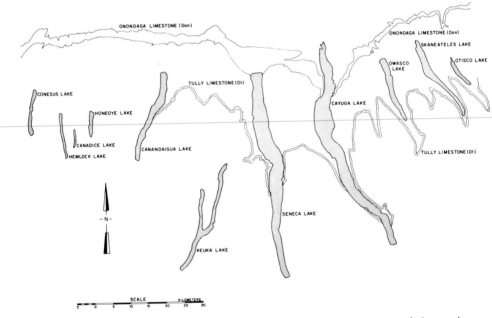

Fig. 3. Location of the Tully and Onondaga limestone escarpments relative to the Finger Lakes (Oglesby *et al.*, 1975).

As the glaciers retreated northward their meltwater formed a series of high-level proglacial lakes bordered by the ice front on the north and the highlands that surround the head basins to the south. In the Cayuga basin, for example, a series of 11 such lakes were formed during the 1000 or so years required for the last ice sheet to retreat from the area. Their emphemeral history is marked by successive lines of lake sediment deposits and even occasional wave cut cliffs. The presence or absence of glacial marine relict species (Dadswell, 1974; Chamberlain, 1975) provides evidence that of the Finger Lakes only Canadice was at an elevation above the level of a postglacial lake that must have been continuous throughout the region.

The retreat of the second glaciation was retarded for several hundred years a short distance south of the heads of the Finger Lakes. The massive end-moraines formed by the deposition of glacial debris insured that, once northern routes to the Mohawk and St. Lawrence drainage were passed by the retreating ice, the direction of flow of the Finger Lakes and their precursors would be to the north rather than south to the Susquehanna.

Soils

Debris from the last glaciation was generally spread over the Finger Lakes region. In addition to the massive end-moraines found at the southern ends of the basins, there is a basal layer of ground moraine, referred to as glacial till, boulder clay, or hardpan, which is normally covered by other types of deposited materials, some of which have maintained their integrity. Postglacial soils have been derived from others.

In general, the northern portions of the basins contain moderately coarse-textured soils with calcareous substrata, giving way southwardly to complex assemblages of more acid, less well-drained types (Fig. 4a,b,c). The figures are derived from a working map (1:250,000) of New York soil series prepared by Cline and Arnold (1970). Their work is continuing and so details will undoubtedly be changed in the future.

Population Distribution

Estimations based on 1970 data indicate that almost 200,000 individuals reside in the Finger Lakes drainage basins. This represents an increase of slightly more than 50% over the last 40 years. (Table 2 presents a breakdown by lake basin.) This figure is reasonably representative of the individual drainage basins, except for Canadice, which has experienced a twofold increase in population. In the case of Canadice this represents an increase of only 196 individuals.

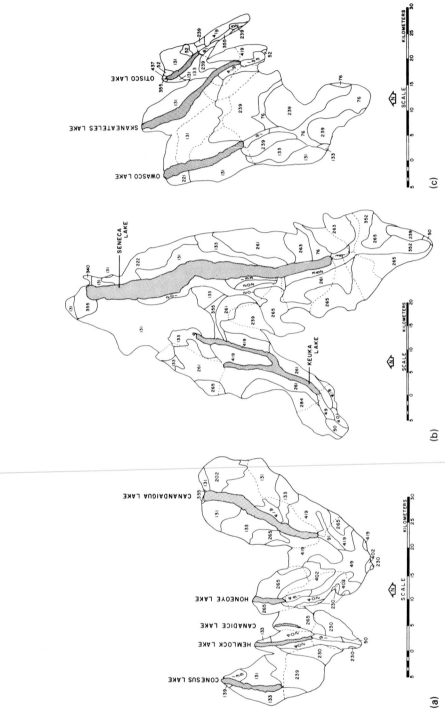

Fig. 4. Soil assemblages in the Finger Lakes region. (a) The five westernmost lakes. (b) Keuka and Seneca. (c) The three easternmost lakes. Soil types are identified in the key to map symbols (from Cline and Arnold, 1970) tabulated on p. 323.

Fig. 4. Map Symbols

Map symbol	Soil association	Other associated soils

I. Areas dominated by medium and moderately coarse-textured mesic soils with sand or gravel substrata

    a. Dominantly well- and moderately well-drained soils not subject to flooding

49	Howard	Phelps, Fredin, Halsey
50	Howard–Chanango	Phelps, Castile, Halsey
52	Palmyra	Atherton, Groton, Homer, Phelps

    b. Dominantly well-drained soils in complex pattern with soils having compact substrata

76	Howard–Langford–Valois	Erie, Halsey, Madrid

    c. Dominantly somewhat poorly drained soils not subject to stream flooding

91	Sloan–Teel	Hamlin, Wayland, Cohoctah

II. Areas dominated by medium and moderately coarse-textured soils with compact loamy substrata

    a. Dominantly well- and moderately well-drained soils with calcareous substrata

131	Honeoye–Lima	Kendaia, Lyons
133	Lansing–Conesus	Appleton, Kendaia, Lyons
139	Ontario–Hilton	Camillus, Madrid, Newstead, Sun
144	Lima–Kendaia	Honeoye, Lyons, Nallis

III. Areas dominated by moderately fine and fine-textured soils with compact, clayey substrata

    a. Dominantly moderately well- and somewhat poorly drained soils with compact substrata

201	Cazenovia–Ovid	Romulus, Alden, Lima
202	Darien	Aurora, Danley, Ilion

    b. Dominantly moderately well- and somewhat poorly drained soils in complex pattern with moderately deep soils over bedrock

221	Cazenovia–Wassaic	Ovid, Romulus, Aurora
222	Darien–Angola	Aurora, Danley, Varick

IV. Areas dominated by medium and moderately coarse-textured mesic soils with fragipans and compact substrata

    a. Dominantly well- and moderately well-drained soils

230	Bath–Lordstown	Mardin, Volusia, Chippewa
239	Langford–Erie	Canaseraga, Ellery, Alden

    b. Dominantly somewhat poorly drained soils

261	Erie–Langford	Ellery, Alden, Papakating
263	Volusia–Lordstown	Chippewa, Mardin, Arnot
265	Volusia–Mardin–Lordstown	Chippewa, Arnot, Tuller

    c. Dominantly well- and moderately well-drained soils in complex patterns with moderately deep and shallow soils over bedrock or with rockland

284	Mardin–Volusia–Lordstown	Bath, Chippewa, Arnot

V. Areas dominated by moderately coarse to fine-textured stone-free soils with silty or clayey substrata

    a. Dominantly well- and moderately well-drained soils with silty substrata

340	Collamer–Dunkirk	Niagara, Canandaigua, Schoharie

(Continued)

Fig. 4. *(Continued)*

Map symbol	Soil association	Other associated soils
b.	Dominantly well- to moderately well-drained soils with clayey substrata	
352	Hudson–Rhineback	Madalin, Fonda, Collamer
355	Schoharie–Odessa	Cayuga, Churchville, Lakemont
c.	Dominantly somewhat poorly and poorly drained soils	
361	Canandaigua–Collamer	Canadice, Niagara, Alden
d.	Dominantly moderately well- to poorly drained soils in complex pattern with soils having compact substrata or rockland	
383	Hudson–Cayuga	Churchville, Rhineback, Madalin
VI.	Areas dominated by moderately deep and shallow soil with bedrock substrata	
a.	Dominantly well-drained, moderately deep soils	
402	Lordstown–Manlius	Mardin, Arnot, Volusia
b.	Dominantly well-drained, moderately deep soils in complex pattern with deep soils having compact substrata	
419	Lordstown–Valousia–Mardin	Arnot, Chippewa, Bath
c.	Dominantly well-drained shallow soils	
437	Benson–Wassaic	Hilton, Appleton, Lyons
VII.	Areas Dominated by organic Soils	
451	Muck and Peat, undifferentiated	Carlisle, Edwards, Wallkill

Population densities, averaged over the whole basin, vary considerably from lake to lake (Table 2). Canadice has the lowest concentration with about 9 individuals/km², and Seneca the highest with about 60 individuals/km². As might be expected, the populations are not uniformly distributed throughout the basins. The highest densities are found in cities and towns, and along the lake shores and tributaries.

Land Use

The major land use categories for each of the lake basins are presented in Table 3. In general, the three westernmost lakes (Hemlock, Canadice, and Honeoye) are dominated by forests (Fig. 5). In the remaining five the northern portions of their basins are in agriculture (Fig. 6), while to the south, drainage areas are primarily dominated by forests (Fig. 5). The major tributaries to all of the lakes pass through this latter region (Fig. 7). There is a transition zone between the forest and agriculturalized areas that contains substantial amounts of land classified as inactive agriculture (Fig. 8). This reflects a pattern of decreasing agricultural usage in the central portion of the Finger Lakes region (Child *et al.*, 1971).

TABLE 2

Total Population (1930 and 1970) and Population Density (1970) in the Various Drainage Basins[a]

	Total no. of individuals		% increase	Population density (individ. km⁻²)
Lake	1930	1970		
Hemlock	836	1186	42	12.3
Canadice	95	291	206	9.2
Honeoye	823	1276	55	13.4
Keuka	9500	12125	28	30.0
Seneca	44561	69559	56	58.9
Owasco	8677	13198	52	28.1
Skaneateles	2515	3931	56	26.0
Otisco	1055	1536	46	16.4

[a] From Anonymous (1932, 1970a,b).

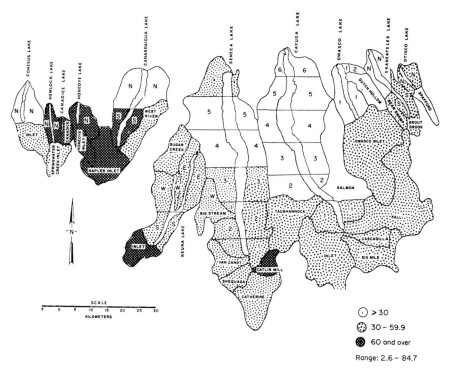

○ > 30

◉ 30 – 59.9

● 60 and over

Range: 2.6 – 84.7

Fig. 5. Percent of total land use in forest in the Finger Lakes region. Numbers 1–5 describe composite watersheds, and are discussed in the source publication (Child et al., 1971).

TABLE 3

Total Drainage Area and That in Various Use Categories (Expressed in km² and as a Percentage of the Total) for the Various Lake Basins[a]

Lake	Active agriculture		Inactive agriculture		Forest		Residential		Total drainage area
	km²	%	km²	%	km²	%	km²	%	km²
Hemlock	24.2	25	8.5	9	62.8	65	0.7	1	96.2
Canadice	4.5	14	1.4	4	25.7	81	0.2	1	31.8
Honeoye	9.1	10	3.6	4	80.7	85	1.6	2	95.0
Keuka	132.2	33	59.7	15	198.9	49	13.8	3	404.6
Seneca	501.5	42	165.5	14	476.5	40	37.1	3	1180.6
Owasco	227.6	48	70.5	15	164.4	35	7.5	2	470.0
Skaneateles	66.6	44	24.3	16	55.2	36	5.3	4	154.0
Otisco	43.4	46	14.7	16	33.6	36	2.1	2	93.8

[a] From Child et al. (1971).

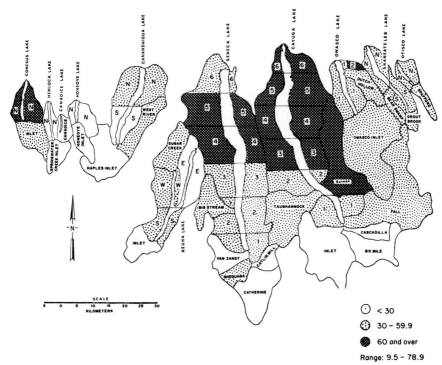

○ < 30

◉ 30 – 59.9

● 60 and over

Range: 9.5 – 78.9

Fig. 6.   Percent of total land use in agriculture in the Finger Lakes region. Numbers 1–
5 describe composite watersheds, and are discussed in the source publication (Child *et al.*,
1971).

As might be expected from the preceding discussion, the economic viability of farm areas exhibits a pattern ranging from a low potential in the
southern part of the region to moderate to high viability in the northern
parts of the basins (Conklin and Linton, 1969). This pattern is a reflection
of the fact that the soils in the northern portions of the drainage basins are
better buffered as a result of underlying lime deposits (Olsen *et al.*, 1969).

Vegetation

The Finger Lakes region was at one time almost completely covered with
forest, and even today this vegetation type is a dominant feature of the landscape in many places. Most of the forests that do exist are in some stage of
recovery from the activities of man during the last 200 years. Much of the
land was farmed, this activity reaching its zenith in the 1880's, and that land
not in agriculture was cut over for its timber. Thus, there are few virgin
stands of forest to be found today (DeLaubenfels, 1966).

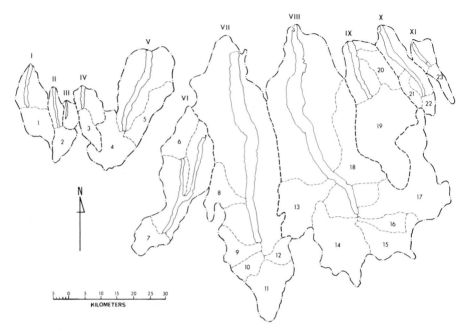

Fig. 7. Location of major stream drainage basins in the Finger Lakes region (adapted from Child *et al.,* 1971). Unnumbered basins are composits of numerous smaller drainage areas. Lakes and stream basins are identified in the tabulation on p. 329.

Four natural vegetation zones are to be found in the region: the northern hardwoods, elm–red maple–northern hardwoods, oak–northern hardwoods, and pine–oak–northern hardwoods (Fig. 9).

As their names imply, these various forest types are best described by comparing them with the northern hardwoods, a group of trees that occupies all of the state except the extreme southeast and the higher areas in the Adirondack Mountains (DeLaubenfels, 1966).

The northern hardwoods are made up of a large, nonuniform group of trees dominated by beech (*Fagus grandifolia* Ehrh.) and sugar maple (*Acer saccharum* Marsh.). Basswood (*Tilia americana* L.), white ash (*Fraxinus americana* L.), and black cherry (*Prunus serotina* Ehrh.) are regular associates in warmer areas, with yellow birch (*Betula lutea* Michx. f.) becoming a codominant in cooler regions. Hemlock [*Tsuga canadensis* (L.) Carr,], white pine (*Pinus strobus* L.), and white cedar (*Thuja occidentalis* L.) are abundant, but not evenly distributed. Hemlock, at times occurring in nearly pure groves, is found on moist, shady slopes and in ravines. This tree was once more common, but was severely cut over for its wood and bark (a source of tannin) in the nineteenth century. White cedar may be

Fig. 7. (Map Symbols)

Lake	Map symbol	Stream basin
Conesus	I-1	Inlet
Hemlock	II-2	Springwater Creek
Canadice	III	
Honeoye	IV-3	Inlet
Canandaigua	V-4	Naples Inlet
	V-5	West River
Keuka	VI-6	Sugar Creek
	VI-7	Inlet
Seneca	VII-8	Big Stream
	VII-9	Van Zandt Creek
	VII-10	Shequaga Creek
	VII-11	Catherine Creek
	VII-12	Catlin Mill Creek
Cayuga	VIII-13	Taughannock Creek
	VIII-14	Inlet
	VIII-15	Six Mile Creek
	VIII-16	Cascadilla Creek
	VIII-17	Fall Creek
	VIII-18	Salmon Creek
Owasco	IX-19	Inlet
	IX-20	Dutch Hollow Brook
Skaneateles	X-21	Bear Swamp Creek
	X-22	Grout Brook
Otisco	XI-23	Spafford Creek

found along with hemlock in abandoned fields and on poorly drained limey soils. Alder (*Alnus* spp.) and larch [*Larix laricina* (Du Roi) K. Koch.] are to be found on wet soils. White pine is an early colonizer of abandoned fields (DeLaubenfels, 1966).

The elm–red maple–northern hardwoods zone is characterized by the relative abundance of elm (*Ulmus americana* L.), where not killed by disease, and red maple (*Acer rubrum* L.). Oak (*Quercus* spp.) and northern hardwoods are present, but in reduced numbers. This assemblage of trees is found on the poorly drained soils of the Ontario Plain north of the Finger Lakes region and in abandoned fields, its presence being due to disturbed edaphic conditions. Oaks would normally be common, but have been reduced because the well-drained soils that they prefer are still in agriculture in the form of crops and pasture (DeLaubenfels, 1966).

The oak–northern hardwoods zone is a transition area where these two groups of trees intermingle or alternate. This assemblage is common in the nonagriculturalized portions of the Finger Lakes region. Slope is important in determining sharp, local variations. South and southwest facing slopes, as a result of receiving more sunlight, may support almost pure stands of oak, which require higher temperatures than the northern hardwoods. Hickory (*Carya* spp.) may be mixed in with the oak. White pine, red cedar (*Juniperus virginiana* L.), white ash, hawthorn (*Crataegus* spp.), and locust (*Robinia pseudoacacia* L.) tend to invade abandoned pastures in this zone. American elm can also be found where the cover is open, and in poorly drained areas along with red maple (DeLaubenfels, 1966).

The white pine–oak–northern hardwoods zone is composed of white pine and oak with northern hardwoods (Stout, 1958). This vegetation type is limited to the southern extent of the Finger Lakes region.

In conclusion, even though the area's forests have been greatly affected, and at one time almost entirely removed by the activities of man, they are now in the process of expanding as marginal farmland is abandoned and gradually invaded by trees. Timber is still harvested in the region but on a selective and nonintensive basis.

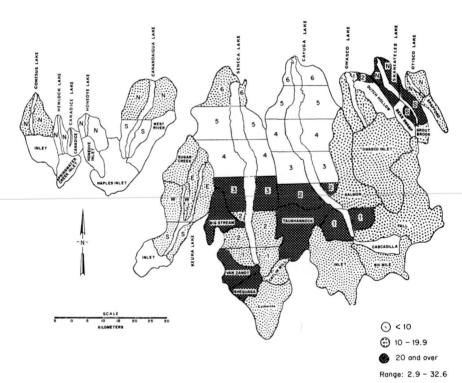

Fig. 8. Percent of total land use in inactive agriculture in the Finger Lakes region. Numbers 1–5 describe composite watersheds, and are discussed in the source publication (Child *et al.*, 1971).

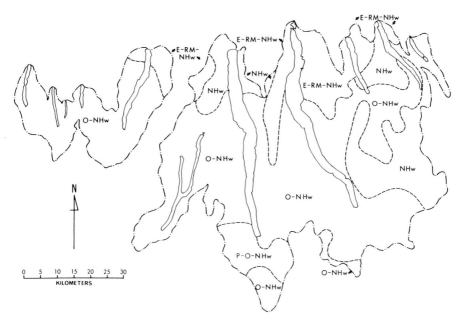

Fig. 9. Major potential vegetation zones in the Finger Lakes region. Redrawn from a map of New York in Stout (1958). Zone symbols are identified in the tabulation below.

Fig. 9.   Map Symbols

Map symbol	Potential vegetation zones
NHw	Northern hardwoods
E-RM-NHw	Elm-red maple-northern hardwoods
O-NHw	Oak-northern hardwoods
P-O-NHw	White pine-oak-northern hardwoods

Point Source Pollutants

Point source pollutants are introduced into several of the lakes being considered. Wineries located in the Keuka Lake basin impose a sizable organic load on some of the streams entering the lake. This is especially so at Hammondsport. Sewage from some shoreline cottages enters the lake directly (Anonymous, 1975a).

In the Seneca Lake basin, Watkins Glen and Mountour Falls contribute domestic waste to Catherine Creek, a major tributary. At the northern end, the City of Geneva's primary sewage treatment plant discharges about 7.6 thousand cubic meters per day into the lake, along with some industrial waste and the contents of storm sewers. Industrial waste and degraded

Keuka Outlet water enter Seneca Lake at Dresden on the western shore (Anonymous, 1975a).

The Owasco Inlet has fairly good quality when it enters the lake, even though it receives treated sewage from the villages of Groton (primary) and Moravia (secondary). There are no significant industrial or municipal discharges directly into the lake. Cottage development is heavy along the shore, with some of the dwellings adding their wastes directly to the lake (Anonymous, 1975a).

Although located in a heavily wooded watershed, Honeoye Lake is used extensively for recreation, boating, bathing, and fishing. Lakefront property is privately owned, and highly developed. As a result, water quality has suffered from failing septic systems, oil from power boats, and other causes (Anonymous, 1975a).

Hemlock and Canadice Lakes have served as water supplies for the City of Rochester for more than 100 years. The city owns all of the lakeshore property, and maintains strict control over the watershed areas.

Otisco Lake receives minimal pollution, but does suffer from shoreline erosion resulting from water withdrawal by the Onondaga County Water Authority. Samples taken in 1965 showed that the bottom of the lake was composed almost entirely of silt. During that year the lake experienced a drop in level of 4 m (13 ft) (Anonymous, 1975a). There is concern that this may be a growing problem. Skaneateles Lake receives no significant discharges.

In the not too distant past (the 1920's) a number of waters within and around the periphery of the Finger Lakes region were strongly affected by point source pollution. Owasco Inlet received cheese factory and milk condensary waste, plus that from a typewriter works, both of which were located in Groton. Sewage from the Village of Groton and waste from a milk shipping station at Locke also found their way into the Inlet (Wagner, 1928).

The Keuka Lake Outlet, which flows into Seneca Lake, received waste fibers and dyes as well as various glues and chemicals from paper and woolen mills along its banks (Claasen, 1928). To this pollution burden was added sewage from Penn Yan, cannery waste, and CS_2 from a carbon disulfide factory (Wagner, 1928).

Stream Flow

Surface runoff values based on long-term estimates for the various drainage basins are presented in Table 4. Honeoye has the least amount (0.292 m yr^{-1}) and Owasco the greatest (0.473 m yr^{-1}). Inputs to the lakes are greatest for Seneca (652 $\times$ 10^6 m^3 yr^{-1}). Canadice receives the least amount of water inflow (10 $\times$ 10^6 m^3 yr^{-1}).

TABLE 4

Surface Runoff in the Various Basins[a]

Lake	m yr^{-1}	$\times$ 10^6 m^3 yr^{-1}
Hemlock	0.330	36.6
Canadice	0.325	10.1
Honeoye	0.292	27.8
Keuka	0.305	148
Seneca	0.356	652
Owasco	0.473	255
Skaneateles	0.432	81.6
Otisco	0.381	33.5

[a] From Knox and Nordenson (1955).

Climatology

The Finger Lakes are located in a climatic region characterized by cold snowy winters and warm dry summers. The area is, as a whole, one of the driest in the state. Nonetheless, the district is reasonably well suited for horticulture, and plants with deep root systems, such as grapes, are widely grown (Carter, 1966).

Pertinent climatological factors, as exemplified by long-term averages (Dethier and Pack, 1963a,b) near the center of the region, are as follows: Annual average solar radiation 97.6 $\times$ 10^2 gm cal cm^{-2}, with a maximum of 172.4 $\times$ 10^2 gm cal cm^{-2} in July. Mean precipitation is 87 cm yr^{-1}, with the smallest amounts falling during the December to March period. Winter snow melt commonly occurs in late March or early April. Air temperature is normally distributed around a July maximum (20.5°C) and a January minimum (-4.4°C).

Precipitation based on long-term averages (Dethier, 1966) is presented in Fig. 10. Two regional trends are apparent. The annual average precipitation decreases from east to west. Maximum precipitation occurs at the eastern extent of the area where the long-term annual average is almost 100 cm yr^{-1}. To the west this decreases to about 75 cm yr^{-1}. There is also a less well-defined north–south trend, with the southern extents of the lake basins receiving somewhat less precipitation.

With the exception of Seneca and Cayuga, all of the lakes regularly freeze over, although the central regions of the deeper lakes may remain open during mild winters. Freezing usually occurs during January or early February. Local records indicate that Seneca Lake has been completely frozen over on several occasions during February and March (Birge and Juday, 1914), i.e., in 1855, 1875, 1885, and 1912.

Crain (1974) has computed potential and actual evapotranspiration for the Western Oswego River drainage basin using the method of

Thornthwaite (1948). The calculations are applicable to two of the lakes being considered here. Thornthwaite's method takes into account air temperature, precipitation, duration of sunlight, and soil moisture content. The major problem with the method is that it is correct only for the site where the data were collected and may vary considerably a short distance away (Crain, 1974). Evapotranspiration estimates for four stations are presented in Table 5.

Potential evapotranspiration is the amount of water that will be lost by evapotranspiration if sufficient water is available at all times from precipitation and soil moisture to satisfy demands. The annual average potential evapotranspiration will vary directly at each station as a function of the mean annual temperature, and increases 2.3 cm per degree Celsius. Mean annual air temperature varies inversely with altitude (Crain, 1974).

Examination of Table 5 shows that computations of potential evapotranspiration do not agree well with those for actual evapotranspiration. This is due to the fact that less precipitation is available at lower altitudes in the region. Thus, evapotranspiration demands are not met during the summer months. For the central Finger Lakes region the following generalizations may be made: (1) In areas that lie below about 275 m, mean annual evapo-

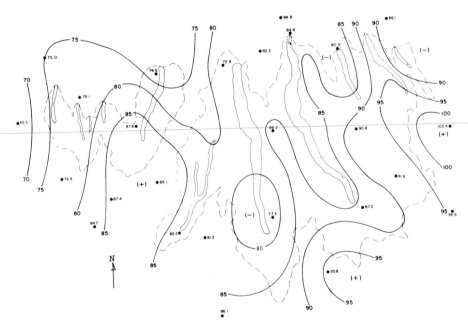

Fig. 10. Regional precipitation based on long-term averages. Individual station averages are located on the map with a filled circle.

TABLE 5

Mean Annual Potential and Actual Evapotranspiration in the Central Finger Lakes[a]

Parameter[b]	Geneva (Alt. 180 m)	Canandaigua (Alt. 220 m)	Penn Yan (Alt. 220 m)	Ithaca (Alt. 290 m)
T°C	9.4	8.9	8.9	8.3
P	83.1	74.9	80.0	87.4
PE	65.5	64.0	64.0	62.7
AE	57.4	54.6	56.1	62.0
SP	25.7	20.3	23.9	25.4

[a] From Crain (1974).

[b] T°C, mean air temperature in °Celsius (long-term average); P, mean precipitation in cm yr^{-1} (long-term average); PE, mean potential evapotranspiration in cm yr^{-1} (computed); AE, mean actual evapotranspiration in cm yr^{-1} (computed); SP, mean water surplus (surplus precipitation) in cm yr^{-1} (computed).

transpiration ranges from 53 to 58 cm. (2) The highest annual water loss occurs at about 290 m where precipitation is in sufficient supply to meet requirements of potential evapotranspiration. (3) At an altitude of about 600 m precipitation meets the needs of potential evapotranspiration, but because of the decrease in temperature due to increased altitude, the mean annual average is only about 51 cm (Crain, 1974). As might be expected from precipitation data, there is an average annual water surplus of from 20 to 50 cm in the central Finger Lakes region (Fig. 17 in Crain, 1975).

LIMNOLOGY

The Finger Lakes were among the first large inland bodies of water in North America to be studied. In 1874–1897 extensive soundings by civil engineering classes from Cornell University provided information for mapping the bathymetry of the six larger lakes. In 1910–1912, and again in 1918 the Wisconsin limnologists Edward Birge and Chancey Juday made brief sampling trips to the Finger Lakes. The lakes were again the sites for scientific study in the late 1920's as part of the work done by groups of scientists carrying out the New York State Biological Survey. It was not until the early years of this decade that comprehensive investigations were again attempted. The following information on Finger Lakes' limnology incorporates the highlights of earlier work but draws most heavily on recent information, much of which is of limited availability in the form of reports and theses.

Physical

Morphometry

Morphometric data were presented in an earlier section in Table 1. Bathymetric maps may be found in Figs. 11–18, and their sources are listed in Table 6. The eight lakes being considered vary widely in size. Canadice has the smallest surface area (2.6 km²) and Seneca the largest (175.4 km²). Honeoye is by far the shallowest with a mean depth of only 4.9 m; Seneca is the deepest of all of the Finger Lakes (mean depth 88.6 m). Seneca Lake's volume is an order of magnitude greater than any of the others (excluding Cayuga), being $15,539.5 \times 10^6$ m³ compared with one-tenth this for Skaneateles, the lake holding the next greatest amount of water. Honeoye has the smallest volume (38.8×10^6 m³).

Water Retention Time

Long-term mean water retention times (Table 7) have been established for the Finger Lakes by a method described in Oglesby et al. (1973). The values thus obtained agree well with annual outflows measured during a year of average precipitation (Oglesby et al., 1975). Honeoye has the shortest retention time (0.8 year) and Seneca the longest (18.1 year).

Temperature Regimes

Seven of the lakes stratify thermally during the warmer months. Honeoye, the shallowest, undergoes transitory stratification during periods of calm. All of the lakes except Seneca are usually stratified from June through September. Stratification in Seneca is usually not well defined until late June or early July, but may persist into November. Mean thermocline

TABLE 6

Sources of Bathymetric Maps

Lake	Surveyed by	Date	Number of soundings	Scale of map
Hemlock	City of Rochester	1849	161	1:14,400
Canadice	City of Rochester	1909	283	1:2,400
Honeoye	N.Y. Conservationist	—	—	1:28,534
Keuka	U.S. Geol. Survey	1884–1888	470	1:40,000
Seneca	Cornell University	1893	405	1:60,000
Owasco	Cornell University	1896–1897	276	1:40,000
Skaneateles	Cornell University	1898	572	1:30,000
Otisco	Cornell University	1897	144	1:24,000

TABLE 7

Water Retention Times (WRT's) for the Finger Lakes

Lake	Drainage area (km^2)	Volume ($\times 10^6$ m^3)	Calculated WRT for long term[a] (yr)	Measured WRT for 1972–1973 water year (yr)	Estimated or measured for long term (yr)
Canadice	31	42.6	4.5	4.0	—
Canandaigua	477	1640.1	7.4	7.8	—
Cayuga	2106	9379.4	9.5	—	12.0
Conesus	231	156.8	1.4	—	—
Hemlock	111	105.9	2.0	1.9	—
Honeoye	95	34.8	0.8	—	—
Keuka	484	1433.7	6.3	6.2	—
Otisco	88	77.8	1.9	—	—
Owasco	539	780.7	—	2.2	3.1
Seneca	1831	15539.5	18.1	—	—
Skaneateles	189	1567.8	17.7	20.3	—

[a] Calculated by assuming that all Finger Lakes have same yield of runoff per unit area of drainage basin as does Owasco, a lake for which over 30 years of good outflow data are available. The following formula is then used:

WRT for lake A = (volume/drainage area) for lake A ÷ (volume/drainage area) for Owasco × WRT for Owasco.

Fig. 11. Bathymetric map of Hemlock Lake. Redrawn from source listed in Table 6 with depth contours in meters.

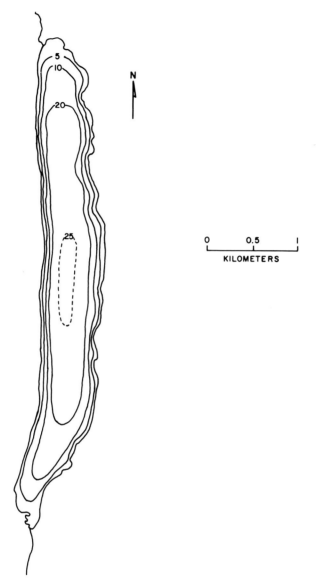

Fig. 12. Bathymetric map of Canadice Lake. Redrawn from source listed in Table 6 with depth contours in meters.

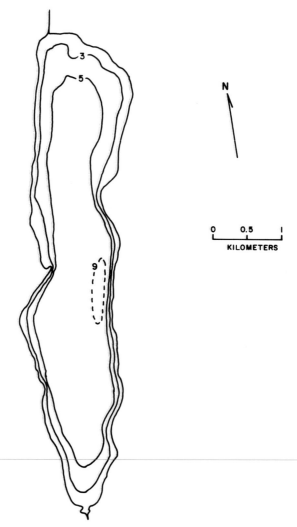

Fig. 13. Bathymetric map of Honeoye Lake. Redrawn from source listed in Table 6 with depth contours in meters.

depths range from 7.7 m (Canadice) to 11.5 m (Seneca) (Table 8). The shallower lakes tend to have warmer bottom waters, and the larger lakes are cooler in the summer and lose heat less rapidly during the winter (Birge and Juday, 1914).

Details of the temporal and spatial distribution of temperatures during 1972 and 1973 were given by Mills (1975) for Hemlock, Owasco, and Skaneateles Lakes. Stewart and Markello (1974) graphically presented the

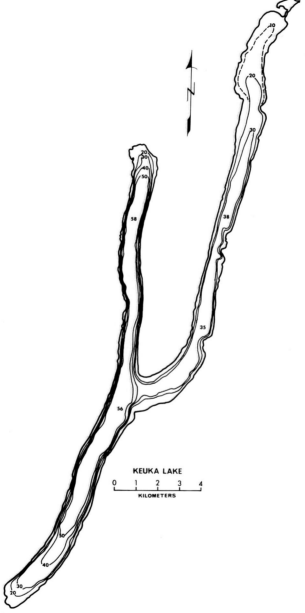

Fig. 14.   Bathymetric map of Keuka Lake. Redrawn from source listed in Table 6 with depth contours in meters.

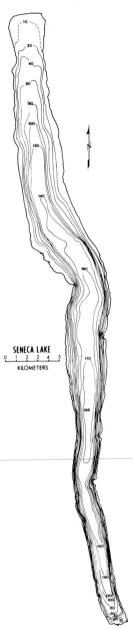

Fig. 15. Bathymetric map of Seneca Lake. Redrawn from source listed in Table 6 with depth contours in meters.

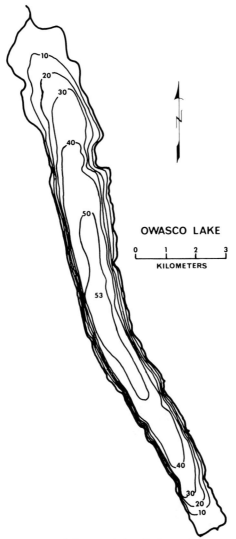

Fig. 16. Bathymetric map of Owasco Lake. Redrawn from source listed in Table 6 with depth contours in meters.

vertical temperature profiles determined for a few occasions during 1969 in Canadice and Honeoye Lakes.

Data presented in Table 9 provide a comparative view for all of the lakes being considered in this publication. The broad interactions of lake depth and size are clearly demonstrated with even this limited amount of information. The August near bottom temperature of Honeoye illustrates the quasi

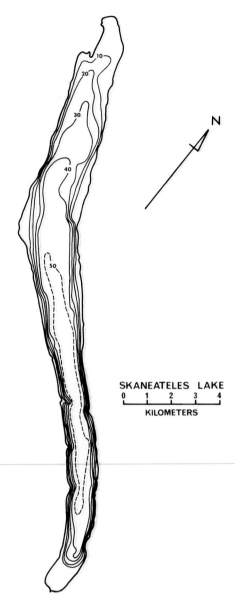

Fig. 17.   Bathymetric map of Skaneateles Lake. Redrawn from source listed in Table 6 with depth contours in meters.

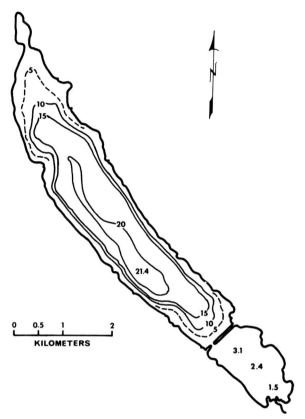

Fig. 18.   Bathymetric map of Otisco Lake. Redrawn from source listed in Table 6 with depth contours in meters.

TABLE 8

Mean Thermocline Depth

Lake	Mean thermocline depth (m)
Hemlock	8.5
Canadice	7.7
Honeoye	Not strat.
Keuka	10.0
Seneca	11.5
Owasco	11.2
Skaneateles	10.7
Otisco	10.0

TABLE 9

Surface and Near Bottom Temperature Measured on a Near Synoptic Basis during April, June, and August of 1973 and Maximum Surface Temperatures for 1972 and 1973[a]

Lake	Calendar day	Temperature (°C)		Calendar day	Max. surface temperature (°C)	
		Surface	Near bottom		1972	1973
Hemlock	114	11.4	7.0	207	27.2	
	199	22.8	9.2	211		26.0
	240	24.2	9.1			
Canadice	114	10.0	6.0		insufficient data	
	199	23.0	8.0			
	240	24.5	8.5			
Honeoye	114	12.5	10.2		insufficient data	
	199	24.0	22.4			
	240	25.0	22.0			
Keuka	115	7.6	4.9		insufficient data	
	199	25.0	6.0			
	240	23.5	5.7			
Seneca	118	5.6	4.0		insufficient data	
	199	23.5	4.0[b]			
	240	23.5	5.5[c]			
Owasco	113	8.5	4.8	202	26.0	
	198	22.8	6.8	190		24.4
	240	23.5	6.4			
Skaneateles	113	6.0	4.5	202	27.0	
	198	22.3	6.0	226		23.5
	240	22.8	4.8			
Otisco	113	11.0	6.5		insufficient data	
	198	23.0	10.0			
	240	23.0	10.5			

[a] As determined by weekly measurements in Hemlock, Owasco, and Skaneateles.
[b] 50 m.

thermal equilibrium between the entire volume of this shallow lake and the atmosphere. Lakes of comparable surface area and length but different depths (e.g., Owasco and Skaneateles, Honeoye and Canadice) illustrate that the deeper member has the lower near bottom temperature. Maximum surface temperatures found for three of the lakes were about 27°C and suggest that the smaller lakes may reach slightly higher temperatures than those that are larger, probably the result of greater evaporative cooling of the latter.

Transparency

Historical transparency data as measured by the Secchi disc are sparse for the early part of this century. The maximum transparencies measured in the summers of 1910 (Birge and Juday, 1914) and 1927 (Muenscher, 1928) were greater than values reported for 1973 (Oglesby, 1974), but there are insufficient measurements that are truly comparable to permit statistical comparisons (Oglesby et al., 1975). Data from the other lakes show no identifiable trends (Table 10).

Mills (1975) examined seasonal transparency trends in three of the lakes (Hemlock, Owasco, and Skaneateles) over the course of a year during 1972–1973 (Fig. 19). The lowest transparencies were usually associated with periods of heavy runoff when large amounts of particulate matter were carried into the lakes. The most noticeable event of this nature occurred during tropical storm Agnes in early June 1972.

TABLE 10

Summer Secchi Disc Transparency[a]

Lake	1910[b]	1927[c]	1971[d]	1972[e]	1973[d,e]
Hemlock	4.7	—	5.0	2.7	2.7
Canadice	4.0	—	—	—	5.2
Honeoye	—	—	—	—	3.0
Keuka	—	—	—	—	7.0
Seneca	8.3	9.1	—	—	3.6
Owasco	—	—	3.4	3.0	2.7
Skaneateles	10.3	—	10.5	5.4	5.1
Otisco	3.0	—	—	—	5.7

[a] In meters.
[b] Birge and Juday (1914); one day in August.
[c] Muenscher (1928); mean of two stations and four sampling dates, June 22–September 8.
[d] Oglesby (1974); mean summer.
[e] Mills (1975); mean summer.

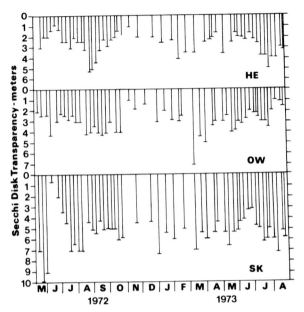

Fig. 19. Seasonal Secchi disc transparencies in Hemlock (HE), Owasco (OW), and Skaneateles (SK). Adapted from a figure in Mills (1975).

Lake Chemistry

The general nature of chemical regimes characterizing the Finger Lakes is determined by the presence of soluble, sedementary rock in their basins. General trends in the distributions of major ions and some trace elements are apparent when the different lakes are compared. Those at higher altitudes have lower dissolved mineral concentrations and Seneca, the bottom of which extends 53 m (174 ft) below sea level, shows the influence of halite deposits. Among the primary nutrients for aquatic plants, silicon is the least variable between lakes, nitrate nitrogen exhibits a distinct east–west gradient of decreasing concentration, and phosphorus is highly variable due to human influences. Of the dissolved gases, critical oxygen levels in the hypolimnion seem to be more dependent on morphology than on biological productivity.

Major Ions

Milliequivalent balances for all of the Finger Lakes are presented in Fig. 20. The most striking features are the high levels of sodium and chloride in Seneca, the abundance of calcium and bicarbonate in all of the lakes, and the substantial variations in levels of ions between the lakes.

The principal sources of sodium and chloride in the largest Finger Lakes have not been identified. Salt (almost pure NaCl) is mined around the southern periphery but the commercial deposits are at a depth substantially below that of the deepest water. Berg (1966) has speculated that Seneca intercepts salt strata. Likens (1974b) has shown that NaCl is derived from within the lake basin proper in the case of Cayuga, a Finger Lake similar to Seneca in NaCl content. The most likely source seems to be the intrusion of saline groundwater. Water with a high salt content is known to surface as springs in the lowlands north of Cayuga (Dudley, 1886), but the location and quantity of any such inflows directly to the lake remain to be verified.

A comparison of present sodium and chloride levels with those given by Berg (1966) for the late 1950's indicates that both ions have increased in abundance in Seneca. Maylath (1973) discussed chloride changes and concluded that an equilibrium concentration of about 120 mg/liter existed from 1920 to 1960. Around 1960–1962 a sudden increase occurred and a new equilibrium at about 160–165 mg/liter was established by the end of the decade. Professor William Ahnsbrak of Hobart College has been conducting research on the origin and flux of chloride in Seneca Lake for the past several years.

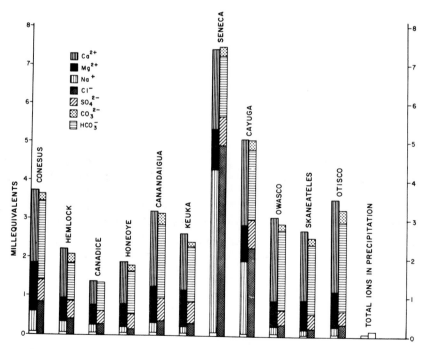

Fig. 20.   Milliequivalent balances for the Finger Lakes (Oglesby *et al.*, 1975).

All of the Finger Lakes contain well-developed calcium bicarbonate buffer systems. The pairing of these two ions and of magnesium and sulfate is apparent in Fig. 20. Calcium and bicarbonate vary substantially between lakes with generally higher values eastwardly. The three major sources of $CaCO_3$–Ca $(HCO_3)_2$ are probably the Tully limestone formation (Fig. 3), calcareous glacial drift deposited in stream channels and as valley head dikes at the southern ends of the basins, and inclusions of calcareous materials in some of the shales and siltstones (see section on Geology). Magnesium and sulfate vary less between systems, with slightly higher values being present in the central lakes. Differences in the calcium to magnesium ratio indicate that substantial variations in the relative amounts of limestone and dolomite probably occur in the various basins.

In general, the higher altitude lakes have lower dissolved solids contents due to a greater degree of rock and soil weathering and perhaps to differences in geological formations within their basins. Water renewal time bears no apparent relation to the levels of dissolved salts. The negative relationship of forest cover ($r = -0.77$) and the positive correlation of population density ($r = 0.85$) with dissolved solids in the lakes provide an interesting reflection on agricultural viability of the soils in the various basins as a function of their buffering capacity (Table 11).

The calcium concentrations in the lakes are highly correlated ($r^2 = 0.86$) with agricultural activities within the drainage basin (Fig. 21). The reason for this is not clear, but several factors can be suggested. In general, agri-

TABLE 11

Percent of Drainage Areas in Forest, Basin Population Densities, and Total Disolved Solids (TDS) in the Various Lakes

Lake	% Forest	Individ. km^{-2}	TDS (mg liter^{-1})
Conesus	34.7	26.4	201
Hemlock	64.4	12.3	127
Canadice	77.9	9.2	76
Honeoye	84.2	13.4	119
Canandaigua	49.7	16.3	187
Keuka	47.7	30.0	165
Seneca	35.1	58.9	270
Cayuga	31.1	42.6	214[a]
Owasco	33.7	28.1	167
Skaneateles	35.2	26.0	148
Otisco	36.6	16.4	183

[a] No synoptic survey data; value represents mean of all stations during summer of 1973.

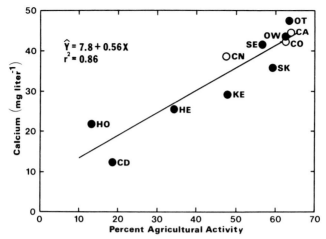

Fig. 21. The relationship between average calcium concentration in the various Finger Lakes and the percentage of the drainage basin in active plus inactive agriculture. The eight lakes being considered in this publication are indicated by filled circles. Key to lake code: CO, Conesus; HE, Hemlock; CD, Canadice, HO, Honeoye; CN, Canandaigua; KE, Keuka; SE, Seneca; CA, Cayuga; OW, Owasco; SK, Skaneateles; OT, Otisco.

culture in New York State tends to be located on well-buffered soils, which leach greater amounts of calcium than those that are less well buffered. In addition, the liming practices that are normally associated with agriculture in the region may tend to increase calcium losses from the landscape. Finally, Likens and Bormann (1972) have shown that disruption of a forest environment by clear cutting increases the export of calcium from the system via streams. Any or all of these factors could conceivably be responsible for the relationship shown in Fig. 21.

Trace Elements

Limited information on trace elements in the Finger Lakes is available from two sources: the U.S. Geological Survey (Anonymous, 1974a, 1975b), which provides data on an extensive array of elements contained in raw water taken for potable supply from some of the lakes, and samples analyzed for iron and manganese, and boron taken by us as part of a special sampling program. The latter study provides the principal information from which variations in time and space can be judged while the former furnishes insights into levels of some of the rarer elements for several of the lakes.

Iron and manganese concentrations were determined on filtered (Reeve–Angel fiberglass filters) water collected during approximately synoptic samplings carried out on three occasions during 1973. The results are shown in Table 12 for surface and near bottom samples. Except for Hemlock,

TABLE 12

Soluble Iron and Manganese Concentrations[a] in the Finger Lakes

Lake	Iron			Manganese		
	4/23–4/28	7/17–7/18	8/28	4/23–4/28	7/17–7/18	8/28
Conesus	40	50	6	6	—	42
Hemlock	50	0	7	8	10	3
Canadice	40	180	13	3	—	1
Honeoye	90	280	13	4	—	1
Canandaigua	20	140	6	3	—	2
Keuka	120	110	15	4	—	4
Seneca	30	10	0	2	—	1
Cayuga	30	200	10	2	—	3
Owasco	20	260	2	3	—	1
Skaneateles	60	120	7	4	—	2
Otisco	110	240	17	2	—	260

[a] In mg m^{-3}.

Keuka, and Seneca, iron concentrations were maximal in April and at a minimum in August. In all cases the August 28 values were very low. Manganese concentrations were at extremely low levels in all lakes at all times except for one moderately high value for Conesus and a very high value for Otisco, both on August 28.

Due to the substitution of borates for phosphates in some household laundry detergents following the June 1, 1973 New York State ban on phosphorus in such products, a survey of boron in selected Finger Lakes was carried out in order to determine background levels and possible short-term changes that might have occurred during the summer of 1973. The results are presented in Table 13.

No consistent pattern of variation between lakes is apparent. Epilimnetic concentrations appeared to be generally lower in all lakes during July and early August, and there was a tendency toward higher levels in waters just off the bottom as the stratified season progressed.

Where reasonably comparable data exist, the boron concentrations given in Table 13 appear to be considerably higher than those reported by the U.S. Geological Survey (Table 14) for the November, 1970 through May, 1973 period. Since changes in detergent use did not seem to produce a short-term response, this variation probably reflects differences in analytical methodologies.

From the data for other elements shown in Table 14, it can be seen that Cu was present at very low concentrations in Keuka and Owasco. Manganese concentrations were low in all of the lakes in agreement with data obtained from the synoptic survey. Iron was the most variable of the ele-

TABLE 13

Boron Concentrations[a] in Some of the Finger Lakes during 1973

Sampling dates	Conesus		Hemlock		Cayuga		Owasco		Skaneateles	
	0–15 m	Bottom	0–15 m	Bottom	0–15 m	Bottom	0–15 m	Bottom	0–15 m	Bottom
4/3–4/4	56	—	52	—	—	—	46	—	48	—
4/17	—	—	—	—	—	—	38	—	—	—
5/14–5/15	—	—	—	—	51	—	26	—	—	—
5/23	—	—	—	—	—	—	53	—	35	40
5/30	53	48	52	42	—	—	49	—	49	—
6/4	42	36	38	18	—	—	46	49	39	36
6/11–6/13	42	40	39	64	25	—	42	79	30	64
6/19	48	48	60	—	—	—	—	—	36	56
6/26	64	—	—	—	—	—	30	48	62	—
7/4	42	—	46	—	—	—	—	—	—	—
7/10	00	00	27	—	—	—	40	—	50	—
7/25	32	—	24	—	—	—	—	—	—	—
7/30	42	54	23	—	—	—	—	—	—	—
8/1	—	—	—	—	24	—	—	—	—	—
8/7	50	—	26	—	—	—	28	—	31	—
8/14	—	—	26	—	—	—	—	—	40	—
8/21–8/22	40	56	48	61	30	49[b]	40	58	40	51

[a] In mg m^{-3}.
[b] 50 m.

TABLE 14

Trace Elements[a] as Determined for Raw Water Samples[b]

Lake		Element						
		Al	Ba	B	Cu	Fe	Mn	Sr
Keuka	avg.	43	16	12	3	190	9	79
(n = 3)	range	16–92	15–16	9–15	2–4	130–280	7–13	72–85
Seneca	avg.	39	25	15	15	112	—	304
(n = 9)	range	10–120	<5–35	9–23	3–34	15–590	<2–<5	270–340
Skaneateles	avg.	50	25	9	11	96	4	70
(n = 30)	range	9–120	20–47	5–15	2–46	6–180	3–7	54–140
Owasco		110	33	11	3	98	14	87
(n = 1)								

[a] In mg m^{-3}.
[b] Anonymous (1974a, 1975b).

ments, and both this variability and the concentrations present agree with the results from the synoptic samplings. The high Sr levels in Seneca are probably associated with salt additions to this lake. Barium and boron were more constant and exhibited smaller differences between lakes than the other elements.

The U.S. Geological Survey reports include data on macronutrients and toxicants as well as a number of trace elements not noted in Table 14. The reason for omitting the latter was that in most instances the other elements were present at concentrations below the level of detection. Since reported levels of detectability varied from sample to sample (probably as a result of differing sample size), it was not possible to express concentrations reported as "less than" in terms of maximum possible values. For example, among those elements of potential biological significance, Mo was reported as being less than values varying from 0.7 to 5 mg m^{-3}.

Nutrients

Detailed information on plant nutrients in the Finger Lakes was first collected for Cayuga in 1968–1969 (Wright, 1969), for Canadice, Conesus, and Honeoye in 1969–1970 (Stewart and Markello, 1974), and for Hemlock, Owasco, and Skaneateles in 1971–1973 (Oglesby et al., 1973, 1975).

Patterns of seasonal changes in the concentrations of soluble reactive phosphorus (SRP), nitrate–N, and soluble reactive silicate–Si (SRSi) in the surface waters of three of the Finger Lakes (Hemlock, Owasco, and Skaneateles) were reported by Oglesby et al. (1975). The results for three of the lakes are presented in Fig. 22a–c. Somewhat more detailed plots of total P and nitrate–N can be found in Stewart and Markello (1974) for Canadice and Honeoye. Oglesby et al. (1975) observed a decline of both SRP and nitrate–N during the summer in the epilimnetic region of each of the lakes. SRP levels were usually less than 3 mg m^{-3}, and were below detection on occasion. The high SRP values occurring in the lakes during late June, 1972 resulted from inputs associated with tropical storm Agnes.

Figure 22a–c shows that the westernmost lake (Hemlock) started out the summer of 1973 with less nitrate–N, i.e., about 400 mg m^{-3}, than the eastern lakes which had about 1800 mg m^{-3}. Apparently as a consequence, nitrate–N concentrations dropped below levels of detection in Hemlock during the summer, whereas in Owasco and Skaneateles about 100–200 mg m^{-3} were still present at the time of fall mixing. During low NO_3–N, E. L. Mills (personal communication) observed increased heterocyst formation coinciding with blue–green maxima, and in some cases, nitrogen fixation was detected by the acetylene reduction method. It has not been determined what effect these low-nitrate concentrations have on limiting the phytoplankton.

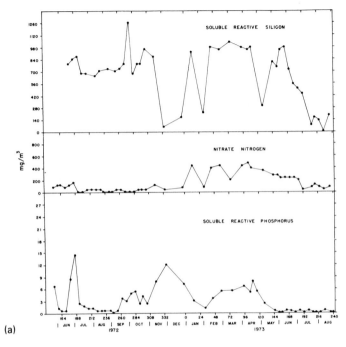

(a)

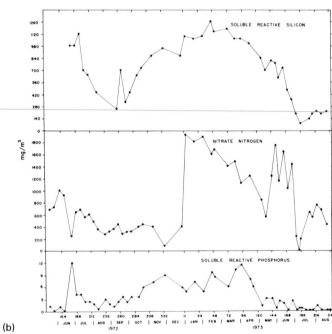

(b)

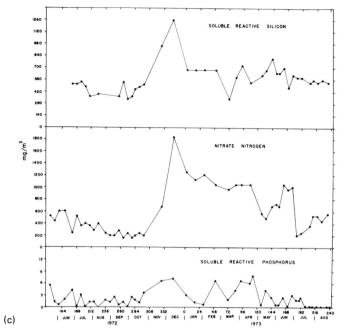

Fig. 22. Seasonal changes in soluble reactive silicate silicon, nitrate nitrogen, and soluble reactive phosphorus in (a). Hemlock, (b) Owasco, and (c) Skaneateles (Mills, 1975).

SRSi concentrations seem to reflect late spring diatom increases in Owasco, and periodic winter diatom peaks in the shallower, western lakes. The lack of any pronounced seasonal change in Skaneateles Lake is probably a reflection of its oligotrophic state. Patterns of dissolved nutrient decrease coincide with increases in phytoplankton standing crops (Mills, 1975).

Nutrient data from a synoptic survey of all of the Finger Lakes are presented in Table 15. All exhibited a decrease in epilimnetic SRP during the summer, but the pattern appeared to differ from lake to lake. No conclusions can be drawn concerning SRP accumulation in the hypolimnion, but more detailed data collected from Hemlock, Owasco, and Skaneateles indicate that this does occur (Mills, 1975).

The total phosphorus concentrations in composite samples (a combination of surface and near bottom water) ranged from a high of 21.7 mg m^{-3} for Honeoye in August to a low of 5.4 mg m^{-3} in Skaneateles during the same month. The highest average concentration for the whole period was measured in Honeoye (18.1 mg m^{-3}); the lowest was from Skaneateles (6.7 mg m^{-3}).

TABLE 15

Nutrient Data from 1973 Synoptic Survey

Lake and calendar day	SRP (mg m⁻³) Epilimnion	SRP (mg m⁻³) Hypolimnion	Total P (mg m⁻³) composite	NO₃⁻-N (mg m⁻³) composite	SRSi (mg m⁻³) composite
Hemlock					
114	6.2		13.5	531	1042
199	0.5	2.5	12.7	74	834
240	0.2	0.8	9.7	229	394
Canadice					
114	5.2		9.2	79	1341
199	1.3	9.3	12.2	49	1401
240	0.2	0.6	8.2	163	710
Honeoye					
114	5.4		16.2	79	709
199	0.9		16.4	49	1957
240	4.2		21.7	79	970
Keuka					
115	7.4		15.1	377	286
199	0.6	3.5	20.4	208	482
240	0.2	1.0	8.0	428	414
Seneca					
118	2.0		17.5	332	153
199	0	6.0	12.8	276	101
240	0	1.4	9.0	243	104
Owasco					
113	6.8		15.6	1756	1057
198	0.4	5.2	11.8	246	156
240	0.2	0.4	8.6	804	811
Skaneateles					
113	4.9		6.9	708	661
198	0.3	1.5	7.7	175	782
240	0.4	0	5.4	511	688
Otisco					
113	3.1		8.4	1258	919
198	0.5	1.2	9.6	859	1440
240	1.8	1.7	9.6	402	2079

Composite SRSi samples ranged from a high of 2079 in Otisco in August to a low of 101 mg m⁻³ in the July sample from Seneca. Some lakes had more SRSi than others, and the concentrations measured in Seneca were consistently lower than those in the rest of the lakes. There is no obvious reason for this variation.

Probably the most striking, and as yet unexplained, feature of the nutrient chemistry of the Finger Lakes is the east–west nitrate-N gradient. Concentrations are lowest in the three western lakes and highest in the three easternmost ones.

Examination of the land use practices in Table 3 does show that there is generally a greater percentage of agriculture in the more easterly basins. Likens (1974a) has reported a significant relationship between the export of inorganic nitrogen in streams and the extent of agricultural activity in the watersheds draining the southern part of the Cayuga Lake basin. However, Conesus, the westernmost lake, has the greatest amount of active agriculture in its basin (Child et al., 1971) but also a very low nitrate–N level (Oglesby et al., 1975). The apparent lack of a firm correlation between nitrate–nitrogen concentrations and either agricultural intensity or population suggests that geochemical sources may be of importance. In reviewing nutrient budgets for other lakes we have not seen this particular possibility considered. However, in the case of the Finger Lakes the many exposed shale formations, highly subject to mechanical disruption by freezing and thawing, would seem to be a potential nitrogen source worth further investigation. A geochemical reservoir is the only one that appears to provide a reasonable explanation for the observed east–west cline.

As stated earlier, there is a paucity of published nutrient data for the Finger Lakes, and with the exception of the work by Stewart and Markello (1974), most studies are represented by only limited sampling. Additional data for several of the lakes can be found in Shampine (1973)—Owasco, Skaneateles, and Otisco; Anonymous (1974b)—Keuka; Anonymous (1974c); Anonymous (1974d)—Seneca; and Stewart and Markello (1974)—Canadice and Honeoye. On the whole these agree well with results of the synoptic survey (Table 15), and of Oglesby et al. (1975).

Dissolved Oxygen

No observable trends toward changing hypolimnetic dissolved oxygen have taken place since 1910 (Birge and Juday, 1914; Wagner, 1928; Oglesby et al., 1975). Lakes with small hypolimnia had, and continue to experience, periods of anoxia in late summer (Fig. 2, Stewart and Markello, 1974). At the other extreme, lakes with large volumes of water beneath the thermocline, e.g., Keuka, Seneca, Owasco and Skaneateles, exhibit smaller reductions in dissolved oxygen during the period preceding fall overturn. In such lakes, variations between years appear to be as much a function of the length of the stratified period and the depth at which the thermocline is established as of the level of primary production. This does not mean that there have been no changes in phytoplankton production per se. In deep bodies of water, such as most of the Finger Lakes, much of the plant material produced may be consumed in the upper region of a lake (Oglesby et al., 1975).

Synoptic survey data on dissolved oxygen concentrations and percentages of air saturation are presented in Table 16. These, together with minima found over two years in the more intensively studied lakes, indicate that

TABLE 16

Surface and Near Bottom Dissolved Oxygen (DO) Measured on a Near Synoptic Basis during April, June, and August of 1973[a]

Lake	Calendar day	Dissolved oxygen				Calendar day	Min. hypolimnetic DO (mg liter^{-1})	
		Surface		Near bottom			1972	1973
		mg liter^{-1}	% sat.	mg liter^{-1}	% sat.			
Hemlock	114	9.7	88	11.3	93	256	2.0 at 9.5°C	
	199	9.1	104	4.9	42	260		3.8 at 9.0°C
	240	9.5	112	5.9	51			
Canadice	114	11.9	105	8.9	71		insufficient data	
	199	8.3	95	6.1	51			
	240	8.9	105	3.6	31			
Honeoye	114	10.9	102	8.7	77		insufficient data	
	199	7.6	89	6.7	77			
	240	10.3	123	8.8	100			
Keuka	115	12.1	101	12.3	96		insufficient data	
	199	8.6	102	7.3	58			
	240	10.0	116	7.9	45			
Seneca	118	12.7	101	12.2	93		insufficient data	
	199	9.2	107	10.2	78[b]			
	240	8.9	103	9.2	73[c]			
Owasco	113	12.9	110	12.9	100	216	6.7 at 6.0°C	
	198	8.1	93	8.0	65	268		6.5 at ~6.4°C
	240	8.4	98	8.9	72			
Skaneateles	113	12.7	102	12.8	99	236	8.3 at 6.0°C	
	198	8.2	94	11.1	89	206		10.4 at ~6°C
	240	8.7	100	10.7	83			
Otisco	113	11.9	107	12.2	99		insufficient data	
	198	7.6	87	3.6	32			
	240	7.3	84	2.3	20			

[a] Minimum near bottom concentrations, and accompanying temperatures as determined by weekly measurements in Hemlock, Owasco, and Skaneateles, are also given.

[b] 50 m

hypolimnetic depletion is likely to pose problems for some of the profundal fauna in Hemlock, Canadice, and Otisco. In addition, Stewart and Markello (1974) showed that bottom waters in Honeoye may become completely anoxic even when only a relatively weak thermal stratification exists.

Stream Chemistry

Stream chemistry can be examined with two purposes in mind: to define the chemical characteristics of the stream itself, or to relate the chemistry of a tributary to that of a receiving lake. Bedrock geology, soil type, land use, population density, industrial wastes, and the hydrologic cycle all affect stream chemistry. In some instances, the influence of one factor will be modified by another. For example, streams in the southern portion of the Finger Lakes region drain the upper shale unit. On the basis of this fact one would expect the waters to be acidic and weakly buffered. In actuality the streams are all highly buffered with calcium carbonate. The reason for this is that many of the stream valleys are oriented in an east–west direction and during the last glacial period were filled with glacial till originating in the limestone units to the north. Stream headwaters pass through this material, dissolving the limestone in the process (Shampine, 1973).

In some instances the relative influences of geology, soil type, and land use cannot be resolved. In central and western New York agriculture tends to occur on soils that are well buffered. Such soils usually contain limestone and tend to lose larger amounts of chemicals than those that are more acidic. This coupling of agriculture, soil buffering capacity, and calcitic rock is such a widespread phenomenon that examples of one of these factors acting alone could not be found.

When stream samples are examined throughout the course of a year, it becomes apparent that not all of the elements and compounds behave in the same fashion. There are indications of both a seasonality and a lack of it. Calcium, magnesium, sodium, chloride, and sulfate concentrations are lowest during spring runoff, and as might be expected there is a negative correlation between concentration and runoff (Likens, 1974a). Potassium, on the other hand, shows no clear seasonal pattern, but concentrations are somewhat elevated at the end of the growing season (Likens, 1974a). Similar results have been reported from the Hubbard Brook Experimental Forest, New Hampshire by Johnson et al. (1969). Likens (1974a) observed that nitrate concentrations in the Cayuga Lake drainage basin varied but noted that there was a seasonal pattern similar to that found in Hubbard Brook (Johnson et al., 1969). Maximum concentrations occurred during the winter with a minimum being observed during the spring growing season (Bouldin et al., 1975; Likens, 1974a). Nitrate concentrations do not appear to change with changing discharge rates during the course of a week or so, requiring the use of monthly means to relate concentration to discharge (Bouldin *et al.*, 1975). Ammonium is also seasonal in pattern, but in this instance high concentrations usually occur during June, July, and August, with low ones being observed during the colder months (Likens, 1974a).

Particulate phosphorus has been considered in some detail by Bouldin *et al.* (1975) in terms of the suspended solids with which much of the phosphorus transported in streams is associated. Suspended solids concentration varies with stream discharge rate in a predictable fashion. For a given rate, the concentration is higher if discharge is increasing (rising limb of the storm hydrograph) than if it is decreasing. Some of the phosphorus associated with this material is desorbable (usually on the order of about 5%, with the maximum being about 10%) (Bouldin *et al.*, 1975). Sediments that are deposited in the stream bed during low flow appear to act as a sink for point source phosphorus introductions during this period. When the sediments are resuspended during periods of high runoff a fraction of this point source phosphorus acts as sorbed phosphorus, equilibrating within hours or days with the surrounding water. Molybdate-reactive phosphorus also varies with discharge and its rate of change, but there is a general decrease in concentration during successive periods of high discharge (Bouldin *et al.*, 1975).

Once the chemistry of the streams in a basin has been characterized, it remains to relate this to lake chemistry. This may be accomplished by simply comparing the dissolved material in a lake with that of its tributaries, or chemical budgets may be computed for the lake. Both approaches require extensive sampling programs because of the variety of variables involved. To date few such studies have been carried out on an adequate basis in the Finger Lakes region or elsewhere. Likens (1974a,b) studied the runoff of water and the chemistry of tributaries in the Cayuga Lake basin during 1970–1971. Bouldin *et al.* (1975) conducted a detailed study of nutrient transport in the streams of the Fall Creek watershed (the major tributary to Cayuga Lake) during 1972–1973. The Environmental Protection Agency obtained estimates of nutrient inputs to Keuka and Seneca Lakes from a limited stream sampling program conducted in 1972–1973 (Anonymous, 1974b,d).

An alternative approach for calculating nutrient input to a lake from the landscape has been suggested by Patalas (1972) based on such parameters as landscape type and population size. Stewart and Markello (1974) applied the method to several New York lakes. Oglesby and Schaffner (1975) have expanded upon the approach using it to calculate phosphorus input to 13 New York lakes including the Finger Lakes. Data on land use, population size, sewage treatment methods, and precipitation are used in the model. Bouldin *et al.* (1975) employed an approach which incorporates surface runoff data. The various methods will be considered again in the nutrient input section.

The tributaries to the Finger Lakes lie within Water Quality Region IIA which includes most of the Appalachian Upland (Shampine, 1973). The region's drainage waters have a relatively low dissolved solids content because of the underlying bedrock which is composed of the comparatively unreactive upper shale unit. The dissolved solids content usually ranges from 90 to 380 mg liter^{-1} during base flow conditions, with an average of about 200 mg liter^{-1}. Waters are all of the calcium carbonate type due to the presence of limestone-containing glacial till. In general, the chemical quality of all streams in the region is good enough for public supply, but since many of the streams contain domestic waste, treatment would be required (Shampine, 1973).

The available sources of data on Finger Lakes tributary chemistry are summarized in Table 17. Most of the information is relatively recent, precluding the estimation of any long-term changes that may have taken place. A lack of compatibility in the parameters sampled, and the apparent nonexistence of tributary chemistry data for some of the lakes, makes basin-to-basin comparisons impossible.

Certain postulates may be made as to the long-term changes that have most likely occurred in the chemistry of streams in the Finger Lakes region. Wholesale changes in stream chemistry should have taken place as the area was settled and cleared for agriculture during the nineteenth century, with such modifications probably resembling those observed by Likens *et al.* (1970) in the Hubbard Brook Experimental Forest, New Hampshire. If this analogy is appropriate, calcium, magnesium, and potassium concentrations should have increased to a considerable extent. Smaller increases in sodium and chloride may also have occured. Sulfate levels should have decreased, but transport of nitrate from the landscape should have been greatly enhanced. Phosphorus losses should also have increased. At present, these trends should be reversed as more and more land goes into inactive agriculture, eventually reverting to woodland. Insight as to what changes actually did take place and at what rates can probably best be obtained by paleolimnological studies of the lakes, an approach that is in its infancy as far as the Finger Lakes are concerned.

If lakes do in fact reflect the chemistry of their tributaries, it should be possible to make several general observations as to stream chemistry. On the average, tributary calcium levels should be higher in lake basins with a high percentage of the land in agriculture. The Finger Lakes indeed exhibit

TABLE 17

Sources of Stream Chemistry Data[a]

Lake	Period	Stream	Frequency	Organization	Parameters measured	Reference
Hemlock	1972	Springwater Creek	6 days during year	Cornell University	pH, CO_3^{2-}, HCO_3^-, CL^-, So_4^{2-}, Ca^{2+}, NO_3^-, SRP	Unpublished
Keuka	1965	10 tributaries	1 day during low flow	USGS[b]	pH, CO_3^{2-}, HCO_3^-, Cl^-, SO_4^{2-}, Ca^{2+}, hardness, conductivity	Crain (1975)
Keuka	1966	6 tributaries	1 day during high flow	USGS	Cl^-, SO_4^{2-}, dissolved solids, hardness, conductivity, NO_3^-	Crain (1975)
Keuka	1972	7 tributaries	10–14 days per tributary station during the year	EPA[c]	NO_3^-, NH_3, Tot. Kjeldahl N, PO_4^{3-}, tot. P	Anonymous (1974b)
Seneca	1965	22 tributaries	1 day during low flow	USGS	Same as Keuka, 1965	Crain (1975)
Seneca	1966	16 tributaries	1 day during high flow	USGS	Same as Keuka, 1966	Crain (1975)
Seneca	1972	13 tributaries	5–14 days per tributary station during the year	EPA	Same as Keuka, 1972	Anonymous (1974d)
Seneca	1972	5 tributaries	variable depending upon the parameter, usually 6–8 days per tributary station during the summer	Hobart College	pH, Cl^-, conductivity, alkalinity, DO, BOD, COD, NO_3^-, NH_3, PO_4^{3-}, acid hydrol. P, tot. P	Anonymous (1972)

Lake	Year	Stations	Sampling frequency	Agency	Parameters measured	Reference
Owasco	1966–1967	Inlet and Dutch Hollow Brook	9/1/66 and 3/29/67	USGS	pH, HCO_3^-, Cl^-, SO_4^{2-}, F^-, Ca^{2+}, Mg^{2+}, Na^+, K^+, dissolved solids, hardness, conductivity, NO_3^-, NO_2^-, NH_4^+, org. N, PO_4^{3-}, tot. PO_4^{3-}, SiO_3	Shampine (1973)
Owasco	1972	Inlet (4 stations) Dutch Hollow Brook, Sucker Brook, Hemlock Creek, Mill Creek	Inlet stations, Dutch Hollow and Sucker; variable, usually 10–28 days during the year, Hemlock and Mill, 1–3 days	Cornell University	Same as Hemlock, 1972	Oglesby et al. (1973)
Skaneateles	1966–1967	Grout Brook	9/1/66 and 3/29/67	USGS	Same as Owasco, 1966–1967	Shampine (1973)
Skaneateles	1972	Grout Brook, Bear Swamp Creek	4–6 days during the year	Cornell University	Same as Hemlock, 1972	Unpublished
Otisco	1966–1967	Spafford Creek	9/1/66 and 3/29/67	USGS	Same as Owasco, 1966–1967	Shampine (1973)

[a] Anonymous (1972).
[b] United States Geological Survey.
[c] Environmental Protection Agency.

a trend in accordance with this observation (Fig. 21). There should also be a
detectable east–west gradient in stream nitrate concentrations, but here
there are confusing aspects to the trend (see section entitled Nutrients).

Although much of the preceding has been of a speculative nature due to
the scarcity of information, some observations can be made from the existing data presented in Tables 18–22. In the Seneca Lake basin (Table 18)
there appears to be a general increase in tributary sulfate concentration
northward (Fig. 23), although there is a considerable amount of scatter in
the data. During low-flow conditions, concentrations ranged from 14 mg
liter^{-1} in Texas Hollow Creek to 85 mg liter^{-1} in Benton Run. This trend is
thought to be related to the bedrock geology as gypsum deposits are known
to occur in the northern portion of the basin.

The impact of human habitation was evident in the stream chemistry of
the Owasco Lake basin, an area studied in some detail by Oglesby *et al.*
(1973) during 1971–1972. The nitrate–N maximum annual range (165–3830
mg m^{-3}) and mean (1309 mg m^{-3}) were observed in the Oswaco Inlet (the
southernmost tributary) below the Village of Moravia. Both parameters
decreased in a regular fashion upstream (Table 22). Sucker Brook, located
at the north end of the lake, had the lowest nitrate–N concentration
encountered during the survey (73 mg m^{-3}), and the lowest mean concentration (840 mg m^{-3}). Dutch Hollow Brook, a highly agriculturalized
watershed with 59% of its drainage area in active agriculture (Child *et al.*,
1971), had a mean concentration of 956 mg m^{-3}, similar to that found in

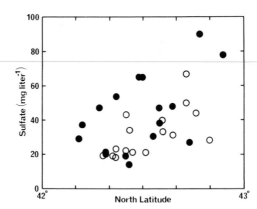

Fig. 23. Changes in the sulfate concentration in streams of the Seneca Lake
drainage basin as a function of latitude. Sampling stations provided the estimate of
latitude. Filled circles indicate measurements made during low and open circles during
high flow.

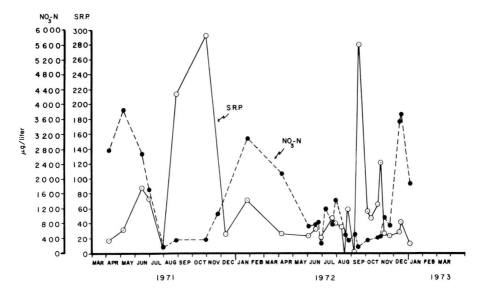

Fig. 24. Seasonal changes in nitrate nitrogen and soluble reactive phosphorus (SRP) in the Owasco Inlet below the Village of Moravia.

the inlet above the Village of Groton, although the range in concentration was more like that observed farther downstream.

Soluble reactive phosphorus concentrations primarily reflected discharges of human wastes. The highest concentrations were in the inlet below Groton (annual range = 12–734 mg SRP m^{-3}; mean = 92 mg SRP m^{-3}). Concentrations decreased downstream, but they were still two times higher below Moravia than above Groton (Table 22). SRP levels were low in Sucker Brook (mean = 20 mg m^{-3}) and Dutch Hollow Brook (mean = 10 mg m^{-3}) relative to the inlet, indicating that agricultural acitvities play a subordinate role relative to the human population in influencing the SRP content of streams in the region.

An example of the seasonal variation in nitrate and SRP is presented in Fig. 24 for samples collected below the Village of Moravia. Nitrate levels were highest during the cooler months, and at a minimum during the growing season. SRP concentrations were highest during late summer–early fall.

Examination of Tables 18 through 22 shows a general agreement when a stream was sampled at different times by different groups. This indicates that the data as presented probably describe the streams reasonably well for conditions existing during the past decade.

Table 18

Chemical Analyses of Samples Collected from Small Streams[a,b]

		Date	SO_4^{2-} (mg liter^{-1})	Cl^- (mg liter^{-1})	NO_3-N (mg m^{-3})	Dissolved solids at 180°C (mg liter^{-1})	Hardness (Ca, Mg) as CaCO$_3$ (mg liter^{-1})	Specific conductance at 25°C (μ S cm^{-1})	Lat.
Seneca									
Beardsley Hollow Creek tributary	HF	2/14/66	19	1.3	40	56	36	84	42°18'/22"
Johns Creek tributary	HF	2/14/66	23	5.5	1200	75	39	113	42/21/46
Glen Creek tributary	HF	2/14/66	18	8.7		—	36	106	42/21/46
Hamilton Creek	HF	2/14/66	19	13		—	—	119	42/20/49
Seneca Lake Tributary 7	HF	2/13/66	43	5.7	1500	159	112	243	42/24/36
Texas Hollow Creek tributary	HF	2/14/66	22	8.3		—	—	158	42/25/21
Seneca Lake Tributary 6	HF	2/14/66	34	3.9	450	91	50	126	42/26/29
Rock Stream tributary	HF	2/13/66	21	1.6		—	28	73	42/27/10
Sawmill Creek tributary	HF	2/14/66	21	2.1	430	—	34	88	42/30/30

Plum Point Creek tributary	HF	2/13/66	33	7.5	5000	137	84	203	42/35/43
Seneca Lake Tributary 8	HF	2/14/66	40	12	2700	186	132	287	42/36/13
Seneca Lake Tributary 4	HF	2/13/66	31	6.6		—	—	206	42/38/32
Seneca Lake Tributary 9	HF	2/13/66	50	9.2	3600	191	142	306	42/43/28
Seneca Lake Tributary 10	HF	2/13/66	44	13		—	153	331	42/46/18
Seneca Lake Tributary 11	HF	2/13/66	28	4.3	1800	—	113	241	42/50/10
Kashong Creek	HF	2/13/66	64	8.6		—	—	390	42/43/27
tributary	HF	2/14/66	70	9.1	8100	247	188	393	
Keuka									
Keuka Inlet tributary	HF	2/13/66	19	1.1	40	—	24	67	42/23/23
Keuka Lake Tributary 8	HF	2/13/66	27	1.9	40	—	21	62	42/24/50
Keuka Lake Tributary 5 tributary	HF	2/13/66	21	5.8		—	—	124	42/35/42
Keuka Lake Tributary 9	HF	2/13/66	18	5.3		—	—	103	42/30/32
Keuka Lake Tributary 10	HF	2/13/66	19	1.7		—	37	90	42/33/36
Keuka Lake Tributary 11	HF	2/14/66	35	7.4	3400	166	124	258	42/37/53

(Continued)

Table 18 (Continued)

		Date	Ca²⁺ (mg liter⁻¹)	HCO₃⁻ (mg liter⁻¹)	CO₃²⁻ (mg liter⁻¹)	SO₄²⁺ (mg liter⁻¹)	Cl⁻ (mg liter⁻¹)	Hardness (Ca, Mg) as CaCO₃ (mg liter⁻¹)	Specific conductance at 25°C (μS cm⁻¹)	pH	Lat.
Seneca											
Catherine Creek tributary	LF	7/2/65	70	280	0	29	9.2	252	500	7.7	42°11′/23″
Catherine Creek	LF	7/2/65	41	135	0	37	5.0	144	306	8.1	42/12/19
Catherine Creek	LF	7/2/65	—	—	—	—	9.6	200	406	—	42/13/56
Catherine Creek	LF	7/2/65	—	—	—	—	3.2	148	295	—	42/13/46
Sleeper Creek	LF	7/1/65	59	217	0	47	4.8	214	436	8.0	42/17/09
Shequaga Creek	LF	7/2/65	40	148	0	21	9.2	143	307	7.6	42/19/02
Mitchell Hollow Creek	LF	7/1/65	49	191	0	20	3.4	167	337	7.9	42/19/32
Cranberry Creek	LF	7/1/65	—	—	—	—	8.4	117	246	—	42/21/14
Catlin Mill Creek	LF	7/1/65	—	—	—	—	12	172	368	—	42/20/38
Glen Creek	LF	7/2/65	51	160	0	54	17	188	415	8.0	42/22/07
Texas Hollow Creek	LF	7/1/65	38	151	0	14	4.0	134	279	7.7	42/25/45
Hector Falls Creek	LF	7/1/65	48	179	0	19	4.6	164	342	7.8	42/25/21
Rock Stream	LF	6/30/65	64	182	0	65	31	228	518	7.8	42/28/52
Club Hollow Creek	LF	7/1/65	20	62	0	30	4.4	80	192	7.3	42/33/21
Big Stream	LF	7/1/65	—	—	—	—	6.6	152	313	—	42/30/04

Big Stream	LF	6/30/65	—	—	—	—	14	175	386	—	42/31/05
Saw Mill Creek	LF	6/30/65	59	176	0	67	5.8	206	423	7.9	42/29/51
Plum Point Creek	LF	6/30/65	46	157	0	47	4.0	165	370	7.9	42/35/26
Mill Creek	LF	7/2/65	39	139	0	38	18	138	368	7.8	42/35/08
Seneca Lake Tributary 4	LF	6/30/65	66	294	0	48	5.8	281	560	7.9	42/39/15
Seneca Lake Tributary 5	LF	7/2/65	—	—	—	—	133	288	928	—	42/44/35
Kashong Creek	LF	7/1/65	58	268	0	27	34	261	557	7.6	42/43/59
Reerer Creek	LF	6/30/65	—	—	—	—	112	340	902	—	42/47/23
Benton Run	LF	7/1/65	84	265	0	85	12	337	614	8.0	42/46/59
Marsh Creek	LF	7/1/65	82	324	—	78	6.0	350	657	7.8	42/53/59
Keuka											
Keuka Inlet	LF	7/2/65	60	219	0	41	13	233	464	7.7	42/22/39
Mitchellsville Creek	LF	7/2/65	11	37	0	20	0.2	44	115	7.4	42/25/21
Keuka Inlet Tributary 2	LF	7/1/65	55	188	0	68	6.6	212	444	7.8	42/23/31
Glen Brook	LF	7/1/65	—	—	—	—	9.4	188	320	—	42/24/34
Keuka Lake Tributary 3	LF	7/1/65	—	—	—	—	12	146	341	—	42/27/11
Keuka Lake Tributary 4	LF	7/1/65	45	96	0	156	26	255	545	7.4	42/29/02
Sugar Creek	LF	7/2/65	60	240	0	42	9.2	244	447	7.9	42/40/30
Keuka Lake Tributary 5	LF	7/2/65	64	208	0	66	27	200	542	7.8	42/35/21
Keuka Lake Tributary 7	LF	7/1/65	—	—	—	—	26	184	450	—	42/33/13
Willow Grove Creek	LF	7/1/65	—	—	—	—	—	14	256	553	42/36/44

[a] Crain (1975).

[b] High overland flow on Feb. 13–14, 1966 designated by HF and low flow on June 30 and July 2, 1965 by LF.

Table 19

Inputs of Selected Allochthonous Materials to Seneca Lake from Five Major Tributaries during the Summer of 1972[a]

Parameter		Marsh Creek	Texas Hollow Creek	Sta. 1[b] Catherine Creek	Sta. 2[c] Catherine Creek	Sta. 3[d] Catherine Creek	Big Stream Creek	Sta. 4[e] Keuka Lake Outlet	Sta. 5[f] Keuka Lake Outlet	Sta. 6[g] Keuka Lake Outlet
DO (mg liter⁻¹)	Mean	4.66	8.23	8.24	7.02	6.39	8.94	8.65	8.37	8.54
	Max.	6.98	9.12	9.19	7.74	7.92	9.78	9.77	8.88	9.19
	Min.	2.80	5.55	7.64	6.00	5.15	8.20	8.25	7.89	8.00
	n	7	7	8	7	8	8	7	7	7
BOD (mg liter⁻¹)	Mean	2.40	0.68	0.67	1.46	2.47	2.14	1.13	1.18	1.47
	Max.	4.44	1.63	1.13	2.24	3.51	7.56	1.45	1.47	1.80
	Min.	1.23	0.03	0.41	1.20	1.37	0.43	0.81	0.91	1.16
	n	7	8	8	6	7	8	7	7	7
COD (mg liter⁻¹)	Mean	34.91	11.97	19.17	15.40	19.83	17.33	12.07	14.92	17.52
	Max.	54.34	22.38	28.97	28.08	33.33	46.25	19.68	47.56	45.47
	Min.	22.75	5.03	8.41	9.36	4.29	4.22	5.50	5.45	7.18
	n	6	7	8	6	8	8	8	8	8
Cl⁻ (mg liter⁻¹)	Mean	75.0	<10	21.7	25.0	44.8	12.5	7.4	7.6	7.9
	Max.	157.9		29.0	54.5	79.7	16.8	8.9	10.0	9.4
	Min.	24.5		17.8	13.6	13.8	6.7	4.8	5.4	7.3
	n	6	8	8	7	8	8	8	8	8
NO₃⁻-N (mg liter⁻¹)	Mean	1.1	0.2	0.5	0.6	0.5	1.3	0.5	0.4	0.6
	Max.	2.0	0.3	1.0	1.0	0.9	1.9	1.4	1.1	2.1
	Min.	0.4	0.1	0.3	0.3	0.3	0.7	0.2	0.2	0.2
	n	7	8	8	7	8	8	8	8	8
NH₃-N (mg liter⁻¹)	Mean	0.5	0.0		0.1		0.0		0.0	
	Max.	1.2			0.4					
	Min.	0.0			0.0					
	n	3	2		3		4		3	

Parameter	Statistic	1	2	3	4	5	6	7	8	9
PO_4^{3-}-P (mg liter⁻¹)	Mean	0.21	3.89	0.03	0.36	0.04	3.19			0.03
	Max.	0.28			1.19	0.06	12.50			
	Min.	0.14	<0.03	<0.03	0.06	0.03	0.06	<0.03	<0.03	<0.03
	n	4	5	5	4	3	4	3	3	3
Acid hydrolyzable P (mg liter⁻¹)	Mean	0.17	11.31	1.07	0.33	0.71	3.80	1.22	0.28	1.31
	Max.	0.26			1.05	1.76	14.95		1.03	5.13
	Min.	0.07	<0.03	<0.03	0.07	0.05	0.04	<0.03	0.03	0.06
	n	3	7	5	4	5	4	5	4	4
Total P (mg liter⁻¹)	Mean	0.49	3.76	0.37	0.38	0.44	4.54	1.81	1.51	1.22
	Max.	1.25		1.00	1.07	1.42	13.26			4.75
	Min.	0.11	<0.03	0.05	0.13	0.07	0.10	<0.03	<0.03	0.03
	n	4	8	3	4	5	3	5	5	5
Specific conductance (μS cm⁻¹ at 250°C)	Mean	815	217	406	398	423	246	205	217	224
	Max.	1088	238	840	838	836	350	232	241	260
	Min.	557	175	140	290	187	112	150	170	158
	n	7	8	8	7	8	8	8	5	5
pH	Mean	7.8	7.7	7.9	7.8	7.6	8.0	8.1	8.0	7.9
	Max.	8.0	7.8	8.0	7.9	7.8	8.2	8.4	8.2	8.0
	Min.	7.6	7.6	7.9	7.7	7.6	7.4	7.7	7.8	7.8
	n	7	8	8	7	8	8	8	8	8
Alkalinity as $CaCO_3$ (mg liter⁻¹)	Mean	241	92	175	144	126	109	76	80	88
	Max.	267	108	217	186	151	138	79	85	100
	Min.	197	68	132	100	101	40	73	77	83
	n	7	8	8	7	8	8	8	8	8

[a] Station numbers increase as a downstream sequence; Anonymous (1972).
[b] Station 1, above Montour Falls.
[c] Station 2, below Montour Falls sewage treatment plant.
[d] Station 3, at Watkins Glen.
[e] Station 4, above Penn Yan.
[f] Station 5, below Penn Yan sewage treatment plant.
[g] Station 6, at Dresden.

TABLE 20

Nitrogen and Phosphorus Concentrations (mg/m^{-3}) in Major Tributaries to Seneca and Keuka Lakes

a. Entering Seneca Lake as Measured during 1972–1973[a]

		Johns Creek	Diversion Channel	Catherine Creek	She-quaga Creek	Glen Creek	Big Stream	Saw-mill Creek	Plumb Point Creek	Keuka Outlet	Kashong Creek	Indian Creek	Hector Falls	Marsh Creek
NO$_3^-$ (mg m^{-3})	Mean	758	706	647	566	231	912	886	1030	694	2685	555	324	1005
	Max.	1320	1220	1580	960	440	1520	1580	2000	2700	5200	1320	640	2130
	Min.	380	60	150	12	92	189	290	20	220	75	19	24	115
	n	13	13	13	13	12	13	14	12	12	12	13	14	13
Tot. Kjeldahl N (mg m^{-3})	Mean	900	649	611	540	575	792	696	545	637	759	519	502	962
	Max.	3900	1470	1200	1760	1890	2000	2100	1540	1760	1800	1380	1800	2000
	Min.	140	160	230	160	<100	220	<100	150	200	<100	130	150	580
	n	13	13	13	13	12	13	14	12	13	13	13	13	14
NH$_3$–N (mg m^{-3})	Mean	32	51	64	26	34	72	38	39	37	91	23	22	138
	Max.	92	189	147	50	78	130	115	80	102	160	64	56	420
	Min.	6	<5	9	<5	12	6	<5	15	12	<5	<5	6	34
	n	13	13	13	13	12	13	14	12	13	13	14	14	14
PO$_4^{3-}$–P (mg m^{-3})	Mean	18	10	14	9	10	87	7	9	32	19	11	8	68
	Max.	21	27	50	20	42	230	18	22	108	56	23	13	198
	Min.	<5	<5	<5	<5	<5	35	<5	<5	10	7	<5	<5	29
	n	13	13	13	13	12	13	14	12	13	13	14	13	14
Tot. P (mg m^{-3})	Mean	40	28	47	21	28	137	10	17	64	33	25	21	166
	Max.	100	90	175	75	75	450	30	45	155	75	95	55	400
	Min.	5	6	<5	<5	10	45	<5	<5	30	7	<5	5	65
	n	13	12	12	12	10	13	13	11	12	11	14	14	14

b. Entering Keuka Lake as Measured during 1972–1973[b]

		Inlet	Unnamed Creek I	Unnamed Creek II	Sugar Creek	Unnamed Creek III	Waneta Outlet	Unnamed Creek IV
NO_3^--N (mg m^{-3})	Mean	754	876	812	680	1310	178	1297
	Max.	1160	1600	1740	1260	1800	350	2900
	Min.	460	770	250	<10	260	32	570
	n	12	11	12	13	14	14	12
Tot. Kjeldahl N (mg m^{-3})	Mean	487	259	738	378	630	821	747
	Max.	750	290	1400	790	1540	2400	2800
	Min.	210	140	170	140	200	340	165
	n	12	10	12	13	14	14	12
NH_3-N (mg m^{-3})	Mean	46	29	38	35	35	50	41
	Max.	92	69	88	84	115	205	170
	Min.	27	9	<5	6	<5	7	<5
	n	12	11	12	13	14	14	12
PO_4^{3-}-P (mg m^{-3})	Mean	17	29	8	8	14	7	18
	Max.	34	56	14	24	33	16	38
	Min.	9	15	<5	<5	<5	<5	6
	n	12	11	12	13	14	14	12
Tot. P (mg m^{-3})	Mean	57	45	17	33	27	36	34
	Max.	130	130	55	120	40	60	80
	Min.	28	20	8	5	10	10	6
	n	11	10	12	13	13	14	12

[a] Anonymous (1974b,d).
[b] Anonymous (1974b).

TABLE 21

Chemical Analyses of Major Tributaries to Owasco, Skaneateles, and Otisco Lakes during 1966–1967[a]

	Date	Ca^{2+}	Mg^{2+}	Na^+	K^+	HCO_3^-	SO_4^{2-}	Cl^-	F^-	Dissolved solids at 180°C	Hardness (Ca, Mg as $CaCO_3$)	Hardness (noncarbonate as $CaCO_3$)	Specific conductance (μ S cm^{-1} at 25°C)	pH
Owasco														
Owasco Inlet at Moravia	8/2/66	—	—	—	—	—	—	19	—	—	—	—	416	—
	9/1/66	56	11	13	1.6	202	25	19	0	238	185	19	414	7.8
	3/29/67	28	4.5	3.0	0.7	80	20	6.5	0.2	118	88	23	196	7.1
Dutch Hollow Brook at Owasco	9/1/66	61	18	6.3	2.0	238	37	10	0.1	270	226	31	450	8.1
	3/29/67	49	9.0	3.0	1.4	144	32	7.0	0.1	195	159	41	322	7.4
Skaneateles														
Grout Brook at Fairhaven	8/2/66	—	—	—	—	—	—	6.9	—	—	—	—	349	—
	9/1/66	43	12	3.0	0.7	158	26	6.5	0	196	157	27	319	7.9
	3/29/67	33	5.1	3.2	1.0	98	18	6.2	0.2	129	104	23	222	7.2
Otisco														
Spafford Creek at Otisco Valley	8/2/66	—	—	—	—	—	—	5.1	—	—	—	—	386	—
	9/1/66	51	16	2.7	1.4	196	30	4.8	0.1	236	193	32	380	7.9
	3/29/67	36	7.2	1.9	0.7	117	26	3.3	0.2	152	122	26	249	7.4

Major ions and physical characteristics

Plant nutrients (mg m^{-3})

	Date	SiO$_2$	NO$_3^-$	NH$_4^+$	NO$_2^-$	Organic N	Total PO$_4$	PO$_4$
Owasco								
Owasco Inlet at Moravia	8/2/66	—	—	—	—	—	—	—
	9/1/66	2.3	40	0	6	0	110	—
	3/29/67	3.6	930	0	0	50	91	30
Dutch Hollow Brook at Owasco	9/1/66	6.1	0	120	3	0	85	—
	3/29/67	3.6	1700	160	3	50	62	—
Skaneateles								
Grout Brook at Fairhaven	8/2/66	—	—	—	—	—	—	—
	9/1/66	5.7	100	0	3	800	0	—
	3/29/67	4.5	1000	310	3	370	46	—
Otisco								
Spafford Creek at Otisco Valley	8/2/66	—	—	—	—	—	—	—
	9/1/66	4.0	90	0	6	720	0	—
	3/29/67	4.6	560	230	3	200	59	20

[a] Shampine (1973). Units are mg liter^{-1} unless otherwise noted.

TABLE 22

Chemical Analyses and pH of Major Tributaries to Owasco, Skaneateles, and Hemlock Lakes during 1971–1972

		Owasco								Skaneateles		Hemlock
		Inlet above Groton	Inlet below Groton	Inlet above Moravia	Inlet below Moravia	Dutch Hollow Brook	Sucker Brook	Hemlock Creek	Mill Creek	Grout Brook	Bear Swamp Creek	Inlet
pH	Mean	8.0	8.0	8.0	8.2	8.2	8.2	8.2	8.2	8.2	8.0	8.2
	Max.	8.3	8.4	8.5	8.8	8.4	8.5	8.4	8.3	8.4	8.5	8.4
	Min.	7.5	7.2	7.0	7.8	8.0	7.8	7.8	8.0	8.1	7.4	7.9
	n	17	17	17	28	14	12	3	3	6	5	6
Ca^{2+} (mg liter^{-1})	Mean	42	46	43	44	46	50	46	44	29	18	42
	Max.	58	59	60	60	76	88	48	47	51	28	46
	Min.	29	34	26	16	6	13	43	40	6	3	38
	n	11	12	13	17	10	10	2	2	4	4	3
CO_3^{2-} (mg liter^{-1})	Mean	5	7	9	8	10	16	6	6	—	—	—
	Max.	14	18	30	24	12	44	8	7	—	—	—
	Min.	0	0	0	0	5	6	5	4	—	—	—
	n	13	13	13	24	12	10	2	3	—	—	—
HCO_3^{-} (mg liter^{-1})	Mean	127	145	128	137	194	195	176	133	116	60	97
	Max.	184	198	200	194	240	298	215	146	146	73	165
	Min.	54	70	49	65	106	79	138	111	86	43	56
	n	13	13	13	24	12	9	2	3	6	6	6
Cl^{-} (mg liter^{-1})	Mean	17	22	17	16	13	14	12	11	13	11	16
	Max.	29	41	33	40	19	29	16	13	22	23	24
	Min.	9	8	5	8	8	6	9	9	9	6	6
	n	15	16	16	27	13	11	2	3	6	6	6
SO_4^{2-} (mg liter^{-1})	Mean	15	20	19	20	23	18	18	14	21	12	30
	Max.	17	21	22	24	27	25		15	24	14	35
	Min.	11	19	18	13	20	5		14	15	6	25
	n	4	5	6	15	4	4	1	2	4	4	5
NO_3-N (mg m^{-3})	Mean	903	1176	1241	1309	956	445	1780	840	664	387	1672
	Max.	2583	2710	3798	3830	3545	1465	2786	1772	1393	830	3798
	Min.	162	165	177	165	77	73	773	241	279	92	165
	n	16	17	17	28	12	12	2	3	6	6	6
SRP (mg m^{-3})	Mean	31	92	73	63	10	20	14	13	30	60	12
	Max.	92	734	589	293	26	45	18	20	114	149	14
	Min.	5	12	0	0	3	2	11	8	8	4	10
	n	17	18	18	29	15	13	2	3	6	6	6

Precipitation Chemistry

The most detailed study of precipitation chemistry in the region was conducted by Likens (1972) during 1970–1971. Samples were collected at six stations in the central Finger Lakes region during this period. The reader is referred to this publication for a description of sampling methods and a more detailed presentation of the data. Chemical analyses of calcium, magnesium, sodium, potassium, ammonium, nitrate, sulfate, chloride, phosphate, pH, and conductivity were made. Hydrogen ion concentrations were estimated from pH. Where possible, comparisons were made with historical data. Some additonal contemporary data collected during 1966–1967 can be found in Shampine (1973).

The annual average concentrations over all six stations, and the range of individual annual station averages, are presented in Table 23. The results are for clean plus slightly dirty (one–five small insects or particles in the collection tubing or reservoir) samples averaged together. Although there is a danger of overestimation if grossly contaminated samples are used, Likens (1972) feels that averaging the results in this fashion gives a more representative measure, especially during periods when only a few clean samples were collected. He also presents the analytical results for only the clean samples, but they will not be considered here. The results of the chemical determinations are for bulk precipitation, a mixture of rain, snow, and dry fallout (Whitehead and Feth, 1964), as the precipitation collectors were continuously exposed to the atmosphere.

Calcium, Magnesium, Sodium, and Potassium. Although somewhat variable from station to station (see range of values in Table 23), the data from the central Finger Lakes region are in close agreement with similar information from the rest of the northeastern United States, and nearby U.S. Geological Survey stations (Likens, 1972; Pearson and Fisher, 1971). Calcium is present in the greatest amount, by weight, with magnesium and sodium being equal, and potassium slightly less than the latter two elements. Historical data do not exist.

Nitrogen and Sulfur. Nitrogen and sulfur are somewhat more interesting in that there are published data on these elements from two of the sampling stations that date back 60 years, which, according to Likens (1972), may represent some of the longest precipitation chemistry records in existence for the United States.

Noticeable changes have taken place in the inorganic nitrogen and sulfur concentration of precipitation since 1915. Samples collected during the first half of the century were characterized by high sulfate (from 4 to more than 8 mg liter^{-1}) and ammonium (up to almost 4 mg liter^{-1} during some years)

TABLE 23

Precipitation Chemistry and Aerial Input during the Period August 1970–July 1971[a]

		Cations											
		Ca²⁺		Mg²⁺		Na⁺		K⁺		H⁺		NH₄⁺	
Station	Precipitation (cm)	Mean conc. (mg liter⁻¹)	Input (gm ha⁻¹)	Mean conc. (mg liter⁻¹)	Input (gm ha⁻¹)	Mean conc. (mg liter⁻¹)	Input (gm ha⁻¹)	Mean conc. (mg liter⁻¹)	Input (gm ha⁻¹)	Mean conc. (mg liter⁻¹)	Input (gm ha⁻¹)	Mean conc. (mg liter⁻¹)	Input (gm ha⁻¹)
Ithaca	96.1	1.55	14896	0.15	1445	0.24	2287	0.08	762	0.10	1003	0.42	4001
Aurora	98.4	1.49	14638	0.22	2208	0.16	1568	0.12	1136	0.12	1217	0.43	4260
Geneva	95.8	0.35	3357	0.06	554	0.06	597	0.04	376	0.10	915	0.36	3419
Canoga	98.8	0.55	5386	0.08	798	0.09	932	0.12	1226	0.10	1009	0.66	6531
Lodi	96.8	0.41	4014	0.07	633	0.07	644	0.15	1450	0.10	978	0.42	4099
Watkins Glen[b]	59.9	2.57	15413	0.54	3212	0.53	3150	0.24	1465	0.11	656	0.40	2394
Mean of stations[b]	97.2	0.87	8458	0.12	1128	0.12	1206	0.10	990	0.10	1024	0.46	4462

	Anions							
	NO₃⁻		Cl⁻		SO₄²⁻		PO₄³⁻	
Station	Mean conc. (mg liter⁻¹)	Input (gm ha⁻¹)	Mean conc. (mg liter⁻¹)	Input (gm ha⁻¹)	Mean conc. (mg liter⁻¹)	Input (gm ha⁻¹)	Mean conc. (µg liter⁻¹)	Input (gm ha⁻¹)
Ithaca	3.02	29005	0.96	9199	6.4	61753	23.0	221.3
Aurora	3.61	35567	1.46	14378	6.5	63503	38.0	374.0
Geneva	1.95	18693	0.42	4052	3.6	34406	15.5	148.7
Canoga	2.13	21035	0.71	6979	4.2	41155	117.9	1165.1
Lodi	2.11	20390	1.78	17269	2.9	27932	68.8	666.0
Watkins Glen[b]	6.36	38105	1.70	10157	9.0	53754	106.8	639.7
Mean of stations[b]	2.56	24938	1.07	10375	4.7	45750	52.6	515.0

[a] Likens (1972).

[b] Sampling did not begin until Nov 1970. Values not included in area means.

and low nitrate concentrations. After 1950 there was a marked reduction in sulfate, and an increase in nitrate.

Likens (1972) is of the opinion that the changes are related to the types of fossil fuels combusted locally. Prior to 1932 coal and wood were the primary source of heat. Natural gas was introduced to the region in 1932, but did not achieve wide usage until the early 1950's. The sulfur content of natural gas is far less than that of coal. On the other hand, oxides of nitrogen are produced by the combustion of natural gas and internal combustion engines, with the latter being the major source of these substances in the atmosphere (Train et al., 1970) of the United States as a whole. Thus, in the Finger Lakes region increases in the nitrate content of rainfall may well be the result of increasing gasoline consumption from about 1947 onward combined with shifts from coal and wood to natural gas.

There have also been changes in the seasonal concentration patterns of sulfate. During the first half of the twentieth century the higher concentrations were observed during winter, which coincided with periods of high coal use. Today, sulfate concentrations are higher during summer with more than 40% of the total annual input to the landscape via atmospheric fallout occurring during June, July, and August (Likens, 1972). A similar pattern has been observed in New Hampshire (Likens and Bormann, 1972).

Inorganic nitrogen concentrations have also been observed to fluctuate seasonally, with both nitrate and ammonium varying in synchrony. The lowest concentrations are observed during mid winter (January and February), with another low occurring during late summer (Likens, 1972). These results are also similar to those obtained in New Hampshire by Likens and Bormann (1972).

Phosphorus. Quantitative measurements of phosphorus are scarce (Likens, 1972), and there are no historical data with which to compare the present results. The annual average concentration for the central Finger Lakes region is 61.7 mg phosphate m^{-3} of precipitation. In contrast to nitrogen and sulfur, only about 8% of the total annual phosphate input to the landscape in the form of atmospheric fallout occurs during the summer months.

Chloride. Likens (1972) points out that the chloride concentrations of precipitation are higher in the central Finger Lakes than at other rural inland stations in the northeast (Pearson and Fisher, 1971), but there is no apparent explanation for this.

Acidity. It is becoming increasingly apparent that precipitation falling throughout the northeastern United States is characterized by a high acidity (e.g., Fisher *et al.*, 1968; Pearson and Fisher, 1971), and the Finger Lakes region is no exception (Likens, 1972). Water from rain and snow is no longer characterized by a weakly acid but highly buffered carbonic acid system. Precipitation is now dominated by stronger, unbuffered, mineral acids which produce greatly lowered pH's. Current measurements of precipitation in the Finger Lakes region indicate that hydrogen ion is the major cation (56% of the cation equivalents), and that sulfate (59%), nitrate (21%), and chloride (20%) are the major anions (Likens, 1972).

Published historical pH data for the region are not available, but Likens (1972) has been able to gain some insight as to what the pH levels of precipitation were by examining bicarbonate information. Collison and Mensching (1932) measured bicarbonate in precipitation samples collected at Geneva during the period 1919–1929. Since bicarbonate could not exist with the strong acids found in present day samples, Likens (1972) feels that this indicates pH's during this earlier period must have been 5.7 or higher. Contemporary values from the central Finger Lakes are on the average about 4.0, with even lower values being measured during the summer.

As Likens (1972) points out, the high-sulfate concentrations reported before 1949 should have produced a concomitant drop in pH as SO_4^{2-} was hydrolyzed to H_2SO_4. Since this apparently did not occur, it appears that much of the sulfate must have been present in an un-ionized particulate form, or neutralized by large amounts of base produced when the coal was combusted. Evidence for such reactions has been reported in the literature (e.g., MacIntire and Young, 1923; Gorham, 1958).

Ground Water Chemistry

The ground waters of the Western Oswego River basin have been studied by Crain (1975), and in general, the results apply to the Finger Lakes region as a whole. Ground water occurs in two basic water-bearing units: (1) bedrock, which in the Finger Lakes is primarily shale, siltstone, and sandstone dating from the Devonian Period, and (2) unconsolidated glacial deposits.

Most of the unconsolidated deposits fall into one of three categories (Crain, 1975): (1) Till formed by direct deposition by the ice sheet, without sorting by standing or running water. These deposits are heterogenous, unsorted mixtures of grain sizes ranging from clay to boulders. (2) Sand and gravel deposits laid down by water from melting snow or glacial ice. (3) Silt and clay deposits that were formed when very fine rock particles (0.005–0.06 mm diameter) settled to the bottom of standing water.

The dissolved solids concentration of the water commonly tapped by wells in the area ranges from 150 to 500 mg liter^{-1}. The median concentration in the shale, siltstone, and sandstone unit is 455 mg liter^{-1}, three times that found in overland runoff from regions underlain by this bedrock unit. This relatively high dissolved solids content is due to the presence of thin beds of carbonate rock, and a matrix of calcium carbonate around the relatively insoluble silicate materials. Thus, the waters are of the calcium bicarbonate type, and have a relatively high mean hardness value (190 mg $CaCO_3$ liter^{-1}).

The dissolved solids concentration of the ground water present in the unconsolidated glacial deposits may be on the order of that found in the bedrock when the former contains fragments of dolomite and limestone transported from the limestone belt to the north of the basins.

Biological

Phytoplankton

A key point to be kept in mind when considering the phytoplankton communities in the Finger Lakes is that the lake ecosystems are both complex and diverse in a comparative sense. This fact requires the input and synthesis of considerable information if one is to study the variations in the structure and dynamics of communities within and between lakes. For example, mean depths differ by more than an order of magnitude, and trophic states, which appear to be related to both lake depth and nutrient input, range from oligotrophic to eutrophic. The herbivore communities differ considerably in the various lakes to the extent that large cladocerans may dominate in one and rotifers in another. Added to this are human-related activites such as land use which may greatly influence the input of a variety of materials, and fishery management programs which can have a considerable effect on the transfer of energy and nutrients through the food web. Because of this complexity the present consideration of the phytoplankton communities is essentially descriptive. Considerations beyond this await a more detailed analysis of the data.

A phytoplankton species checklist for the eight lakes based on samples collected during 1972–1973 is presented in Table 24. A total of 257 taxa are listed, including 107 genera and 232 species, with the remainder being subspecies and varieties. Some of the forms listed are not considered as being planktonic, but since they were collected in the limnetic portions of the lakes, they are included. Such species were usually represented by only a few individuals per count, and did not greatly influence the biomass estimates presented below.

Mills (1975) recently completed a detailed study of the phytoplankton of four of the Finger Lakes (Conesus, Hemlock, Owasco, and Skaneateles). Data were collected over 13 months, including two stratified periods, during

TABLE 24

Phytoplankton Species Checklist for 1972–1973[a]

Taxa	Hemlock	Canadice	Honeoye	Keuka	Seneca	Owasco	Skaneateles	Otisco
Chlorophyta								
Actinastrum gracilimum						X		
Actinastrum hantzschii	P					X		
Actinastrum hantzschii var. *fluviatile*						X		
Ankistrodesmus falcatus	X	X	X	X	X	X	X	X
Ankistrodesmus falcatus var. *acicularis*	P							
Ankistrodesmus spiralis	X	X	X		X	X	X	X
Apiocystis brauniana				X				
Arthrodesmus incus var. *extensus*	X					X		
Arthrodesmus ralfsii						X		
Asterococcus spinosus							X	
Botryococcus sudetica								
Carteria cordiformis	X	X	X	X	X	X	X	X
Carteria klebsii						X	X	
Chlamydomonas sp.	P						X	
Chroococcus sp.	P						P	
Closteriopsis longissima	X	X		X	X	X	X	
Closterium acerosum	P				X			
Closterium acutum						X	X	
Closterium dianae	X							
Closterium moniliforme						X	X	
Coccomonas orbicularis						X	X	
Coelastrum microporum	X		X	X	X	X	X	X
Colacium vesiculosum						P		
Cosmarium bioculatum	X						X	

Species										
Cosmarium botrytis	X		X	X	X			X		
Cosmarium paramense	X			X	X	X		X	X	
Cosmarium pseudopyramidatum								X	X	X
Cosmarium punctulatum	X		X	X	X	X		X	X	
Cosmarium pyramidatum	X				X			X		
Cosmarium reinforme	X							X		
Cosmarium retusum	X			X	X	X		X		X
Crucigenia quadrata	X							X		
Crucigenia rectangularis	X				X			X		
Crucigenia tetrapedia	X		X					X		X
Dictyosphaerium pulchellum	X			X	X	X				
Echinosphaerella limnetica	X					X				
Elakatothrix gelatinosa	X					X				X
Eudorina elegans	P									
Franceia ovalis	X				X			X		
Franceia droescheri	X									
Franceia tuberculata	P							X		
Gleocystis major								P		
Golenkinia paucispina	X		X	X	X			X	X	
Golenkinia radiata	X			X		X		X	X	
Kirschneriella lunaris	X			X	X	X				
Kirschneriella lunaris var. *dianae*	X							X		
Kirschneriella lunaris var. *irregularis*	X		X	X	X			X	X	
Lagerheimia ciliata	X				X			X	X	
Lagerheimia quadriseta	X							X	X	
Lagerheimia subsalsa	X									
Micractinium pusillum	X							X	X	
Micractinium quadrisetum				X				X		
Micrasterias foliacea	X							X		
Mougeotia sp.	X		X		X			X		
Nephrocytium agardhianum	X					X		X		
Nephrocytium limnetica	X			X	X		X	X		X

(Continued)

TABLE 24 (Continued)

Taxa				Lake				
	Hemlock	Canadice	Honeoye	Keuka	Seneca	Owasco	Skaneateles	Otisco
Oocystis borgei	X						X	
Oocystis lacustris	X	X	X	X	X	X	X	X
Oocystis parva							X	
Oocystis pusilla							X	
Ophiocytium capitatum var. *longispina*	X							
Pandorina morum	X		X	X		X	X	X
Pediastrum biradiatum	X					P		
Pediastrum boryanum	X				X	X	X	
Pediastrum boryanum longicorne					X			
Pediastrum duodenarium	X							
Pediastrum duplex	X			X			X	X
Pediastrum duplex var. *clathratum*	X							
Pediastrum duplex var. *gracilimum*	X					X		
Pediastrum duplex var. *reticulatum*	X				X	X		
Pediastrum obtusum	X							
Pediastrum sculpatatum	X					X		
Pediastrum simplex	X					X		X
Pediastrum simplex duodenarium	P							
Pediastrum tetras						X		
Pediastrum tetras var. *tetraodon*	X					X		
Planktosphaeria gelatinosa						X		
Quadrigula closteroides							X	
Quadrigula lacustris	X	X	X	X	X	X	X	X
Scenedesmus abundans	X			X	X		X	X
Scenedesmus abundans var. *asymetrica*	P							

Species							
Scenedesmus abundans var. longicauda	X						
Scenedesmus acumunatus						X	X
Scenedesmus arcuartis						X	X
Scenedesmus arcuartis var. capitatus	P			X		P	
Scenedesmus armatus	X	X		X	X	X	
Scenedesmus armatus var. bicaudatus	X	X		X	X	X	X
Scenedesmus bijuga	X	X	X	X	X	X	X
Scenedesmus bijuga var. alternans	X	X			X	X	
Scenedesmus dimorphus	X					X	
Scenedesmus obliquus	X					X	
Scenedesmus opoliensis	X			X		X	
Scenedesmus protuberans	X					X	
Scenedesmus quadricauda	X			X	X	X	X
Scenedesmus quadricauda var. longispina	X			X	X	X	X
Selenastrum gracile	X		X			X	
Selenastrum minutum	X	X	X	X	X	X	X
Selenastrum westii	X		X	X			
Sphaerocystis schroeteri	X	X	X	X	X	X	X
Staurastrum arctison	X						
Staurastrum brevispina	P						
Staurastrum chaetoceras	P						
Staurastrum crenulatum	P						
Staurastrum dejectum	X						
Staurastrum eustephanum	X						
Staurastrum grallatorium	X						
Staurastrum natator var. crassum	X		X	X	X	X	
Staurastrum paradoxum	X					X	P
Staurastrum polymorphum	X					X	

(Continued)

TABLE 24 (Continued)

	Lake							
Taxa	Hemlock	Canadice	Honeoye	Keuka	Seneca	Owasco	Skaneateles	Otisco
Staurastrum rotula	X							
Staurastrum rugulosum	P							
Staurastrum striolatum	X							
Stylosphaeridium stipitatum	X	X	X	X	X	X	X	
Tetraedron hastatum var. *palatinum*	X							
Tetraedron minimum	X	X		X	X	X	X	
Tetraedron regulare var. *incus* fa. *longispina*	X							
Tetraedron regulare var. *torsum*	X					X		
Tetraedron trigonum	X					X		
Tetraedron trigonum var. *gracile*								
Tetraedron tumidulum					X			
Tetraspora sp.							P	
Treubaria setigerum	X				X	P	X	
Euglenophyta								
Euglena sanguinea	X					X	X	
Phacus acuminatus	X					X	X	
Phacus curvicauda	X							X
Pyrrophyta								
Ceratium hirundinella	X	X	X	X	X	X		X
Glenodinium borgei	X							
Glenodinium gymnodinium	X	X	X	X	X	X	X	X
Glenodinium pulvisculus	X		X			X	X	
Glenodinium quadridens	X	X	X	X	X	X	X	
Gymnodinium caudatum	X	X		X	X	X	X	X
Gymnodinium limneticum	X					X	X	
Gymnodinium palustre								
Peridinium cinctum	X	X	X	X	X	X	X	
Peridinium pusillum	X					X	X	X

Cryptophyta

Cryptomonas erosa	x	x		x		x	x	x	x
Cryptomonas nasuta	x	x		x	x	x	x	x	x
Cryptomonas ovata	x	x	x	x	x	x	x	x	x
Cryptomonas pusilla	x	x	x	x	x	x	x		

Chrysophyta—Chrysophyceae, Xanthophyceae

Biocoeca socialis	x	x	x			x	x		
Chromulina globosa	x	x				x	x		
Chromulina ovalis	x	x	x	x	x	x	x	x	
Chrysamoeba radians	x								
Chrysosphaerella conradi	x	x							
Chrysosphaerella longispina	x	x						x	
Cladomonas fruticulosa	x	x	x	x	x	x	x		
Diceras phaseolus	P	x	x						
Binobryon bavaricum	x	x	x		x	x	x		
Dinobryon borgei	x								
Dinobryon cylindricum	x								
Dinobryon divergens	x					x	x		
Dinobryon eurystoma	x	x	x	x	x	x	x	x	
Dinobryon sertularia	x	x	x	x	x	x	x	x	
Dinobryon sociale	x				x	x	x	x	
Dinobryon sociale var. americanum	x					x			
Hymenomonas roseola	x	x		x		x	x	x	x
Mallomonas akrokomos	x	x	x		x	x	x	x	
Mallomonas dentata	x	x	x	x	x	x	x	x	
Microglena cordiformis	P	x							
Ochromonas sp., 4–6 µm	x	x	x	x	x	x	x	x	x
Ochromonas granulosa						x	x		
Ochromonas simplex							x		
Rhyzochrysis planctonica	x	x			x	x			
Sphaeromantis tetragona				x	x				

(Continued)

389

Table 24 (Continued)

Taxa	Lake							
	Hemlock	Canadice	Honeoye	Keuka	Seneca	Owasco	Skaneateles	Otisco
Synura uvella	P	X	X	X				X
Trachelomonas sp.				X		X	X	
Uroglenopsis americana	P				X		X	
Chrysophyta—Bacillariophyceae								
Centrales								
Coscinodiscus sp.	X				X	X	X	
Cyclotella 5–6 μm	X	X	X	X	X	X	X	X
Cyclotella 10 μm	X	X	X	X	X	X	X	X
Cyclotella 15 μm	X	X	X	X	X	X	X	
Cyclotella 20–25 μm	X	X	X	X	X	X	X	X
Melosira granulata	X	X	X	X	X	X	X	X
Melosira italica	X							
Stephanodiscus astrea			X	X		X	X	X
Pennales								
Achnanthes sp.	X							
Achnanthes clevei						X	X	
Amphiprora ornata	X			X				
Amphiprora paludosa	X							
Amphora ovalis	X						X	
Asterionella formosa	X	X	X	X	X	X	X	X
Cocconeis placentula	X	X	X	X	X	X	X	X
Cymbella lanceolata	X	X		X	X		X	
Cymbella ventricosa	X	X	X	X	X	X	X	
Diatoma tenue var. elongatum	X	X	X	X	X	X	X	X
Diatoma vulgare	X	X		X	X	X	X	X
Diploneis elliptica	X	X			X	X	X	X
Eunotia arcus var. bidens	X							
Eunotia quaternaria	X							

Taxon						
Eunotia sudetica	X					
Fragilaria brevistriata	X			X	X	X
Fragilaria capucina			X		X	X
Fragilaria crotonensis	X	X	X	X	X	X
Fragilaria virescens	X	X	X		X	X
Fragilaria virescens var. virescens	P			X	X	X
Gomphoneis herculeana					P	
Gomphonema sp.				X		
Gomphonema olivaceum	X	X	X	X	X	X
Gyrosigma fasciola	X			X	X	X
Gyrosigma macrum	X	X				
Gyrosigma obtusatum	P					
Hantzschia amphioxys	P					
Mastoglia smithii var. lacustris	P					
Meridion circulare	X	X		X	X	X
Navicula capitata	X			X	X	X
Navicula capitata var. hungarica				X		
Navicula cryptocephala	X	X		X	X	X
Navicula cuspidata				X	X	
Navicula elginensis	X					
Navicula exigua	X			X	X	X
Navicula lanceolata	X			X	X	X
Navicula minima	X	X		X	X	X
Navicula schroeteri escambia	P			X	X	
Navicula subtilissima	X	X		X	X	
Navicula viridula var. linearis		X				
Navicula viridula var. rostellata	X			X	X	X
Nedium affine var. affine	X					
Nitzschia acicularis	X	X		X	X	
Nitzschia brebessonii	X			X	X	
Nitzschia nyassensis	P					
Nitzschia palea	X	X		X	X	X

(Continued)

391

Table 24 *(Continued)*

Taxa	Lake							
	Hemlock	Canadice	Honeoye	Keuka	Seneca	Owasco	Skaneateles	Otisco
Nitzschia sigmoidea	X		X					X
Nitzschia vermicularis	X	X	X	X	X	X	X	X
Pinnularia sp.	X					X		
Pinnularia brebessonii	P		X		X			
Pinnularia torta var. *torta*								
Rhizosolenia eriensis	X	X	X	X		X	X	X
Rhizosolenia eriensis var. *brevispina*	X	X				X		
Rhoicosphenia curvata					X			
Stauroneis anceps	X				X	X		
Surirella linearis	X	X		X	X	X	X	X
Surirella splendida					X			
Synedra acus	X	X	X	X	X	X	X	
Synedra cyclopum	X			X			P	
Synedra delicatissima							X	
Synedra delicatissima var. *delicatissima*	X	X	X	X				X
Synedra radians	X	X	X	X	X		X	X
Synedra rumpens	X	X	X			X	X	
Synedra ulna	X	X	X	X	X	X	X	
Synedra ulna var. *danica*	X							X
Tabellaria fenestrata	X	X	X	X	X	X	X	
Cyanophyta								
Anabaena affine	P							
Anabaena flos-aquae	X	X	X	X	X	X	X	X
Anabaena scheremetievi	X							
Aphanizominon flos-aquae	X			X	X	X	X	X
Coelosphaerium pallidum	X			X		X	X	X

Species								
Dactylococcopsis acicularis	X	X						
Gomphosphaeria lacustris	X	X		X	X	X	X	X
Gomphosphaeria lacustris var compacta	X	X	X	X	X	X	X	X
Lyngbya contorta	X	X		X	X	X	X	X
Lyngbya limnetica	X	X	X	X	X	X	X	X
Merismopedia tenuissima	X	X	X	X	X	X	X	X
Merismopedia trolleri	X			X	X	X		
Microcystis aeruginosa	X	X	X	X	X	X	X	
Oscillatoria prolifica	X	X	X	X	X	X		
Phormidium tenue	P				X	X	X	
Stichosiphon regularis	X	X	X	X	X	X	X	X

a The species that were counted denoted by X, species observed but not included in counts by P.

1972–1973. Three of the lakes are considered in a much abbreviated, but nonetheless somewhat complex, fashion in the following sections.

Details of Community Composition in Three Lakes. *Hemlock.* Forty-six percent of the taxa observed in this lake were chlorophytes. Most were summer species, although a few forms did occur year-round, e.g., *Ankistrodesmus falcatus, Coelastrum microporum, Cosmarium bioculatum, Scenedesmus bijuga, Selenastrum minutum,* and *Tetraëdron minimum.* The cryptomonads *Cryptomonas erosa, C. ovata,* and *C. pusilla* were common throughout the year, as were the dinoflagellates *Gymnodinium caudatum* and *Peridinium cinctum.* Blue-green algae were most evident during the summer, the major representatives being *Anabaena flos-aquae, Gomphosphaeria lacustris, Lyngbya limnetica, Merismopedia tenuissima,* and *Microcystis aeruginosa.* The Chrysophyceae were represented by *Mallomonas* sp., *Chromulina* sp., and *Ochromonas* sp., and the diatoms by such forms as *Asterionella formosa, Cocconeis placentula, Cyclotella* spp., *Fragilaria crotonensis, Melosira granulata, Nitzchia acicularis,* and *Synedra rumpens.*

Cell number in the euphotic zone ranged from 3×10^3 cells ml^{-1} to almost 100×10^3 cells ml^{-1}. Maxima were observed during the spring and summer, and minima during the winter months. Maximum concentrations were usually to be found at the 2- and 5-m depths. Ultraplanktonic forms were the most numerous size component, at times exceeding 90% of the total cells counted. Nannoplankters were next in importance (Mills, 1975).

Phytoplankton biomass (based on cell volume measurements) in the upper 10 m of the water column ranged from less than 1 gm m^{-3} to over 7 gm m^{-3} (Mills, 1975), with the higher concentrations occurring during the warmer months (Fig. 25). Seasonal averages ranged from 0.82 gm m^{-3} (winter, 1972) to about 3 gm m^{-3} (summer, 1973). Diatoms and dinoflagellates were the major contributors to the biomass, followed by the cryptomonads, blue-greens, and chlorophytes. The Chrysophyceae and Xanthophyceae were of least importance, accounting for only 5% of the biomass during the 1972–1973 study period. Diatoms accounted for 38%, with the seasonal averages ranging from 25 to 52% (summer, 1972 and spring, 1973, respectively). Diatom biomass varied from 0.27 gm m^{-3} during the winter of 1973 to 0.88 gm m^{-3} the following summer. Dinoflagellates made up 23% of the biomass during the survey period, with a range of 11 (spring, 1973) to 45% (winter, 1973). The winter maximum was due to a bloom of *Gymnodinium caudatum* and *G. limneticum* (Mills, 1975). Of intermediate importance were chlorophytes, blue-greens, and cryptomonads which accounted for 11–12% of the biomass when averaged

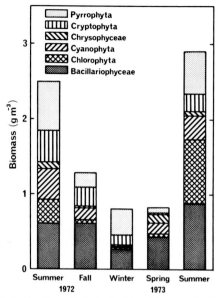

Fig. 25. Seasonal distribution of phytoplankton biomass in Hemlock Lake, and the relative contribution of the various taxonomic groups (data from Mills, 1975).

over the whole survey. The Chlorophyta were of equal importance to the diatoms during the summer of 1973.

Netplankton was the dominant size category in terms of biomass in Hemlock Lake during four out of five seasons, the one exception being spring 1973, when ultraplankton accounted for 43% of the biomass versus 27% for netplankton (Fig. 26). Averaged over all seasons the net plankton accounted for 48% of the biomass, ranging from 27 to 70% (winter, 1973). Ultraplankton and large nannoplankton were somewhat less important, accounting for 24 and 21%, respectively. On a seasonal basis large nannoplankton ranged from 9 to 33%, and ultraplankton from 13 to 43%.

Plant pigments, another index of phytoplankton biomass, were also measured by Mills (1975). Annual ranges are summarized in Table 25, and seasonal concentrations of chlorophyll *a* and phaeo pigments are plotted in Fig. 27. Chlorophyll concentrations were usually higher during the warmer months, but reasonably high concentrations did persist until December 1972. Concentrations during the spring and summer of 1972 were higher than in 1973.

Owasco. In terms of species number, chlorophytes dominated during 1972–1973, accounting for 43% of the forms observed. Most occurred dur-

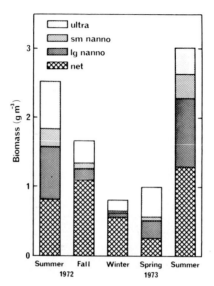

Fig. 26. Seasonal distribution of phytoplankton biomass in Hemlock Lake, and the relative contribution of the different size categories. (data from Mills, 1975). Key to size categories: net, netplankton (cells or colonies with maximum dimension 70 μm); lg nanno, large nannoplankton (maximum dimension 20–70 μm); sm nanno, small nannoplankton (maximum dimension 10–20 μm); ultra, ultraplankton (maximum dimension 10 μm).

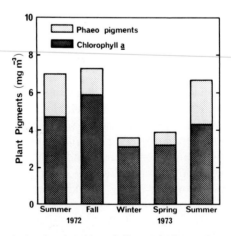

Fig. 27. Seasonal changes in chlorophyll *a* and phaeo pigments in Hemlock Lake (data from Mills, 1975).

TABLE 25

Annual Range of Chlorophyll a and Phaeo Pigments in Hemlock, Owasco, and Skaneateles[a]

	Chlorophyll a		Phaeo pigments	
Lake	1972	1973	1972	1973
Hemlock	Trace—16.7	Trace—11.5	Trace—7.7	Trace—15.2
Owasco	Trace—8.3	Trace—11.5	Trace—6.1	Trace—6.5
Skaneateles	Trace—4.0	Trace—3.7	Trace—5.4	Trace—2.4

[a] Concentrations are in mg m^{-3}.

ing the summer months, with the common species being *Quadrigula lacustris, Scenedesmus bijuga, Staurastrum gracile,* and *S. natator* var. *crassum.* Cryptomonads, *Dinobryon* spp., *Chromulina* spp., *Peridinium cinctum,* and *Gymnodinium caudatum,* were some of the common flagellated forms. *Cladomonas fruticulosa* and *Cryptomonas nasuta* commonly occurred during the winter, and were occasionally detected in the deep, cold waters during the summer. *Asterionella formosa, Cyclotella* sp., *Fragilaria crotonensis,* and *Synedra rumpens* were important components of the winter–spring diatom community.

Cell numbers ranged from about 1×10^3 cells ml^{-1} to about 13×10^3 cells ml^{-1}. Net- and ultraplankton were alternately dominant (Mills, 1975). The former were important during the stratified season, with a shift to the smaller forms occurring during late summer–early fall. The nannoplankton varied in importance.

Phytoplankton biomass in the upper 10 m of the water column varied by an order of magnitude from about 0.5 to 5 gm m^{-3}. Concentrations were usually higher during the warmer months, but a maximum of over 5 mg m^{-3} was observed during late winter 1973 (Mills, 1975). This increase was primarily due to a population of the dinoflagellate *Peridinium cinctum* which had an estimated biomass of 4.4 gm m$^{-3.}$

Netplanktonic forms usually made up a substantial portion of the biomass, accounting for from 55 to 76% of the total (Fig. 28). Averaged over the whole survey they accounted for 68% of the biomass. Small nannoplankters were most evident during the fall of 1972 (26%). The ultraplankton reached a maximum of 70% in early June 1973, but never exceeded 40%, and were usually much less thereafter (Mills, 1975). Seasonally, ultraplankton accounted for from 6 to 12% of the biomass.

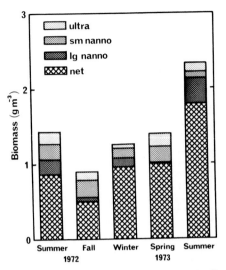

Fig. 28. Seasonal distribution of phytoplankton biomass in Owasco Lake, and the relative contribution of the different size categories (data from Mills, 1975). Key to size categories: net, net plankton (cells or colonies with maximum dimension 70 μm); lg nanno, large nannoplankton (maximum dimension 20–70 μm); sm nanno, small nanno-plankton (maximum dimension 10–20 μm); ultra, ultra plankton (maximum dimension 10 μm).

Diatoms, dinoflagellates, and cryptomonads were the dominant taxonomic groups in terms of biomass. In general, the diatoms dominated during the warmer months and the latter two during the cooler periods (Fig. 29). Diatoms were dominant, comprising at least 40% of the biomass, except during the winter of 1973 when they decreased to very low levels. The dinoflagellates and cryptomonads were essentially equal in biomass when averaged over the whole survey (25% and 21%, respectively), but tended to be quite variable seasonally (Fig. 29). The dinoflagellates ranged from 1 (fall, 1973) to 72% (winter, 1973), and the cryptomonads from 6 (summer, 1973) to 38% (fall, 1972). The three remaining taxonomic groups usually accounted for only a small percentage of the biomass, exceptions being the summer of 1972 when the nondiatom chrysophytes comprised 12% of the biomass, and again in the summer of 1973 when they represented 13%. During the latter period, blue–greens accounted for 18% of the total phytoplankton biomass.

Plant pigment annual ranges are summarized in Table 25, and seasonal change in chlorophyll *a* and phaeo pigments are plotted in Fig. 30. During 1972 relatively high pigment concentrations persisted in the upper 10 m of the water column through December. A sharp increase occurred in May of

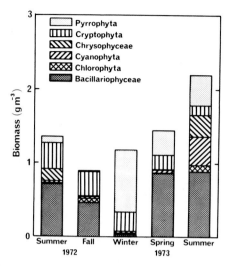

Fig. 29. Seasonal distribution of phytoplankton biomass in Owasco Lake, and the relative contribution of the various taxonomic groups (data from Mills, 1975).

1973, and pigment concentrations appeared to be somewhat higher that summer than in 1972.

Skaneateles. Skaneateles, the most oligotrophic of the Finger Lakes, has the fewest species, lowest cell numbers and biomass, and the smallest seasonal range of cell numbers and biomass. Chlorophyte and diatom species were about equal in number, and accounted for 67% of the forms

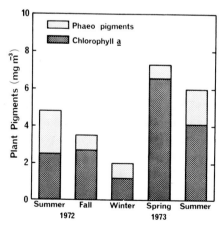

Fig. 30. Seasonal changes in chlorophyll *a* and phaeo pigments in Owasco Lake (data from Mills, 1975).

observed. The green algae were generally common during the summer, being represented by *Oocystis lacustris, Quadrigula lacustris, Scenedesmus bijuga, Selenastrum minutum,* and *Tetraëdron minimum.* Diatoms such as *Cyclotella* sp., *Gomphonema olivaceum, Navicula minima, Synedra rumpens,* and *Rhizosolenia eriensis* were commonly found during the winter and spring. The Chrysophyceae were most conspicuous during the warmer months and included such genera as *Dinobryon, Chromulina, Mallomonas,* and *Ochromonas.* The dinoflagellates *Gymnodinium caudatum* and *Peridinium cinctum* were frequently observed, as were the blue–greens *Gomphosphaeria lacustris, G. lacustris* var. *compacta, Merismopedia tenuissima,* and *Phormidium tenue.*

Netplankton and ultraplankton dominated in terms of cell numbers which ranged from about 9×10^3 cell ml^{-1} in June 1972 to 1×10^3 cell ml^{-1} in January 1973. Ultraplanktonic forms were most important during the fall and winter of 1972–1973, followed by a shift in size distribution to netplankton in the spring of 1973. Large nannoplankters made up a significant portion of the total cell numbers during July and August of 1973 (Mills, 1975).

Phytoplankton biomass in the upper 10 m of the water column rarely exceeded 1.7 gm m^{-3} during Mills' 1972–1973 survey, one exception being early August 1972 when a maximum of slightly more than 3 gm m^{-3} was observed. In general, the three smaller size categories dominated, at times comprising over 90% of the biomass, although the netplankton did on occasion account for at least 60%. On a seasonal basis phytoplankton biomass ranged from 0.2 (spring, 1973) to 1.1 gm m^{-3} (summer, 1972). Net plankton accounted for from 14 to 30% of the total, large nannoplankton 14 to 58%, small nannoplankton 5 to 38%, and ultraplankton 11 to 37% (Fig. 31). Averaged over the whole study period the four size categories accounted for 21, 38, 15, and 25% of the biomass, respectively.

The relative biomasses of the different taxonomic groups varied. In general, blue–greens were dominant in early summer 1972 followed by the nondiatom chrysophytes in August, when *Dinobryon sertularia* alone reached a maximum concentration of 2.8 gm m^{-3}. The cryptomonads were important during the cooler months, with the diatoms gradually increasing in importance up until June 1973 when they comprised almost 90% of the biomass. Important species contributors were *Gomphosphaeria lacustris, Merismopedia tenuissima, Phormidium tenue, Mallamonas* spp., *Cryptomonas* spp., *Asterionella formosa, Fragilaria crotonensis, Synedra rumpens, Cyclotella* sp., and *Peridinium cinctum* (Mills, 1975).

The seasonal dominance by the various taxonomic groups underwent considerable variation. The most noticeable involved members of the Chrysophyta during the summers of 1972 and 1973. In 1972 members of the

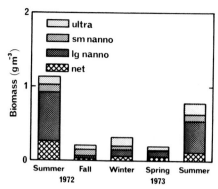

Fig. 31. Seasonal distribution of phytoplankton biomass in Skaneateles Lake, and the relative contribution of the different size categories (data from Mills, 1975). Key to size categories: net, net plankton (cells or colonies with maximum dimension 70 μm); lg nanno, large nannoplankton (maximum dimension 20–70 μm); sm nanno, small nanno-plankton (maximum dimension 10–20 μm); ultra, ultra plankton (maximum dimension 10 μm).

Chrysophyceae and Xanthophyceae accounted for 50% of the biomass, whereas the diatoms made up only 6%. In 1973 the diatoms accounted for 67% and the other chrysophytes 13%. The reason for the shift in taxa is not known but is obviously related to the *Dinobryon sertularia* bloom mentioned in the previous paragraph. Averaged over the whole survey period, the diatoms accounted for 45% of the phytoplankton biomass, the cryptophytes 21%, nondiatom chrysophytes 12%, dinoflagellates 9%, cyanophytes 7%, and chlorophytes 5%. Seasonal ranges were as follows: diatoms, 6–84%; cryptomonads, 8–61%; other chrysophytes, 0–50%; dino-flagellates, 0–16%; blue–greens, 3–15%; and greens, 2–8% (see also Fig. 32).

Plant pigment concentrations in Skaneateles Lake were consistently low (Table 25). Chlorophyll *a* plus phaeo pigment rarely exceeded 4 mg m^{-3}. Seasonal concentrations are summarized in Fig. 33.

A number of important species found in Skaneateles have been associated with unproductive waters (Mills, 1975). *Synedra rumpens* and *Oocystis lacustris* are found in large, deep, oligotrophic lakes (Hutchinson, 1967), with the latter species also being common in unenriched Canadian Shield lakes (Schindler and Nighswander, 1970; Kling and Holmgren, 1972). *Rhizosolenia eriensis* has been found in nutrient-poor offshore waters of Lake Michigan (Holland and Beeton, 1972) and in the more unproductive of the Muskoka lakes (Michalski *et al.*, 1973). Some of the forms appear to be more eutrophic, e.g., *Dinobryon* is often associated with oligotrophic waters but may also be found in more productive lakes (Hutchinson, 1967). The cryptomonads do not appear to be related to trophic status (Michalski *et al.*, 1973), while *Fragilaria crotonensis* appears to prefer conditions

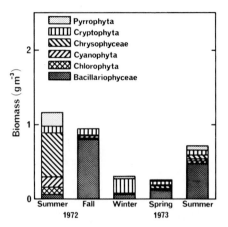

Fig. 32. Seasonal distribution of phytoplankton biomass in Skaneateles Lake, and the relative contributions of the various taxonomic groups (data from Mills, 1975).

approaching eutrophy (Vollenweider and Nauwerck, 1961; Hutchinson, 1967; Michalski *et al.*, 1973; Davis, 1964).

Synoptic Survey. During 1973 all of the Finger Lakes were sampled on three different occasions, once each in April, June, and August (at 0, 5, and 10 m, and just off the bottom), in an effort to gain some understanding of similarities and differences. Only the three upper depths were used for present purposes. Cell counts are summarized in Fig. 34, plant pigments in Fig. 35, and percent composition of the various taxonomic groups in Table 26. Important species and their concentrations are listed in Table 27.

Cell concentrations varied considerably from lake to lake, ranging from 0.8×10^3 cells ml^{-1} in Skaneateles in June to 17×10^3 cells ml^{-1} in Hemlock during the same month, and in some instances within a given lake, e.g., Hemlock had a concentration of 2.5×10^3 cells ml^{-1} in April and 17×10^3 cells

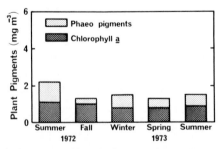

Fig. 33. Seasonal changes in chlorophyll *a* and phaeo pigments in Skaneateles Lake (data from Mills, 1975).

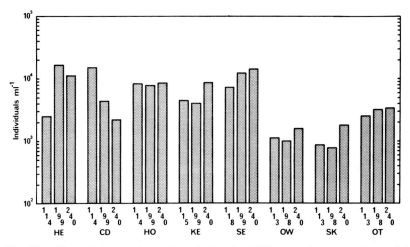

Fig. 34. Phytoplankton cell numbers during the 1973 synoptic survey. Key to lake code: HE, Hemlock; CD, Canadice; HO, Honeoye; KE, Keuka; SE, Seneca; OW, Owasco; SK, Skaneateles; OT, Otisco.

ml^{-1} in June. Canadice exhibited a similar range over the survey period (Fig. 34). Canadice showed a definite spring peak in cell numbers, but in the other lakes there was a general increase as the growing season progressed. Cell numbers changed little in Honeoye.

Examination of the plant pigment data produces an entirely different picture in some instances. The spring peak is still evident in Canadice, but

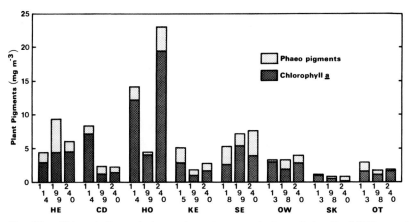

Fig. 35. Chlorophyll *a* and phaeo pigments concentrations during the 1973 synoptic survey Key to lake code: HE, Hemlock; CD, Canadice; HO, Honeoye; KE, Keuka; SE, Seneca; OW, Owasco; SK, Skaneateles; OT, Otisco.

TABLE 26

Relative Abundance of Phytoplankton Taxonomic Groups during the 1973 Synoptic Survey in Terms of Cell Numbers[a]

Lake	Calendar day	Percent[b]					
		Chlorophyta	Cyanophyta	Bacillariophyceae	Chrysophyceae	Cryptophyta	Pyrrophyta
Hemlock	114	11	16	43	17	12	<1
	199	2	64	25	5	3	<1
	240	15	58	16	7	3	<1
Canadice	114	<1	54	4	40	1	<1
	199	3	34	11	47	4	<1
	240	11	23	44	17	4	<1
Honeoye	114	5	<1	68	12	16	<1
	199	1	77	4	12	6	<1
	240	2	79	9	8	5	<1
Keuka	115	7	12	50	12	19	<1
	199	3	40	22	24	10	<1
	240	7	68	7	17	2	<1
Seneca	118	14	30	50	1	4	<1
	199	5	86	3	4	2	<1
	240	8	56	5	27	3	<1
Owasco	113	1	0	89	3	7	<1
	198	5	<1	37	29	27	2
	240	20	16	21	9	34	<1
Skaneateles	113	5	53	18	6	18	<1
	198	10	15	52	4	17	1
	240	7	45	26	7	13	1
Otisco	113	1	18	49	9	22	<1
	198	10	48	14	11	16	<1
	240	5	48	13	19	15	<1

[a] Members of the Xanthophyceae are included with Chrysophyceae.
[b] Percentages are in terms of cell numbers.

there are also spring and late summer peaks of considerable size in Honeoye. This does not appear to be due to a shift in the major taxonomic groups as the diatoms dominated in April (68% of cell numbers) and blue–greens in June and August (77% and 79% of cell numbers, respectively) (Table 26).

During the April survey *Asterionella formosa* was the dominant species in Honeoye, accounting for 60% of the total cell numbers. By June *Gomphosphaeria lacustris* had become dominant, and *A. formosa* had decreased to a very low concentration. In August, a second blue–green species, *Anabaena flos-aquae,* became abundant replacing *G. lacustris. Fragilaria crotonensis, Lyngbya bergei,* and *Chromulina ovalis* were also present in moderate numbers (Table 27).

In general, blue–green algae tended to be the dominant taxonomic group when averaged over the entire sampling period, accounting for from 37 to 57% of cell numbers. An exception to this was Hemlock Lake in which blue–greens accounted for only 5% while the diatoms made up 49%. In each of the other lakes, except Canadice, diatoms were the second most important group, contributing from 19 to 32%. The nondiatom chrysophytes were somewhat more important in Canadice. There was a general trend for blue–greens to increase in number and diatoms to decrease during the course of the study, but exceptions were noted, e.g., the reverse being true in Canadice.

The species composition as described in Table 27 varied enough between the other lakes to warrant at least a brief consideration at this point. Those lakes which have already been considered in this respect will not be included in the present discussion, and the reader is referred to the previous paragraphs concerning them. *Synura uvella* and *Merismopedia tenuissima* were the dominant species in Canadice Lake in April, reaching average concentrations of 6.0×10^3 and 7.9×10^3 cells ml^{-1}, respectively. Other important species were *Cryptomonas pusilla, Oscillatoria prolifica, Cyclotella* sp., and *Chromulina ovalis.* Cell concentrations were considerably lower in June. *S. uvella* had dropped below detection and was replaced by *Dinobryon bavaricum,* but in fewer numbers (1.5×10^3 cells ml^{-1}). *Merismopedia tenuissima* was second in importance, and *Asterionella formosa* and *D. sertularia* were fairly numerous, as were the four remaining April species. Cell concentrations again declined in August, being half those of June. *Merismopedia tenuissima* was still present (0.5×10^3 cells ml^{-1}), but *Dinobryon* was replaced by *Cyclotella* sp. (0.6×10^3 cells ml^{-1}). *Chromulina ovalis* and *Selenastrum minutum* were the only other forms present in any abundance. *Oscillatoria prolifica* had by this time disappeared from the surface waters, but was detected in large numbers (2.9×10^3 cells ml^{-1}) in the bottom waters.

In Keuka Lake *Cyclotella* spp. (1.9×10^3 cells ml^{-1}), *Cryptomonas*

406

TABLE 27

Phytoplankton Abundance[a] Determined during the 1973 Synoptic Sampling of Lakes

	Lake and calendar day											
	Hemlock			Canadice			Honeoye			Keuka		
Taxa	114	199	240	114	199	240	114	199	240	114	199	240
Cyanophyta												
Anabaena flos-aquae			18					648	4880			
Anabaeba scheremetievi			52									
Coelosphaerium pallidum			142									
Gomphosphaeria lacustris		167	88					6783			5181	
Lyngbya birgei									738			
Lyngbya limnetica	130	116										
Merismopedia tenuissima	532	9026	3569	7922	754	502		133		516	1600	626
Oscillatoria prolifica				147								
Phormidium tenue												
Chlorophyta												
Carteria cordiformis							214					
Cosmarium bioculatum			164									
Cosmarium botrytus			73									
Crucigenia quadrata												
Kirchneriella lunaris												
Oocystis lacustris			178									
Scenedesmus bijuga	242	84	1052									

Taxon											
Selenastrum minutum		54			97							178
Sphaerocystis schroeteri		80	60									
Stylosphaeridium stipitatum					52				192			
Chrysophyceae												
Biococca socialis	170	273										
Chromulina ovalis		442	760	140	236	133	1069	1114	502	531	922	1364
Cladomonas fruticulosa	1						176					
Dinobryon bavaricum	82				1488							
Dinobryon sertularia					203							
Ochromonas sp.	74	44	52	140	110	81			354			
Synura uvella				5995								
Uroglenopsis americana												
Bacillariophceae												
Asterionella formosa	715	840	154		216		6455		132			
Cyclotella spp.		938	1726		140	579	361			1895	900	
Fragilaria crotonensis			266				328	610				
Fragilaria virescens												
Melosira granulata	96							257				
Nitzschia acicularis	120											
Rhizosolenia brevispina		192										
Synedra radians					59							
Synedra rumpens												
Cryptophyta												
Cryptomonas pusilla	302	457		192	176		1567		811			
Cryptomonas ovata			332		73		240	198		398		

(Continued)

407

Table 27 *(Continued)*

Taxa	Seneca 118	Seneca 199	Seneca 240	Owasco 113	Owasco 198	Owasco 240	Skaneateles 113	Skaneateles 198	Skaneateles 240	Otisco 113	Otisco 198	Otisco 240
Cyanophyta												
Anabaena flos-aquae												127
Anabaeba scheremetievi												
Coelosphaerium pallidum												
Gomphosphaeria lacustris	1631	143	3794				88	15	389			
Lyngbya birgei												
Lyngbya limnetica	841		241									
Merismopedia tenuissima		10574	3908							472	1416	1446
Oscillatoria prolifica												
Phormidium tenue								177				
Chlorophyta												
Carteria cordiformis												
Cosmarium bioculatum												
Cosmarium botrytus												
Crucigenia quadrata						177						
Kirchneriella lunaris							45		59			
Oocystis lacustris			294									

	1	2	3	4	5	6	7	8	9	10	11	12
Scenedesmus bijuga	1088											
Selenastrum minutum	317	420	221									
Sphaerocystis schroeteri												
Stylosphaeridium stipitatum							44					
Chrysophyceae												
Bicoeca socialis												
Chromulina ovalis	295	435				81				177	339	575
Cladomonas fruticulosa												
Dinobryon bavaricum				416								
Dinobryon sertularia		170		88		29						
Ochromonas sp.												
Synura uvella		3407										
Uroglenopsis americana												
Bacillariophceae												
Asterionella formosa			167	370						579		
Cyclotella spp.	3400	214	756	870	14	162	81	125	427	545	405	361
Fragilaria crotonensis												
Fragilaria virescens												
Melosira granulata												
Nitzschia acicularis												
Rhizosolenia brevispina												
Synedra radians							81	649				
Synedra rumpens												
Cryptophyta												
Cryptomonas pusilla	236	441	66	243	236		140	125	132	295	472	472
Cryptomonas ovata									66			

[a] Units are cells ml^{-1}.

pusilla (0.8 × 10³ cells ml⁻¹), *Chromulina ovalis* (0.5 × 10³ cells ml⁻¹), and *Merismopedia tenuissima* (0.5 × 10³ cells ml⁻¹) were the dominant species during the April survey. In June four of these were still important, but their relative positions had changed: *M. tenuissima* had tripled in concentration to 1.6 × 10³ cells ml⁻¹; *C. ovalis* increased to a lesser extent (0.9 × 10³ cells ml⁻¹); and *Cyclotella* spp. and *Cryptomonas pusilla* both decreased to 0.9 × 10³ and 0.4 × 10³ cells ml⁻¹, respectively. By August another blue–green, *Gomphosphaeria lacustris,* had appeared and was by far the dominant species with a concentration of 5.2 × 10³ cells ml⁻¹. *Chromulina ovalis* (1.4 × 10³ cells ml⁻¹) and *M. tenuissima* (0.6 × 10³) were of lesser importance.

Cyclotella spp. (3.4 × 10³ cells ml⁻¹), *Gomphosphaeria lacustris* (1.6 × 10³ cells ml⁻¹), and *Lyngbya limnetica* (0.8 × 10³ cells ml⁻¹) were major phytoplankters in Seneca during the April survey. An additional species, *Scenedesmus bijuga,* was also present in reasonable numbers (1.0 × 10³ cells ml⁻¹), and is of interest because of the fact that although the green algae only accounted for 9% of the total number of cells it was due almost entirely to this one species. Blue–greens completely dominated the phytoplankton by June, comprising 85% of the total. *Merismopedia tenuissima* was of major importance with a concentration of 10.6 × 10³ cells ml⁻¹, while *G. lacustris* played a very minor role (0.1 × 10³ cells ml⁻¹). The latter species increased in August to 3.8 × 10³ cells ml⁻¹, and was codominant with the chrysophyte *Uroglenopsis americana* (3.4 × 10³ cells ml⁻¹) and *M. tenuissima* (3.9 × 10³ cells ml⁻¹). At this point, the blue–greens accounted for 56% of the phytoplankton cell numbers, and the chrysophytes 27%.

Diatoms were the dominant taxonomic group in Otisco Lake in April, being represented by *Asterionella formosa* (0.6 × 10³ cells ml⁻¹) and *Cyclotella* sp. (0.5 × 10³ cells ml⁻¹). *Cryptomonas pusilla* (0.3 × 10³ cells ml⁻¹) and *Merismopedia tenuissima* (0.5 × 10³ cells ml⁻¹) were present in similar concentrations. The diatoms were replaced in importance by the blue–greens in June, with *M. tenuissima* being the dominant species. *Cryptomonas pusilla* (0.5 × 10³ cells ml⁻¹), *Cyclotella* sp. (0.4 × 10³ cells ml⁻¹), and *Chromulina ovalis* (0.3 × 10³ cells ml⁻¹) were the other major forms. August differed little from June as to the relative composition of the various taxonomic groups, and the same species that dominated in June did so in August at about the same concentration.

The foregoing may well be interesting in a descriptive sense, but so far no attempt has been made to relate the differences that have been observed to the dynamics of the lake ecosystems. Unfortunately, such an attempt is beyond the scope of the present publication, and any elucidation of these phenomena awaits further analysis and experimentation.

Species Richness. Of the four lakes studied by Mills (1975) during 1972–1973, Hemlock was the richest in taxa with 203 forms, including 89 genera and 187 species. Owasco was second with 151 forms, 82 genera, and 140 species. Skaneateles had 129 forms, 71 genera, and 127 species.

When only the samples from the synoptic survey were considered, Hemlock still had the richest flora with 92 forms, 56 genera, and 88 species. Seneca was a very close second with 90 forms, 57 genera, and 84 species. Honeoye (61 forms, 46 genera, and 61 species) and Skaneateles (66 forms, 48 genera, and 65 species) had the least diverse planktonic flora.

Historical Comparisons.   There is a dearth of historical phytoplankton data, and those survey results that do exist (Birge and Juday, 1914, 1921; Muenscher, 1928) do not permit accurate quantitative comparison with contemporary results. The first published survey of the Finger Lakes phytoplankton was conducted by Birge and Juday (1914) in 1910, when net samples were collected from all but Honeoye. In general, they found that three different groups of algae were well represented: the Chlorophyceae by *Staurastrum;* the diatoms by *Melosira, Cyclotella, Tabellaria, Fragilaria, Synedra, Asterionella,* and *Navicula;* and the blue-greens by *Anabaena, Aphanizominon, Lyngbya, Oscillatoria, Coelosphaerium, Clathrocystis, Gloeocapsa,* and *Aphanocapsa.*

In terms of individual lakes, Otisco was dominated by blue-greens (*Clathrocystis* and *Coelosphaerium*) as was Owasco (*Clathrocystis*). Diatoms (predominantly *Asterionella, Fragilaria,* and *Tabellaria*) were the most important forms in the other lakes. Other species of importance were *Ceratium,* found in all the lakes, and *Dinobryon,* which occurred in seven out of ten, but was abundant only in Owasco. *Mallomonas* was restricted to the metalimnion of Canadice and Otisco, reaching a maximum of 2.1 cells ml^{-1} in the former.

Birge and Juday (1921) sampled the net- and nannoplankton of Seneca on August 1, 1918. The most important forms were *Cryptomonas,* "monads," *Aphanocapsa, Oocystis, Stephanodiscus,* and *Synedra.* The mean concentration in the 0 to 10 m layer was 263.8×10^6 individuals m^{-3}. Seneca had a concentration of 14.2×10^9 cells m^{-3} in August 1973, but direct comparison between the two values is not possible since it is not clear from the data of Birge and Juday whether *Aphanocapsa* individuals represent colonies or cells, a factor that could effect cell numbers by two orders of magnitude. With taxa that permit direct comparison (Table 28) the 1973 counts appear to be considerably higher.

Muenscher (1928) sampled the phytoplankton of Seneca Lake at two stations (north and south) on four occasions in June, July, August, and September 1927. *Asterionella* was the most common diatom, being abundant in the surface waters early in the stratified season. *Fragilaria* was present toward the latter part of the season. Trace amounts of *Tabellaria* were found near the surface. The blue-green *Anabaena* was also found during the

TABLE 28

Abundance of Dominant Phytoplankton Taxa in Seneca Lake during August 1918 and 1973[a]

Taxa	1918 individuals ml^{-1}	1973 cells ml^{-1}
Chlorophyta		
Oocystis	10	
Oocystis lacustris		294
Scenedesmus bijuga		420
Selenastrum minutum		221
Pyrrophyta		
Ceratium	61 × 10^{-3}	
Cryptophyta		
Cryptomonas	20	
Cryptomonas pusilla		441
Chrysophyceae		
Dinobryon	2 × 10^{-3}	
Monads	102	
Chromulina ovalis		435
Ochromonas sp.		170
Uroglenopsis americana		3407
Bacillariophyceae		
Cyclotella spp.		756
Melosira	2 × 10^{-3}	
Stephanodiscus	16	
Asterionella	135 × 10^{-3}	
Fragilaria	17 × 10^{-3}	
Synedra	1	
Tabellaria	8 × 10^{-3}	
Cyanophyta		
Anabaena	2 × 10^{-3}	
Aphanocapsa	115	
Gomphosphaeria lacustris		3794
Lyngbya limnetica		241
Microcystis	8 × 10^{-3}	
Merismopedia tenuissima		3908
Total	264	14087

[a] The 1918 counts are from Birge and Juday (1921). Counts represent average concentrations in upper 10 m of water column.

latter part of the season in the surface waters, and on August 20 *Microcystis* was detected in all but the deepest waters. *Ceratium* was also an August species. *Dinobryon* occurred irregularly in the samples. Members of the Chlorophyceae were rare, but *Actinastrum*, *Sphaerocystis*, and *Staurastrum* were found. Several nannoplankters were also observed in reasonable numbers. *Gleotheca* and *Gomphosphaeria* were the most abundant cyanophytes, occurring in large numbers during August and September from 0 to 25 m. *Scenedesmus* and *Oocystis* were the most abundant greens. *Scenedesmus* was rather common in July, August, and September, but *Oocystis* was not found during July and August. Table 28 lists some of the other more commonly observed species.

A rough estimate of cell numbers can be obtained from figures presented in Muenscher (1928), and these are listed in Table 29. The counts are in individuals per liter, and range from 21×10^3 to 164×10^3 with the highest concentration being measured in September. These numbers are of the same order of magnitude as those reported by Birge and Juday (1921), but considerably below present day levels.

Counts of the dominant genera present in Keuka (Anonymous, 1974b), Seneca (Anonymous, 1974d), and Owasco (Anonymous, 1974c) during April, July, and October 1972 are listed in Table 30. The concentrations measured in Keuka are similar, but those for Seneca are about five times lower than those measured by us in 1973.

Zooplankton

Problems of similar complexity to those encountered in studying the phytoplankton must also be contended with when considering the zooplankton. Lake depth affects both habitat availability and suitability (e.g., through hypolimnetic oxygen depletion). Major predators may differ considerably in the various lakes. For example, the alewife is present in some lakes but absent in others. In those lakes where it does occur the period over which it has been present ranges from a few to hundreds of years. As a result, the present treatment of zooplankton communities is primarily descriptive in nature.

A survey of the zooplankton in the Finger Lakes was conducted in April, June, and August 1973 (Chamberlain, 1975). A species checklist for the lakes being considered here is presented in Table 31. The 61 species collected were distributed among the major taxonomic groups as follows: calanoid copepods, 7; cyclopoid copepods, 6; cladocerans, 15; malacostracans, 3; protozoans, 6; and rotifers, 24. Twelve of the species were not found in the eight lakes being considered in this document, i.e., *Limnocalanus macrurus* (only in Canandaigua), *Cyclops scutifer* (Cayuga), *Ceriodaphnia laticaudata* (Cayuga), *Gammarus lacustris* (Cayuga), *Codonella cratera* (Cayuga), *Thecacineta cothurnoides* (Cayuga), *Chromogaster ovalis* (Cayuga), *Gastropus hyptopus* (Cayuga), *Hexarthra mira* (Cayuga), and *Notholca acuminata* (Cayuga).

Hemlock had the greatest number of species (29), and its neighbor Canadice the fewest (20); Canadice and Honeoye only had two copepod species, whereas the other lakes had from five to seven.

TABLE 29

Seneca Lake Phytoplankton Data (Muenscher, 1928) Averaged for 0–10 m[a]

Location	Date	Plankton concentration						
		Netplankton			Nannoplankton			
		Protozoa	Algae	Diatoms	Protozoa	Algae	Diatoms	Total
South end of lake	6-24	—	—	500	—	—	20,000	21×10^3
	7-16	<100	—	250	6025	60,000	50,000	116×10^3
	8-20	1000	1700	250	100–1000	100,000	20,000	123×10^3
	9-6	1500	100–500	100–200	2000	140,000	20,000	164×10^3
North end of lake	6-22	<100	—	700	<100	50,000	20,000	71×10^3
	7-18	100–500	—	100–200	3000	60,000	50,000	113×10^3
	8-18	1500	1700	600	2500	60,000	12,500	79×10^3
	9-7	1000	100–500	400		120,000	22,500	144×10^3

[a] Categories are as presented in Muenscher. Important genera listed below. Counts in individuals liter^{-1}.

Important Genera

Net

Protozoa

Ceratium: Absent in June, rather abundant in August (0–15 m).

Dinobryon: Irregular occurrences.

Algae

Anabaena: Near surface, latter part of season.

Microcystis: Found in all but deepest samples in August, only traces in September.

Actinastrum, *Sphaerocystis*, *Staurastrum*: Were found, but Chlorophyceae were generally rare.

Diatoms

Asterionella: Most common diatom, abundant near surface early in season.

Fragilaria: Found near latter part of season.

Tabellaria: Only traces near surface.

Nanno

Algae

Gloeothece and *Gomphosphaeria*: Two most abundant blue–greens, found in large numbers August and September (0–25 m).

Microcystis: Fairly common to north in August.

Gloeocapsa, *Aphanocapsa*, and *Merismopedia*: Last two species in large numbers at south end, few to the north.

Scenedesmus and *Oocystis*: Most abundant greens, *Scenedesmus*, common in July, August, and September with trace amounts in June, *Oocystis* occurred in July and August.

Characium, *Coelastrum*, *Crucigenia*, *Gloeotaenium*, *Pediastrum*, *Quadrigula*, *Elacatothrix*, *Cosmarium*, *Closterium*, *Eudorina*, and *Teträdron*: Also observed.

TABLE 30

Dominant Phytoplankton Genera in Keuka, Seneca, and Owasco during April, July, and October 1972[a,b]

	Keuka			Seneca			Owasco		
	5/27	7/21	10/14	5/16	7/23	10/14	5/28	7/24	10/12
Chlorophyta									
Kirchneriella									
Oocystis			679						
Schroederia		778		705	80				
Pyrrophyta									
Peridinium				126					
Crytophyta									
Cryptomonas	550			136	120	211			
Chrysophyta—Chrysophyceae									
Dinobryon		2296		931	80	532	585	530	402
Chrysophyta—Bacillarophyceae									
Centrales									
Cyclotella		271			141		120		
Melosira									502

Pennales								
Asterionella	282	1212		362		482		
Fragilaria	521	832					548	743
Navicula	166						42	
Rhizosolenia								
Synedra							48	
Cyanophyta								
Anabaena	181					211		
Chroococcus			3321					
Merismopedia								
Oscillatoria			679		74			
Polycystis (Microcystis)			679				253	110
Rhaphidiopsis							386	60
Unidentified								
"Flagellates"			906			412	277	231
Other genera	341	452	2642	281	142	642	164	663
Total	2041	5841	8906	2541	647	2490	1627	2651

[a] Anonymous (1974b,c,d).
[b] All counts in number ml^{-1}.

TABLE 31

Zooplankton Species Checklist for 1972–1973[a]

Taxa	Hemlock	Canadice	Honeoye	Keuka	Seneca	Owasco	Skaneateles	Otisco
Calanoid Copepods								
Diaptomus minutus	x	x	—	—	—	x	x	x
Diaptomus oregonensis	—	—	—	x	x	—	—	x
Diaptomus sicilis	—	—	—	x	x	x	x	x
Epischura lacustris	—	—	—	x	—	x	x	x
Epischura nordenskiöldi	—	—	—	—	—	x	x	x
Limnocalanus macrurus	—	—	—	—	—	—	x	—
Senecella calanoides	—	—	—	—	x	x	x	—
Cyclopoid Copepods								
Cyclops bicuspidatus	x	—	—	x	x	—	x	—
Cyclops scutifer	—	—	—	—	—	—	—	—
Cyclops varicans	x	—	—	—	—	—	—	—
Cyclops vernalis	x	x	x	—	x	x	x	x
Mesocyclops edax	x	—	x	x	—	—	—	x
Tropocyclops prasinus	—	—	—	x	x	—	—	—
Cladocerans								
Bosmina longirostris	x	x	x	x	x	x	x	x
Ceriodaphnia laticaudata	—	—	—	—	—	—	—	—
Ceriolaphnia pulchella	x	—	—	—	x	—	x	—
Ceriodaphnia quadrangula	—	—	—	x	x	—	—	—
Ceriodaphnia reticulata	—	—	—	x	—	—	—	—
Chydorus sphaericus	x	x	x	—	—	—	x	—
Daphnia dubia	—	—	—	x	—	x	—	x
Daphnia galeata mendotae	—	—	x	x	x	x	x	x
Daphnia longiremis	x	x	x	—	—	x	—	x
Daphnia parvula	—	—	—	x	x	—	—	x

418

Taxon	1	2	3	4	5	6	7	8
Daphnia pulex	x	x	—	x	—	—	—	x
Daphnia retrocurva	—	x	x	x	x	x	x	x
Diaphanosoma leuchtenbergianum	x	—	x	—	x	x	x	x
Leptodora kindtii	x	x	x	x	x	x	x	x
Polyphemus pediculus	—	—	—	x	—	—	—	—
Malacostracans								
Gammarus lacustris	—	—	—	—	—	—	—	—
Mysis relicta	—	x	x	x	x	x	x	x
Pontoporeia affinis	—	—	x	x	—	—	—	—
Protozoans								
Codonella cratera	—	—	—	—	—	—	—	—
Difflugia oblonga	—	x	x	—	—	x	—	—
Difflugia lebes	x	x	x	x	x	x	x	x
Difflugia tuberculata	—	—	—	—	—	—	—	x
Glaucoma scintillans	x	x	x	x	—	—	—	—
Thecacineta cothurnoides	—	—	—	—	—	—	—	—
Rotifers								
Asplanchna priodonta	x	x	x	x	x	x	x	x
Ascomorpha sultans	x	x	—	—	—	—	x	—
Brachionus calyciflorus	—	—	—	—	—	—	—	x
Chromogaster ovalis	x	x	x	x	x	x	x	—
Chonochilus uricornis	—	—	x	x	x	x	—	x
Filinia longiseta	—	—	—	—	—	—	—	—
Gastropus hyptopus	—	—	—	—	—	—	—	—
Gastropus minor	x	—	—	—	—	—	—	—
Hexarthra mira	—	—	—	—	—	—	—	—
Kellicottia longispina	x	x	x	x	x	x	x	x
Keratella cochlearis	x	x	x	x	x	x	x	x
Keratella hiemalis	—	—	—	—	—	—	—	—
Keratella quadrata	x	—	x	x	x	x	x	x
Monostyla quadridentata	x	x	x	x	x	x	x	x

(Continued)

TABLE 31 (Continued)

Taxa	Lake							
	Hemlock	Canadice	Honeoye	Keuka	Seneca	Owasco	Skaneateles	Otisco
Notholca acuminata	—	—	—	—	—	—	—	—
Ploesoma hudsoni	x	—	—	x	x	x	—	—
Ploesoma lenticulare	x	x	—	—	—	—	—	—
Ploesoma triacanthum	—	—	—	—	x	x	—	—
Ploesoma truncatum	x	x	—	x	x	x	x	—
Polyarthra euryptera	x	—	x	x	x	x	—	—
Polyarthra vulgaris	x	x	x	x	x	x	x	x
Synchaeta stylata	—	x	—	—	x	—	—	—
Trichocerca longiseta	—	—	x	—	—	—	—	—
Trichocerca multicrinis	x	x	x	x	x	x	x	—

[a] Presence indicated by x.

Historical Comparisons. Historical comparisons of a sort can be made with the data of Birge and Juday (1914) (Table 32), Hall and Waterman (1967) (Table 33), and Chamberlain (1975) (Table 31). Exact comparisons are not possible because in most instances the organisms were identified in the earliest study only to genus level.

Birge and Juday (1914) list the occurrence of 20 forms (Table 32). During their survey they found four calanoid copepods, at least two cyclopoids (reported as *Cyclops* spp.), eight cladocerans, and seven rotifers. Birge and Juday reported *Daphnia hyalina* as occurring in all but Seneca and Honeoye, but the species was not observed by either Hall and Waterman (1967) or Chamberlain (1975). The rotifers *Anuroes* and *Notholca* reported in the 1910 survey were not found by Chamberlain in his 1972–1973 samples. Hall and Waterman (1967) and Chamberlain (1975) list a greater number of taxa than Birge and Juday (1914) even when the various forms are only considered to genus level.

The overall results of Hall and Waterman (1967) and Chamberlain (1975) are roughly comparable (Tables 31 and 33), but some differences can be noted. Three of the species found by the former were not observed by Chamberlain, i.e., *Ergasilus chautauguaensis* and *Daphnia ambigua* in Otisco, and *Holopedium gibbernum* in Keuka. Of those species observed in both surveys, Hall and Waterman reported eight occurrences in specific lakes not reported by Chamberlain: one in Seneca and two each in Canadice, Skaneateles, and Otisco; whereas Chamberlain reported 17 not listed by Hall and Waterman: one each in Honeoye and Owasco, two each in Canadice and Otisco, three in Skaneateles, and four each in Hemlock and Seneca.

Chamberlain (1975) has made an historical comparison of zooplankton abundance between the 1910 samples (Birge and Juday, 1914) and those that he collected in 1972–1973 in Hemlock, Owasco, and Skaneateles. To do this, he converted the August 1910 counts to average number of individuals per cubic meter in the water column. Contemporary estimates were made from samples taken during the period from about two weeks before the 1910 sampling date and extending two weeks beyond. This involved the pooling of data from four or five sampling dates. The results are summarized in Table 34.

In general, the standing crops were greater in 1972–1973 than in 1910. Chamberlain (1975) points out that the 1973 data for the small plankters are probably more comparable to those from the Birge and Juday (1914) study because a smaller mesh net (no. 20) was used than in 1972 (no. 10). Chamberlain infers from the number of nauplii and small rotifers taken in 1910 that a no. 20 or 25 mesh net was used.

TABLE 32

Zooplankton Species Reported in August 1910[a,b]

Taxa	Lake							
	Hemlock	Canadice	Honeoye	Keuka	Seneca	Owasco	Skaneateles	Otisco
Calanoid Copepods								
Diaptomus minutus	x	x	—	x	x	x	x	x
Diaptomus sicilis	—	—	—	—	x	—	—	—
Epischura	—	—	—	x	—	x	—	—
Senecella calanoides[c]	—	—	—	—	x	x	x	—
Cyclopoid Copepods								
Cyclops spp.	x	x	—	x	x	x	x	x
Cladocerans								
Bosmina	x	—	—	x	x	x	x	—
Ceriodaphnia	—	—	—	x	x	x	—	—
Daphnia hyalina	x	x	—	x	—	x	x	x
Daphnia pulex	—	x	—	—	—	—	—	—
Daphnia retrocurva	x	—	—	—	x	—	—	—
Diaphanosoma	x	x	—	—	—	x	x	x
Leptodora kindtii[d]	x	x	—	x	—	x	x	x
Polyphemus pediculus[d]	—	—	—	—	x	—	—	—
Rotifers								
Anuraea spp. (*Keratella* spp.)	x	x	—	x	x	—	—	x
Asplanchna	—	—	—	x	x	x	x	—
Conochilus	x	—	—	x	x	x	x	—
Notholca longispina (*Kellicottia longispina*)	x	x	—	x	x	—	—	x
Ploeosoma	—	—	—	—	x	—	—	—
Polyarthra	x	x	—	x	x	x	x	x
Triarthra (*Filinia*)	x	—	—	x	—	—	—	—

[a] Birge and Juday (1914).

[b] Presence indicated by x. Honeoye was not sampled. Revised taxonomic names are placed within parentheses.

[c] Originally reported as *Limnocalanus* (Juday, 1925).

Examination of Table 35 shows that cladocerans were more numerous in 1972–1973. The same is also generally so for copepods. Rotifers exhibited the greatest increase in number, especially in Hemlock where the 1973 average was 20 times that for 1910. Hemlock had a 4.8-fold increase in numbers of total comparable genera and a 7.4-fold increase when all genera were considered. Differences were less pronounced in Owasco and Skaneateles.

Relative Abundances.   Chamberlain (1975) conducted a detailed study of the zooplankton in Conesus, Hemlock, Owasco, and Skaneteles Lakes during 1972–1973. Two stratified seasons were included in the survey, but a change in net mesh size (no. 10 to no. 20) and sampling format just prior to the second season requires that the data be considered separately. During both stratified seasons, maximum zooplankton concentrations occurred in Hemlock with lower levels being observed in Owasco and Skaneateles, in that order (Fig. 36 and 37). The 1972 maxima were Hemlock, 2.1×10^5 individuals m^{-3}; Owasco, 1.2×10^5 individuals m^{-3}, and Skaneateles, 0.7×10^5 individuals m^{-3}. Hemlock had a similar maximum concentration in 1973 (2.5×10^5 individuals m^{-3}), but lowered abundances were reported

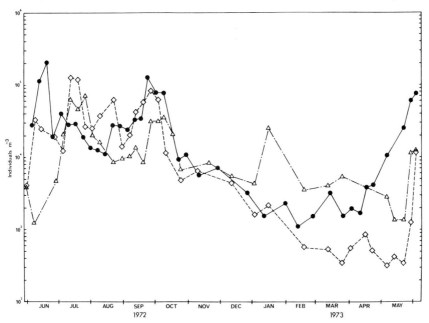

Fig. 36. Total zooplankton volumetric abundance in Hemlock (filled circles), Owasco (open diamonds), and Skaneateles (open triangles) during the period June 1972–May 1973.

TABLE 33

Zooplankton Species Reported in 1965[a,b]

Taxa	Lake							
	Hemlock	Canadice	Honeoye	Keuka	Seneca	Owasco	Skaneateles	Otisco
Calanoid Copepods								
Diaptomus minutus	x	x	—	—	—	x	x	x
Diaptomus oregonensis	—	—	—	x	x	—	—	x
Diaptomus sicilis	—	—	—	x	x	x	x	x
Epischura lacustris	—	x	—	x	—	x	x	x
Senecella calanoides	—	—	—	—	x	x	—	—
Cyclopoid Copepods								
Cyclops bicuspidatus thomasi	—	—	—	x	x	—	—	—
Cyclops vernalis	x	x	x	—	—	x	x	x
Ergasilus chautauguaensis	—	—	—	—	—	—	—	x
Eucyclops agilis	—	—	—	—	—	—	—	x
Eucyclops speratus	x	—	x	x	—	—	—	x
Mesocyclops edax	x	—	—	x	x	—	x	x
Tropocyclops prasinus	—	—	—	x	x	—	—	—

424

Cladocerans

Species						
Bosmina longirostris	x	x	x	x	x	x
Ceriodaphnia pulchella	—	x	x	—		x
Ceriodaphnia quadrangula	—	—	x	—		—
Ceriodaphnia reticulata	—	—	x	—		—
Chydorus sphaericus	x	—	—	—		x
Daphnia ambigua	—	—	—	—		x
Daphnia dubia	—	x	x	—	x	—
Daphnia galeata mendotae	x	x	x	x	x	x
Daphnia longiremis	—	—	—	—		—
Daphnia parvula	—	x	x	x	—	x
Daphnia pulex	—	—	—	—		x
Daphnia retrocurva	x	x	x	x	x	x
Diaphanosoma leuchtenbergianum	—	—	x	x	—	x
Holopedium gibberum	—	x	—	—	—	—
Leptodora kindtii	x	x	x	—	x	x

[a] Hall and Waterman (1967).
[b] Presence indicated by x.

425

TABLE 34

Comparison of Total Zooplankton[a] in the Water Column for August 1910 with Mean Standing Crops for August 1972 and 1973[b]

Lake	1910	1972	1973
Hemlock	31,146	20,468	229,476
Owasco	15,483	37,391	33,765
Skaneateles	5,683	13,409	10,561

[a] Units are individuals m^{-3}.
[b] Chamberlain (1975).

for Owasco and Skaneateles (0.6×10^5 and 0.2×10^5 individuals m^{-3}, respectively).

The most noticeable difference between the two years was the increase in the number of nauplii caught. For example, during 1972 the maximum concentration in Hemlock Lake was estimated at 500 individuals m^{-3} compared with 4.0×10^4 individuals m^{-3} in 1973. An order of magnitude increase was also noted in Skaneateles, but differences were considerably less in Owasco. A second factor which may have affected the estimation process was the change in vertical towing depths. In Skaneateles this was increased from 45 m to 54–55 m, and in Hemlock from 15 to 16–17m. The

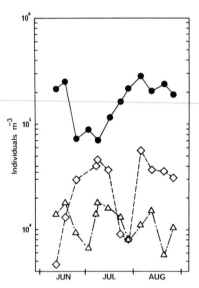

Fig. 37. Total zooplankton volumetric abundance in Hemlock (filled circles), Owasco (open diamonds), and Skaneateles (open triangles) during the period June–August 1973.

TABLE 35

Comparison of Zooplankton Abundances in Hemlock, Owasco, and Skaneateles Lakes: 1910 vs 1972–1973[a,b]

	Lake								
	Hemlock			Owasco			Skaneateles		
Taxa	Aug 23 1910	Mean 1972	Mean 1973	Aug 13 1910	Mean 1972	Mean 1973	Aug 15 1910	Mean 1972	Mean 1973
Copepods									
Cyclops	4,015	8,330	16,145	498	114	169	98	0	27
Diaptomus	4,908	0	0	7,240	12,915	94	2,728	3,776	4,154
Epischura	—	—	—	20	0	0	—	—	—
Senecella	—	—	—	53	127	47	43	485	82
Cladocera									
Bosimina	191	1,904	8,447	486	566	4,014	161	29	34
Daphnia	3,788	2,890	2,727	222	3,353	0	65	1,097	286
Diaphanosoma	367	238	0	172	4,674	28	—	—	—
Rotifers									
Asplanchna	—	—	—	253	1,638	780	50	0	0
Conochilus	298	0	25,790	39	13,919	10,735	8	6,740	1,068
Filinia	848	0	0	—	—	—	—	—	—
Kellicottia	1,494	0	636	—	—	—	—	—	—
Keratella	142	0	33,811	—	—	—	—	—	—
Polyarthra	1,629	170	34,602	125	0	2,275	644	0	551
Nauplii	13,466	68	28,632	6,375	85	949	1,886	43	2,570
Totals (comparable general)	31,145	13,600	150,791	15,483	37,391	19,091	5,683	12,170	8,772
Totals (all species)	31,145	20,468	229,476	15,483	37,391	33,765	5,683	13,409	10,561

[a] From Chamberlain (1975).
[b] Counts in individuals m^{-3}.

maximum depths of the lakes are Hemlock, 27.5 m; Owasco, 54 m; and
Skaneateles, 90.5 m.

During the winter months concentrations were lower in Owasco
(minimum of 340 individual m^{-3}) than either Hemlock (1.1 × 10^3 individuals m^{-3}) or Skaneateles (1.3 × 10^3 individuals m^{-3}). A sharp increase in
concentration was observed during January in Skaneateles (25 × 10^3 individuals m^{-3}), and was due almost entirely to a pulse of *Diaptomus minutus*
(Chamberlain, 1975).

In terms of number per unit area (Fig. 38), Owasco, with its relatively
high zooplankton concentration and greater towing depth, had a greater
maximum than Hemlock in 1972 (5.6 × 10^6 individuals m^{-2} during July and
3.1 × 10^6 individuals m^{-2} during June, respectively). Skaneateles reached a
level of 3.1 × 10^6 individuals m^{-2} in July. During the following year, Hemlock had the maximum concentration (4.6 × 10^6 individuals m^{-2}), followed
by Owasco (3.0 × 10^6 individuals m^{-2}), and Skaneateles (1.4 × 10^6 individuals m^{-2}) (Fig. 39).

Details of Community Structure and Seasonal Changes. *Hemlock.*
Crustaceans were the dominant taxonomic group in Hemlock during the
summer of 1972, and again from October 1972 through April 1973 (Chamberlain, 1975). At times, they made up over 90% of the total zooplankton
during these periods. From May through August 1973 crustaceans
decreased in importance, usually accounting for less than 30% of the total,

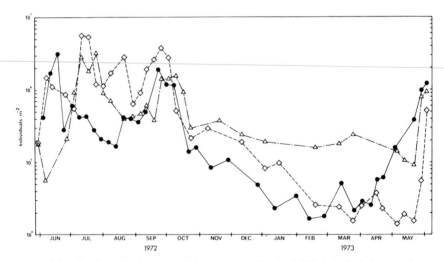

Fig. 38. Total zooplankton areal abundance in Hemlock (filled circles), Owasco (open
diamonds), and Skaneateles (open triangles) during the period June 1972–May 1973.

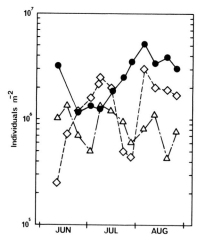

Fig. 39. Total zooplankton areal abundance in Hemlock (filled circles), Owasco (open diamonds), and Skaneateles (open triangles) during the period June–August 1973.

except for a peak of 45% in July. During this time, rotifers were the dominant zooplankters.

Cladocerans were generally the most important of the Crustacea, attaining a maximum of 70% of the total zooplankton count in early April 1973. *Daphnia longiremis* was the dominant species in 1972, reaching a maximum concentration of 10×10^3 individuals m^{-3} in July. *Bosmina longirostris,* a species that was encountered on all but one sampling date, was present but at lower concentrations (mean of 1.2×10^3 individuals m^{-3}). *Bosmina* increased somewhat during late fall and early winter, reaching a peak of 3.7×10^3 individuals in m^{-3} in November. *Daphnia longiremis* was considerably more numerous the following summer, peaking at 15×10^3 individuals m^{-3} in June, and again at 6.5×10 individuals m^{-3} in August. *Ceriodaphnia pulchella* was present less frequently, but did reach sizeable concentrations in September 1972, and during July and August 1973 when it was the dominant crustacean.

Cyclopoid copepods were also represented in reasonable numbers, at times accounting for 60% of the total zooplankton. *Cyclops bicuspidatus* was observed on 47 out of 50 sampling dates, and was most abundant during July 1972 when it was the dominant crustacean. It was also well represented in August and September of that same year, and peaked again in June 1973. *Mesocyclops edax* was present from July through September 1972, and also peaked in June 1973 at a maximum for the study period. *Cyclops vernalis* was generally present, but was only important in June and August 1973. Calanoid copepods were only minimally represented, and were absent from the 1972 samples.

Rotifers were a numerically important component of the Hemlock Lake zooplankton community during June 1972, and again in September–October. They were the numerically dominant zooplankters during the summer of 1973, accounting for 50–80% of the total number. *Conochilus unicornis* was very abundant on all three occasions. *Asplanchna priodonta* was also common, occurring on 35 of 50 sampling dates. Three peaks were observed during the fall and summer of 1972. The species reached a maximum level in May 1973, and was also present in reasonable numbers in August. *Kellicottia longispina* was scarce in 1972, but reached high concentrations in May–June 1973. This peak was followed by a rapid decline in numbers. *Polyarthra vulgaris,* also rare in 1972, was an important species during the summer of 1973, increasing in number after the decline of *K. longispina;* it reached a maximum concentration in August.

Owasco. Crustaceans were an important but not always dominant component of the Owasco Lake zooplankton community. Percent contribution for the 1972–1973 study period ranged from less than 10 up to 100%. During the 1972 stratified period they alternated in relative importance with the rotifers. The latter organisms assumed a more dominant role from June through August 1973. Crustaceans were relatively more abundant during the cooler months when total numbers of zooplankton were low.

Calanoid copepods were usually the dominant crustaceans, at times accounting for almost all of the Crustacea. *Diaptomus sicilis* was the most important species during 1972, reaching a maximum concentration of 29×10^3 individuals m^{-3} in September. *Diaptomus minutus* was also prevalent at this time, but not in as great numbers (maximum of 7.7×10 individuals m^{-3} in early August). Both of these species were only minimally represented after January 1973. *Cyclops vernalis* was intermittently present at low concentrations.

Cladocerans were present during August 1972, but did not exceed 25% of the total community in number. Their relative importance tended to increase from October 1972 on, culminating in June 1973 when they comprised more than 90% of the zooplankton. *Bosmina longirostris* was the dominant species, peaking in June–July 1973, and again in August.

Rotifers were periodically important during 1972, sometimes accounting for 80–90% of the zooplankton. They dominated during July and August 1973. *Conochilus unicornis* was at times prominent during the summer and early fall of 1972 and reached a maximum concentration of 117×10^3 individuals m^{-3}. The species peaked twice in 1973 during June and August but at levels considerably below the previous year. *Asplanchna priodonta* was present during 1972 but did not exceed 4.2×10^3 individuals m^{-3}. It occurred again in 1973, reaching a maximum concentration of 12×10^3 individuals m^{-3} in late June. *Kellicottia longispina* was intermittently present but at low concentrations.

Skaneateles. The zooplankton in Skaneateles was dominated by crustaceans except for two periods, once during July 1972, and again in October 1972 when rotifers were the most numerous taxonomic group. Calanoid copepods were by far the most prevalent zooplankters, *Diaptomus minutus* being the most common species. This organism was relatively abundant throughout the year, and peak concentrations were observed during late September–October 1972 (23 × 10^3 individuals m^{-3}), late January 1973 (25 × 10^3 individuals m^{-3}), and June–July 1973 (9.2 × 10^3 individuals m^{-3}). *Diaphanosoma leuchtenbergianum* was present during part of the sampling period (July–November 1972 and July–August 1973), but never exceeded 6 × 10^3 individuals m^{-3}. *Senecella calanoides, Epischura lacustris,* and the cyclopoid *Cyclops vernalis* were periodically present in low numbers.

Cladocerans increased from July through September 1972, but never exceeded 40% of the total community. A smaller pulse was noted in the summer of 1973. *Daphnia retrocurva* was the dominant species (maximum of 2.4 × 10^3 individuals m^{-3}), followed by *D. galeata mendotae. Bosmina longirostris* was a minor component of the group.

Chonochilus unicornis was the only rotifer to occur in any numbers, and had peak concentrations ranging from 17 × 10^3 to 117 × 10^3 individuals m^{-3} on five occasions during the study period. Concentrations were highest during June–October 1972.

Biomass of Zooplankton in Hemlock, Owasco, and Skaneateles. The average biomass of the various taxonomic groups and total zooplankton biomass for Hemlock, Owasco, and Skaneateles are summarized on a seasonal basis in Fig. 40. The relative importance of the various taxonomic groups differs considerably between lakes. In Hemlock, cyclopoid copepods were always the dominant group, accounting for 36–73% of the total biomass. Cladocerans were also numerous throughout the sampling period, making up 22–40% of the biomass. Nauplii and rotifers appeared to be far more abundant in 1973, but Chamberlain (1975) attributes this to the change in net mesh size rather than to actual population shifts. Calanoid copepods were the dominant taxonomic group in Owasco, and were a major contributor to the biomass (64–90%). Cladocerans were also important, but to a much lesser extent, making up 3–26% of the total biomass. Cyclopoid copepods were also present but played only a minor role, accounting for 1–4% of the biomass. Nauplii were present in slightly greater amounts in 1973 but did not show the great increase in mass that they did in Hemlock. Rotifers were always of minor importance during the survey. Skaneateles Lake was also dominated by calanoids in terms of biomass (70–100%), with cladocerans being of secondary importance (0–29%). Cyclopoid copepods and rotifers were of no consequence, although they were encountered from

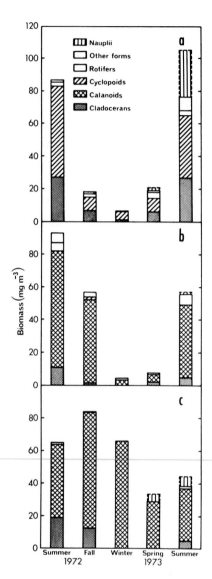

Fig. 40. Seasonal changes in zooplankton biomass in (a) Hemlock, (b) Owasco, and (c) Skaneateles, and the relative contribution of the various taxonomic groups (data from Chamberlain, 1975).

time to time in the samples. Nauplii were considerably more numerous in 1973, contributing to 12–14% of the total zooplankton biomass.

Hemlock had the highest average seasonal biomass concentration at 105 mg m^{-3} during the summer of 1973. Owasco was second with 93 mg m^{-3} (summer, 1972), and Skaneateles last with 83 mg m^{-3} (fall, 1972). Skaneateles was unusual in that it maintained a relatively high biomass level during the colder months, i.e., 66 mg m^{-3} (winter, 1973). By this time of year Hemlock had decreased to 5.7 mg m^{-3}, and Owasco to 4.5 mg m^{-3}. The explanation for this is not known.

The highest individual biomass measurement (318 mg m^{-3}) occurred in Owasco during July 1973. The maximum for Hemlock was 222 mg m^{-3} (July, 1972), and for Skaneateles 189 mg m^{-3} (October, 1972).

Rooted Vegetation

The morphometry of the Finger Lakes dictates that, in general, rooted macrophytes will be confined to narrow, inshore bands along the north–south axis, the vicinity of deltas, and the extreme ends of the lakes. Considerable extents of bottom that are completely free of macrophytes occur where steep drop offs and rock bottoms combine to produce an unsuitable habitat.

Although restricted in area, macrophyte growths are dense in places and are regarded as a major nuisance in some of the lakes. The plant producing the greatest interference with water-based recreation is milfoil, *Myriophylium spicatum*. Limited historical data suggest that the relative, and perhaps absolute, abundance of this macrophyte had increased during the last few decades, e.g., Oglesby *et al.* (1976) for the southern end of Cayuga and the 1927 and 1974 descriptions of Seneca Lake reviewed below. *Chara*, *Nitella* and the potamogetons seemed to have decreased during this same period.

Historical Descriptions. Eaton (1928) provides a brief, general description of the rooted vegetation in the Finger Lakes of the Oswego River system based on data collected during the 1927 biological survey. Extensive beds of eel grass or wild celery (*Vallisneria*), pondweeds (*Potamogeton*), hornwort (*Ceratophyllum*), ditch grass (*Elodea*), and other submerged forms were primarily restricted to the ends of the lakes. These species usually occurred on a muddy bottom in quiet water at a depth of 3–5 m (10–17 ft), with some plants being observed at depths up to 7.6 m (25 ft). Growth was poor along wave-swept shores.

The alga *Chara foetida* (musk grass) was universally distributed on all of the lake bottoms up to a depth of at least 6–9 m (20–30 ft), and when conditions were favorable up to about 12.8 m (40–44 ft). Although it lay close to the bottom, *C. foetida* was frequently uprooted by the south swell which is characteristic of the lakes. *Nitella* and other species of *Chara* were often found in association with *C. foetida*.

Chara attracted coots, redheads, and other species of ducks when wild celery and sago pondweed (*Potamogeton pectinatus*) were scarce, not an uncommon occurrence in the Finger Lakes (Eaton, 1928). These algae are also the favorite food of the golden shiner.

The flora of Seneca Lake has been described in somewhat more detail by Muenscher (1928). There were only two localities in the lake where extensive beds of macrophytes were found. The largest and most productive were at the northern end, especially in the northwest corner. These covered a total area of less than 2.5 km² (1 square mile), and were usually limited to waters from 1.5 to 4.6 m (5–15 ft) deep. The vegetation was very dense in the shallow waters close to shore. Numerous species were present, but the flora was dominated by the potamogetons.

Species of *Chara* and *Nitella* were to be found in the deeper waters. Almost pure stands of algae covered extensive areas, although there were occasions when *Potamogeton gramenius* var. *graminifolius* was found in association with them. Beds of *Chara* were found from a depth of a few feet up to about 6 m (20 ft). *Nitella* was rarely found in water less than 3 m (10 ft) deep, being more common at a depth of 4.6–7.6 m (15–25 ft).

The only other macrophyte bed of any extent was located between Dresden and Long Point along the western shore of the lake. There were several fairly dense aggregations composed of *Potamogeton* with minor species intermingled. The beds were mostly located near shore, but, in a few instances, they extended out more than 100 m (300 yd). The total area of the beds was estimated at probably covering less than 1.3 km² (0.5 square mile).

Scattered macrophyte beds were located among the eastern and western shores. A few were also noted at the southern end of the lake in the southeastern corner near the inlet. Muenscher (1928) estimated that the total area of lake bottom covered by these plants was about 5.2 km² (2 square miles).

Cattail marshes were found to occur to some extent at the southern end of the lake; *Typha angustifolia* was the dominant species. Other emergent plants such as *Scirpus acutus* and *S. americanus* were found along the outer margins of the cattail marshes, covering extensive areas of shallow water (Muenscher, 1928).

Baston and Ross (1975) surveyed five of the lakes being considered here in 1974. The reader is referred to their publication for a description of sampling methods and analytical procedures. A summary of their observations is given as follows.

Keuka.   The most extensive macrophyte beds in Keuka Lake were found at the northern ends of the two arms and at its southern end. Other beds occurred along the lake perimeter to the north of prominent points, spits, and deltas and various man-made structures such as piers and docks.

The rooted plant community at the northern end of the western arm was somewhat influenced by a spit that extends 90 m outward from the western shore. The lagoon created by the spit was weed free due to the 12-m water depth, but north of the hole the water becomes shallower, and macrophyte growth occurs. On the landward side of the spit the cover was 60% *Myriophyllum spicatum,* 30% *Heteranthera dubia,* 5% *Potamogeton* sp., and 5% *Najas flexilis. Myriophyllum spicatum* could also be found north and south of the spit, and in patches along the eastern shore. The eastern side of the spit was intermittently covered with *Najas flexilis* (55–75%), *Chara* sp. (5–25%), and *Myriophyllum spicatum.* Growth in this region was restricted to a narrow band along the shore due to the rapid increase in depth lakeward.

The northern end of the eastern arm was heavily covered with *Myriophyllum spicatum* (60–80%), with occasional occurrences of *Potamogeton crispus* and *P. amplifolius.* A rich growth of *Najas flexilis* occurred in about 5 m of water. *Nitella* sp. could be found south of this bed in about 6 m of water. *Chara* sp. was found to the west.

Well-established macrophyte beds were found at the southern end of the lake. Diminutive growth forms of *Valisneria americana* and *Myriophylium spicatum* were near shore in about 1 m of water. Baston and Ross (1975) felt that this reduced size may have been a response to turbid water conditions. Northward, in 1.8–3.0 m of water, *Myriophyllum spicatum* (10%) and *Heteranthera dubia* (40%) covered the bottom. Beyond this area, in water about 4.5 m deep, *M. spicatum* occurred in a pure stand. A rapid increase in depth limited any further plant growth. Macrophyte development in the southwestern corner was inhibited by the rocky substrate composed of large cobble.

Seneca.   Dense growths of aquatic angiosperms were found in Seneca Lake where depth and substrate permitted. Three-quarters of the shoreline were fringed with beds from 9 to 400 m wide. The only places where growth did not occur were areas of rocky bottom or where depth increased rapidly. A continuous band of rooted macrophytes extended around the northern end of the lake. *Myriophyllum spicatum* was the dominant species, with the plants averaging 3 m in height. The bottom was of variable composition, being a mixture of sand (43–75%), silt (15–28%), and clay (15–28%). *Myriophyllum spicatum* covered 99% of the bottom, the remaining 1%

being made up of *Potamogeton pectinatus* and *P. richardsonii.* Extensive beds of *M. spicatum* occurred in the northeast corner of the lake along the Department of Transportation channel. Cover varied from 50 to 100%. These beds occupied almost three square kilometers. Closer to shore the bottom became rock, and the beds ceased. The *M. spicatum* that occurred in Glass Factory Bay was unusual in that it grew on a very rocky substrate, the roots of the plants being embedded in the small amounts of silt between the large stones. These roots differed from the diffuse systems found in other areas in that they were stubby, short, and woody in appearance. The same species was found growing along the shoreline of the bay which has a sand (81%) bottom. Proceeding south along the western shore the milfoil bed extended 270 m out from shore, then narrowed to 90 m at Dresden. The sharp increase in depth south of Brewer Point terminated the weed bed. On the eastern side of the lake a bed approximately 180 m wide extended south from Pastime Park to Willard State Hospital. From Willard to Lamoreaux Landing the bed tapered to 90 m. Only isolated beds occurred south of there.

The flora at the southern end of the lake was somewhat more diverse. The southeastern corner of the lake bottom was covered by white sand in about 2.4–3.6 m of water. The rooted plant community was made up of 50% *Myriophyllum spicatum,* 40% *Heteranthera dubia,* 5% *Elodea canadensis,* 1% *Potamogeton crispus,* 1% *Chara* sp., and 1% *P. perfoliatus.* In deeper waters there were patches of *M. spicatum* and *H. dubia.* About 450 m offshore water depth increased from 3 to 15 m, limiting further plant growth. A rocky bottom and deep water precluded plant growth in the southwest corner of the lake, but occasional patches of *E. canadensis* were found close to the shore.

Owasco. The perimeter of Owasco Lake was essentially weed free because of a rocky bottom and relatively deep water close to shore. Exceptions to this were the shallower northern and southern ends. A sand bar extended along the northwest shore of the lake. The dominant species in this area was *Chara* sp. which covered 75% of the bottom. *Heteranthera dubia* (20%) and *E. canadensis* were also found. In the northeast corner near Sucker Brook the substrate is primarily heavy sand; *Elodea canadensis* (80%) and *Valisneria americana* (10%) were the dominant species. *Myriophyllum spicatum* dominated in the outlet, covering 60% of the bottom. A second sand bar occurred in the Poplar Cove area, extending out 270 m from the shore, and following its contour. *Chara* sp. was the dominant form, but the cover was quite variable. Occasional patches of *E. canadensis* and *M. exalbescens* were found close to shore. A locally abundant stand of *Potamogeton amplifolius* growing on sandy substrate occurred to the north of Bucks Point. South of there the bottom was composed of large rocks and cobble. Patches of *Chara* sp. (5–40% cover) were to be found, but most of the bottom was devoid of vegetation. The most highly concentrated macrobenthic population in the lake was in the southeast corner growing on a mucky substrate. *Heteranthera dubia* was the dominant specied (30% cover) while *V. americana* and *P. crispus* were both common (10% cover). *Ceratophyllum demersum* was found on occasion. The substrate in the southwest corner is composed of small cobble and unconsolidated shale fragments; *E. canadensis* (80%) and *Chara* sp. (20%) were dominant. Isolated patches of *H. dubia* were present.

Skaneateles. Baston and Ross (1975) state that Skaneateles Lake contained an ideal composition and density of macrobenthic plants. *Chara* sp. and *Elodea canadensis,* the dominant forms, were low-growing submerged types which in their opinion struck a proper balance with the common pondweeds and associated species, i.e., *Valisneria americana, Heteranthera dubia,* and *Ceratophyllum demersum.* The northern perimeter of the lake has a shale bottom which prevented plant growth close to the shore. Lakeward the major species were *Potamogeton richardsonii, V. americana, Najas flexilis,* and *P. canadensis.* These species also occurred intermittently along the eastern shore. There was a heavy concentration of *E. canadensis* along the western shore where the water was 3–6 m deep, and the bottom was covered with a shallow layer of sandy silt.

The bottom in the southeast corner of the lake varied from sand to rock. *Elodea canadensis* was the dominant offshore species, covering 85% of the bottom. *Valisneria americana* (10%), *Chara* sp., *P. crispus,* and *P. pectinatus* (the latter three species totaled 5% cover) were also found. The water in this area was about 3.6 m deep. Moving closer to shore (22 m) in shallower water (3 m), *Chara* sp. dominated with about 80% bottom cover. *Valisneria americana* was also common (20% cover). The bottom in this region was covered with about 10 cm of muck. *Chara* was also present in the southwest corner covering 100% of the sandy clay bottom. This stratum was about 70 cm deep. At greater depths, i.e., 7.5 m. the bottom was covered with small rock fragments, and devoid of any rooted plant growth.

Otisco. The extensive shallow areas at the northern and southern ends of Otisco Lake would be expected to provide extensive areas suitable for macrophyte growth. Such a situation was not observed by Baston and Ross (1975) in 1974. The perimeter of the lake was generally free of dense rooted plant beds except for a few isolated locations. At the northern end of the lake a weed bed was becoming established along the west shore near Laders Point in Glen Cove. The dominant species, *Heteranthera dubia,* covered 80%

of the bottom of the bay. *Najas flexilis* was also present (5% cover). The water depth was about 1.8 m. Bottom sediment was 48% sand, 45% silt, and 7% clay. *Valisneria americana* was found floating on the surface but was not sampled in the growing condition. *Polygonum amphibium* grew along the shore in about 1 m of water. *Polygonum amphibium* was dominant in|Turtle Bay on the eastern shore. A band of rooted macrophytes, reaching out about 24 m from shore, extended north from the bay to the Narrows, and south for about 800 m. *Heteranthera dubia* accounted for about 25% cover in this band, and was the dominant species. *Elodea canadensis* and *Chara* sp. were present in smaller amounts, covering approximately 5% of the area.

The bottom at the southern end of the lake, north of the causeway, was composed of shale fragments and devoid of weeds. The shallow pool south of the causeway would seemingly be an excellent habitat for rooted plant growth but was almost completely free of weeds at the time of the survey. *Polygonium amphibium* was the only species found, and that only along the fringe of the marsh. It would appear that the lack of plant growth could have been due to poor light penetration. Secchi disc transparency was about 0.6 m, while the mean depth of the pool was 1.2 m. The cause of the turbidity was not known. The sediment was composed of 28% sand, 2% silt, and 70% clay, a good substrate for macrobenthic plant growth.

The general absence of well-developed macrophyte beds in the shallow areas was surprising, as copper sulfate had in the past been used to control plant growth in such locations (Baston and Ross, 1975). Baston and Ross (1975) suggested that the dearth of vegetation may have been a response to the lake level having been lowered an estimated 3.9 to 4.2 m by the fall of 1965, when a severe drought occurred in the northeastern United States. (Part of the water for the Syracuse and Onondaga systems is supplied by the lake.) That winter unseasonally heavy snows covered the region. When the lake returned to its normal level in 1967 the weed problem no longer existed, and for the past decade the lake has been essentially weed free.

Fish

In general, fish communities of the Finger Lakes are of the cold water, salmonid type characteristic of deep, well-oxygenated, north temperate latitude lakes in the Western Hemisphere. However, littoral zones and marsh areas assure a variety of habitats for shallow water species, many of which are typically cited as being of the warm water variety. The morphometry of the two shallowest lakes, Honeoye and Otisco, results in environments that preclude some of the cold water stenotherms from their communities.

All of the Finger Lakes support well-developed sport fisheries. Commercial fishing is prohibited. The principal management tools that have been employed are catch regulation, habitat improvement (for anadromous salmonids), and, most heavily emphasized, stocking, including the introduction of new species.

The only systematic, comprehensive study of the fish and fisheries in the Finger Lakes was that conducted during 1927 by the New York State Conservation Department (now the Department of Environmental Conservation). Subsequently, this agency has carried out various management programs accompanied by limited study of the fish communities. Some further information is also available from nonscientific documents and several special purpose studies.

Fish Communities and Fisheries in the Past.   The zoogeography of fish in the Finger Lakes has not yet been studied in any systematic way. However, Dadswell (1974), in his investigation of how glacial relict invertebrates and cottids are distributed in eastern North America, concluded that "The community probably dispersed from the Genesee Valley glacial lakes into the rest of the Finger Lakes basins. . . during the Lake Hall phase about 13,000 B.P." It might therefore be expected that fish species present at that time were shared in common with the Great Lakes, except in the case of Canadice, where altitude probably resulted in an early isolation from other waters.

Except for a reference to abundant salmon and eels in Cayuga during the mid-seventeenth century (Raffieux, 1671–1672), descriptions of Finger Lakes' fish communities are not available until after a considerable period of human influence had taken place. Major impacts on fish community structure were probably exerted by the construction of the Erie barge canal and its accompanying locks in the early 1800's, introductions of nonnative species by fish culturists beginning in the latter third of the nineteenth century, construction of numerous dams on lake tributaries for water power, and sewage and industrial wastes discharged into surface waters in the basins.

The first comprehensive survey of fish in the Finger Lakes was carried out in the summer of 1927 (Eaton, 1928; Greeley, 1928). A variety of gear was used to sample the various habitats represented in those Finger Lakes that are part of the Oswego drainage. The qualitative description of fish communities which emerged from this study is presented in Table 36. Canandaigua and Cayuga are included for comparative purposes. The latter, because of its proximity to Cornell University, is also the most extensively studied of the Finger Lakes. It is logical to suppose that some of the rarer species, such as the sturgeon, listed for Cayuga also occurred in Seneca and possibly some of the other larger lakes. Common names of fish were generally used by Greeley (1928) and Eaton (1928). We have grouped

TABLE 36

Fishes Found in the Central and Eastern Finger Lakes

| Common name | Fish | | |
	Family	Genus	Species
Lake (sea) lamprey	Petromyzonidae	*Petromyzon*	*marinus*
Lake sturgeon	Acipenseridae	*Acipenser*	*fulvescens*
Long-nosed gar	Lepisosteidae	*Lepisosteus*	*osseus*
Bowfin	Amiidae	*Amia*	*calva*
Sawbelly (alewife)	Clupeidae	*Alosa*	*pseudoharengus*
Gizzard shad		*Dorosoma*	*cepedianum*
Rainbow trout	Salmonidae	*Salmo*	*gairdneri*
Brown trout		*Salmo*	*trutta*
Lake trout		*Salvelinus*	*namaycush*
Lake cisco		*Coregonus*	*artedii*
Common whitefish		*Coregonus*	*clupeaformis*
Smelt	Osmeridae	*Osmerus*	*mordax*
Mud minnow	Umbridae	*Umbra*	*limi*
Northern pike	Esocidae	*Esox*	*niger*
Chain pickerel		*Esox*	*lucius*
Lake chub	Cyprinidae	*Couesius*	*plumbeus*
Carp		*Cyprinus*	*carpio*
Black-nosed dace		*Rhinichthys*	*atratulus*
Long-nosed dace		*Rhinichthys*	*cataractae*
Fallfish		*Semotilus*	*corporalis*
Horned dace (creek chub)		*Semotilus*	*atromaculatus*
(Blackchin shiner)		*Notropis*	*heterodon*
Spottail minnow (shiner)		*Notropis*	*hudsonius*
Satin-fin minnow (shiner)		*Notropis*	*spilopterus*
Common shiner		*Notropis*	*cornutus chrysocephalus*
Common shiner		*Notropis*	*cornutus frontalis*
Silverfin (emerald shiner)		*Notropis*	*atherinoldes*
Cut-lips minnow		*Exoglossum*	*maxillingua*
Golden shiner		*Notemigonus*	*chrysoleucas*
Silvery minnow		*Hybognathus*	*nuchalis*
Blunt-nosed minnow		*Pimephales*	*notatus*
Sturgeon (longnose) sucker	Catostomidae	*Catostomus*	*catostomas*
Common (white) sucker		*Catostomus*	*commersoni*
Chub sucker		*Erimyzon*	*sucetta*
Hog sucker		*Hypentelium*	*nigricans*
Red-fin (redhorse) sucker		*Moxostoma*	sp.

during the Summer of 1927[a,b]

			Occurrence			
Canandaigua	Keuka	Seneca	Cayuga	Owasco	Skaneateles	Otisco
✓	—	✓	✓	—	—	—
—	—	—	✓	—	—	—
—	—	—	✓	—	—	—
—	—	—	✓	—	—	—
—	✓	✓	✓	—	—	—
—	—	—	✓	—	—	—
✓	✓	✓	✓	✓	✓	—
✓	✓	✓	—	—	✓	✓
✓	✓	✓	✓	✓	✓	—
✓	✓	✓	✓	✓	✓	✓
✓	✓	✓	✓	—	✓	—
✓	—	—	—	✓	✓	—
✓	✓	✓	✓	✓	—	—
—	—	—	✓	—	—	—
✓	✓	✓	✓	✓	✓	✓
—	—	—	—	✓	✓	—
✓	✓	✓	✓	✓	—	✓
✓	—	✓	✓	✓	—	—
✓	—	✓	✓	✓	✓	—
✓	—	✓	✓	—	—	—
✓	—	✓	✓	✓	✓	✓
—	✓	—	✓	—	—	—
✓	✓	✓	✓	✓	—	—
—	✓	✓	✓	—	—	—
—	—	—	✓	—	—	—
✓	✓	✓	✓	✓	✓	—
—	✓	—	✓	—	—	—
—	—	—	—	✓	✓	—
✓	✓	✓	✓	✓	—	✓
✓	—	✓	✓	✓	—	✓
✓	✓	✓	✓	✓	✓	✓
—	—	—	—	✓	—	—
✓	✓	✓	✓	✓	✓	✓
—	—	—	✓	—	—	—
—	—	—	✓	—	—	—
—	—	—	✓	—	—	—

(Continued)

TABLE 36

Common name	Fish		
	Family	Genus	Species
Short-headed red-fin sucker		*Moxostoma*	*macrolepidotum*
Spotted (channel) catfish	Ictaluridae	*Ictalurus*	*punctatus*
Common (brown) bullhead		*Ictalurus*	*nebulosus*
Yellow bullhead		*Ictalurus*	*natalis*
Stonecat		*Noturus*	*flavus*
Tadpole stonecat (madtom)		*Noturus*	*gyrinus*
Margined (brindle?) stonecat		*Noturus*	*miurus*
Eel (American eel)	Anguillidae	*Anguilla*	*rostrata*
Killifish	Cyprinodontidae	*Fundulus*	*diaphanus*
Eel-pout (burbot)	Gadidae	*Lota*	*lota*
Nine-spined stickleback	Gasterostidae	*Pungitius*	*pungitius*
Brook stickleback		*Culoea*	*inconstans*
Trout perch	Percopsidae	*Percopsis*	*omiscomaycus*
White bass	Percichthyidae	*Morone*	*chrysops*
Rock bass	Centrarchidae	*Ambloplites*	*rupestris*
Common sunfish (pumkinseed)		*Lepomis*	*gibbosus*
Bluegill		*Lepomis*	*machrochirus*
Smallmouth bass		*Micropterus*	*dolomieui*
Largemouth bass		*Micropterus*	*salmoides*
Calico bass (black crappie)		*Pomoxis*	*nigromaculatus*
Yellow perch	Percidae	*Perca*	*flavescens*
Sauger		*Stizostedion*	*canadense*
Yellow pike (walleye)		*Stizostedion*	*vitreum*
Zebra darter (lag perch)		*Percina*	*caprodes*
Tesselated (Johny) darter		*Etheostoma*	*nigrum*
Fantail darter		*Etheostoma*	*flabellare*
Sheepshead	Scianidae	*Aplodinotus*	*grunnieus*
Skipjack (brook silverside)	Atherinidae	*Labisdesthes*	*sicculus*
(Mottled) sculpin	Cottidae	*Cottus*	*bairdi*
(Slimy) sculpin		*Cottus*	*cognatus*
Total number of taxa	23	44	66

[a] Greely (1928) and Eaton (1928).
[b] Presence indicated by checkmark (√).

(*Continued*)

			Occurrence			
Canandaigua	Keuka	Seneca	Cayuga	Owasco	Skaneateles	Otisco
—	—	—	√	—	—	—
—	—	—	√	—	—	—
√	√	√	√	√	√	√
—	—	—	√	—	—	—
—	—	—	—	—	√	—
—	—	—	√	√	—	—
—	√	—	—	—	—	—
√	—	√	√	√	—	—
√	√	√	√	√	√	√
√	—	—	√	—	—	—
√	—	—	—	—	—	—
√	√	√	√	√	—	—
√	—	√	√	—	—	—
—	—	√	√	—	—	—
√	√	√	√	√	√	√
√	√	√	√	√	—	√
—	—	—	√	—	—	—
√	√	√	√	√	√	√
—	√	√	√	√	—	√
—	—	—	√	—	—	—
√	√	√	√	√	√	√
—	—	—	√	—	—	—
√	—	√	√	√	—	—
√	√	√	√	√	—	—
√	√	√	√	√	—	—
—	—	—	—	√	—	—
—	—	—	√	—	—	—
—	—	—	√	—	—	—
√	√	√	√	—	—	—
√	—	√	√	—	√	—

fish by families; genera and species have been identified where possible by reference to Scott and Crossman (1973) and the American Fisheries Society (Anonymous, 1966) list of common and scientific names.

The western Finger Lakes, those in the Genesee River drainage, were apparently not investigated systematically. However, the following occurrences were noted by Greeley (1927): Hemlock—common whitefish, cisco, lake trout, chain pickerel, smallmouth bass, largemouth bass, and yellow perch; Canadice—chain pickerel; Honeoye—chain pickerel, largemouth bass, walleye, and yellow perch. The muskalonge, *Esox masquimongy,* was noted as having been introduced to Honeoye, but no specimens were taken.

From the investigation cited above and patterns of distribution given by Scott and Crossman (1973), it seems probable that native species shared by most of the Finger Lakes and still fairly common to abundant in 1927 included yellow perch, lake trout, smallmouth bass, cisco, white sucker, pumpkinseed, chain pickerel, brown bullhead, rock bass, and killifish. Those species formerly present in one or more of the lakes but declining or extirpated included Atlantic salmon, eel, bowfin, lake sturgeon, and the long-nosed gar. The fate of the first two of these was probably dictated by downstream dams and pollution, while for the last three destruction of spawning grounds has most likely been the determining stress. Introduced species listed in Table 36 include carp, rainbow trout, and brown trout. Smelt and alewives represent species whose origin is uncertain for particular lakes and which have become more widely distributed than formerly.

The primary purpose of the 1927 survey was to develop fish stocking policies for the stream and lake waters of the basins (Moore, 1928). Eaton (1928) stated that 150 years previously there had been a plentiful supply of lake fish for the 20,000 or so Seneca and Cayuga Indians who inhabited the Oswego watershed, but the white man's invasion of the region has altered these conditions, generally for the worse. Eaton's assessment of the fisheries and fishery potentials of the Finger Lakes in 1927 is summarized in the following observations.

Most of the streams entering the lakes appeared to have been seriously affected by the destruction of forest cover and modifications of the landscape resulting from agricultural activities. These streams provided little encouragement for the suckers and minnows that used to breed in them successfully.

The weed beds in the lakes were in general poorly distributed, exceptions being the ends of the lakes and some sheltered bays. This resulted in only a relatively small area being available for weed inhabiting species.

Keuka was considered to be the epitome of a lake with high sport fishing potential. It contained neither lampreys nor burbot, and the alewife was abundant as a forage fish. Food for bottom feeders was plentiful in the form of *Pontoporeia* and *Chironomus.* The weed beds near Branchport and Penn Yan provided satisfactory breeding sites for perch, bullhead and largemouth bass. The inlets at Hammondsport and Branchport were good streams for rainbow trout to spawn in. The cisco was present in sufficient numbers to supplement the diet of trout. More lake trout were taken in this lake in a single week by fishermen than were taken during the entire season in Owasco, Canandaigua and Skaneateles, the population being maintained by proper stocking and protection of the spawning grounds. The whitefish was scarce, but could probably be increased in number by restocking.

Seneca Lake had an abundant supply of alewives as a forage species, but the cisco existed only in reduced numbers. The lamprey was present, but not the burbot. Lake trout, small-mouth bass and yellow perch were the predominant food species. Perch were abundant, with many in the 1 to 2 pound class. The whitefish was scarce and possibly extinct, and the eel was fast disappearing. Walleye had been stocked in recent years, but were scarce, being restricted to the head of the lake and Dresden Bay. Eaton felt that the lake could support a much larger lake trout population that it did, and that this could be accomplished by proper stocking and lamprey control.

Owasco was naturally adapted to the lake trout, rainbow trout, walleye and smallmouth bass. The lamprey and burbot were missing, but forage species for the trout were scarce. Almost all of the trout that were taken during the survey were in an emaciated condition. Cisco and smelt had been recently introduced, but were still too few in number to feed the trout. Of the trout taken some contained smelt, cisco and sculpin, but no fish had more than two in its stomach. Most had been feeding on *Mysis* and other small organisms. The omnivorous walleye appeared to find sufficient food in the form of insect larvae and small fish to grow successfully. This was the only species taken in any number by sportsmen during the survey. Eels were present, but decreasing rapidly in number. Carp were abundant at the head of the lake, and were thought to seriously interfere with spawning of the walleye, perch and bullhead, and it was thought that they should be controlled. Suckers were still plentiful as there were still some streams in which they could spawn successfully. It was in Owasco Lake that the only sturgeon sucker taken in the Finger Lakes during the survey was found.

The cool clear waters of Skaneateles were well suited to lake trout, rainbow trout, steelhead, whitefish and cisco. The shallows provided black bass with favorable breeding and feeding grounds, and conditions were adequate for the perch and sucker. Eaton felt that both the rainbow and steelhead should be encouraged because there were a number of streams in which they could spawn successfully. He also suggested the stocking of lake trout, whitefish and cisco in great numbers, but does not comment on the presence of an adequate food supply for all of the species.

Otisco was the shallowest, warmest and weediest of the lakes surveyed. The deeper waters were not suited for fish during the summer due to low dissolved oxygen levels. The lake was definitely not a lake trout lake, but walleye, black bass, pickerel, sunfish, bullhead and suckers were thought to be well adapted to such conditions. Eaton suggested the introduction of blue-gill, crappie and catfish. The number of cottages and fishermen were large relative to the size of the lake, and there were complaints as to the scarcity of fish.

The foregoing summary of Eaton's observations indicates the importance he and his colleagues placed on determining the availability of suitable food in assessing the potential of a lake to produce various gamefish. Of the fish sampled during the summer of 1927, stomach contents of 2500 specimens were studied (Eaton, 1928).

Current Status of Fish and Fisheries. Personnel of the New York State Department of Environmental Conservation have periodically sampled the fish communities in the Finger Lakes since the 1927 survey. A variety of sampling gear has been used with gill netting being emphasized. Species taken during 1965–1975 are summarized in Table 37.

Recent samplings cannot be directly compared with the intensive survey of 1927 as evidenced by the absence of the rarer forms recorded in the latter but not in Table 36. Nevertheless, the presence of some of the species taken during 1965–1975 but not during the earlier study may indicate their spread into or through the Finger Lakes either by natural means or as a result of stocking. For example, the alewife was taken in Canandaigua, Owasco, and Otisco during the recent surveys, and a local resident has told one of the authors that this species is now in Skaneateles. Sportsmen recognize the alewife as being a preferred forage species for the lake trout. Current rumor suggests that in at least one instance, Owasco Lake, fishermen have clandestinely introduced *A. pseudoharengus* with the hope of stimulating the production of lake trout.

This and other possible additions to the fish fauna of the different lakes are indicated in Table 38. Gamefish added to the community of a particular lake almost certainly reflect stocking by the State. Mechanisms for the apparent spread of other fish, such as the bluegill and the white perch, are unknown.

Thomas Chiotti, Conservation Biologist, New York State Department of Environmental Conservation, has kindly provided the following description of fisheries and fishery management activities for three of the Finger Lakes, and some complementary information can be found in Parker (1975).

Otisco Lake was in the last ten years a very excellent producer of brown trout. However, the unknown source of introduction of the white perch has caused poor returns of stocked brown trout. This has resulted in the cessation of brown trout stocking by the State and also consideration of stocking salmonids in the future. Onondaga County continues to stock the lake with trout annually, however. The return is insignificant. The intensity of the white perch takeover is shown somewhat in Table VI-B of Appendix 201-A (*not included*). Long range future plans are indefinite, with the only consideration at present being that if culture techniques of walleye rearing allow the production of large numbers of fingerling walleyes, they may be recommended for stocking. There are presently no fish species stocked by D.E.C. in Otisco Lake. The sportfishery is declining and supported only by occasional large walleyes, smallmouth bass and largemouth bass.

Owasco Lake is presently a very good producer of lake trout. Present stocking policy calls for 10,500 yearling equivalents of these fish. Since the state has adopted a yearling policy recently, future stocking will be yearlings only rather than the combination of fall fingerlings and spring yearlings. At present, we are seeing the effects of a 1972–74 stocking policy of 24,000 lake trout yearling equivalents. Fishing remains good, however preliminary analysis indicates that growth rate and condition factor of fish sampled has declined from 1975 to 1976.

Rainbow trout and brown trout are being stocked at the rate of 5,000 yearlings each per annum. These fish are shorter lived and allow fisheries management personnel to make much more rapid changes in management strategy than do the longer lived lake trout. The use of these fish is considered an experimental management technique at this time.

The alewife has only recently become established in the lake, though smelt populations have been high since the early 1940's. The alewife's origin is unknown. It is believed to be responsible for the increase in production of salmonids, particularly lake trout, since changes in growth and condition factor are well documented, both before and after the appearance of the alewife.

The cisco has declined in Owasco Lake in recent years. It was apparently declining prior to the establishment of the alewife. However, it is possible that the increased pressure on the cisco population from the alewife may have caused further decline. If this is correct, the cisco may be suffering from what is known as the Willer–Hile syndrome in which a species is self-regulating as to its density. However, an outside species such as the alewife may act to do this interspecifically in which case the cisco population may never recover to its former abundance.

It should also be noted that Owasco Lake also possesses a *native* (italics inserted by us) strain of rainbow trout in addition to native lake trout. These rainbows spawn primarily in the Owasco Inlet system to perpetuate their particular strain.

Recent netting data has shown a relative decline in the abundance of alewives and smelt which is felt to be a result of heavy predation by the salmonid population. Therefore, the lake will be monitored quite closely to keep the predator–prey relationship in the best balance.

Skaneateles Lake has a quite stable and abundant population of purely native lake trout which are also slow growing. Production of young may actually be to the point that a more liberal season would increase the average size of the fish harvested. There are no smelt or alewives present in the lake at this time. Yellow perch is the primary forage.

The lake also has a population of wild strain rainbow trout which are spawned in Grout Brook. In addition, 10,000–15,000 domestic rainbows are stocked annually by the Onondaga County hatchery system which produce a lake fishery for this species.

Insight into current fishery management practice for Hemlock and Canadice is provided by Abraham (1976a,b). It appears that in the former lake natural reproduction of lake trout is low, and stocking must be carried out in order to provide a viable fishery. Once established, lake trout grew well on an alewife–smelt forage base. The available forage seems to be underutilized and increased salmonid stocking was recommended. In the autumn of 1975 an experimental stocking of 6000 Atlantic Salmon (*Salmo salar*) was carried out, and it was suggested that this be continued for at least three years (Abraham, 1976a).

Canadice contains a lake trout population that reproduces so successfully that stocking has now been discontinued. Growth rates of lake trout indicated their production was in balance with the availability of prey, principally the alewife. Prior introductions of brook trout and splake failed but that of yearling rainbows was successful.

Benthos

The benthos of the Finger Lakes has received relatively little attention, but some work has been done on Skaneateles by Harman and Berg (1970), who reported 13 species of molluscs, not including the Sphaeriidae. Three substrate types are found in Skaneateles: loose, wave-washed channery at the shore; cobbles, sand, and bedrock covered with silt just below this; and

TABLE 37

Fishes Taken by the New York State Department of Environmental Conservation[a] in Samplings during the 1965–75 Period[b]

Fish						Occurrence					
Family	Common name	Hemlock	Canadice	Canandaigua	Keuka	Seneca	Cayuga	Owasco	Skaneateles	Otisco
Clupeidae	Alewife	✓	✓	✓	✓	✓	✓	✓	—	✓
Salmonidae	Lake trout	✓	✓	✓	✓	✓	✓	✓	✓	—
	Rainbow trout	✓	✓	—	✓	✓	+	✓	✓	✓
	Brown trout	✓	—	✓	✓	✓	+	✓	—	✓
	Brook trout	—	✓	—	—	—	—	—	—	—
	Splake	—	✓	—	✓	—	—	—	—	—
	Cicso	✓	✓	—	✓	—	✓	✓	✓	✓
	Lake whitefish	✓	✓	—	✓	—	—	—	—	—
Osmeridae	Smelt	✓	—	—	—	—	+	—	—	—
	Rainbow smelt	—	—	—	✓	—	✓	✓	—	—
Esocidae	Great northern pike	✓	✓	—	✓	✓	+	✓	✓	✓
	Chain pickerel	✓	✓	—	✓	✓	+	—	✓	✓
Cyprinidae	Carp	✓	—	—	—	—	+	✓	✓	✓
	Spottail shiner	—	—	—	✓	—	+	—	—	✓
	Bluntnose minnow	—	—	—	✓	—	+	—	—	—
	Golden shiner	✓	✓	—	—	—	+	✓	—	✓

Family	Species											
Castostomidae	Common sucker	√	√	√	√	√	√	√	√	√	√	√
	Bigmouth buffalo	—	—	—	—	—	√	—	—	—	—	—
	Hog sucker	—	—	—	—	—	—	—	—	√	—	—
	Longnose sucker	—	—	—	—	—	—	—	—	√	√	√
Ictaluridae	Brown bullhead	√	√	√	√	√	√	√	√	√	√	√
	Black bullhead	—	—	—	—	—	√	—	—	—	—	—
	Stonecat	—	—	—	—	—	—	—	—	√	—	—
Cyprinodontidae	Banded killifish	—	√	—	—	—	—	—	—	—	—	—
Percichthidae	White perch or white bass(?)	—	—	—	—	—	√	—	√	—	√	√
	White perch	—	—	—	—	—	—	—	—	—	—	—
Centrarchidae	Rock bass	—	√	√	√	√	—	√	√	√	—	√
	Pumpkinseed	√	√	√	√	√	√	√	√	√	√	√
	Largemouth bass	√	√	√	√	√	√	√	√	√	—	√
	Smallmouth bass	√	√	√	√	√	√	√	√	—	√	√
	Bluegill	—	√	√	√	√	√	√	√	—	—	√
	Black crappie	√	√	√	√	√	√	√	√	—	—	√
Percidae	Yellow perch	√	√	√	√	√	√	√	√	√	√	√
	Walleye	—	—	—	—	—	—	—	+	—	—	√
	John darter	—	—	—	—	—	—	+	—	+	—	—
	Sampling devices	G,T,E	G,T,E	G,T,E	G,T,E	G,T,E	G,T, E,S	G,T,E	G,T,S	G,T,E	G,E	G,T,E

[a] Chiotti (1976) based on samplings during 1965–1975. (presence indicated by √). Others have been added for Cayuga based on personal communication with Professor D. A. Webster of Cornell University (presence indicated by +).

[b] Sampling devices are indicated by G, for gill net; T for trap net, E, for electrofishing; and S, for seine.

TABLE 38

Occurrences of Species (Plus Signs) Found during 1965–1975 in the Oswego Drainage Lakes (Table 37) but Not Recorded for the 1927 Survey (Table 36)

Fish	Lakes for which occurrence newly recorded						
	Canandaigua	Keuka	Seneca	Cayuga	Owasco	Skaneateles	Otisco
Alewife	+				+	?	+
Rainbow trout							+
Brown trout				+	+		+
Cisco				+			
Great northern pike		+	+		+		
Spottail shiner		+	+				+
Golden shiner						+	+
Bigmouth buffalo (*Ictiobus cyprinellus*)		+					
Hog sucker			+		+		
Longnose sucker						+	
Black bullhead (*Ictalurus melas*)		+					
Stonecat			+				
White perch				+	+		
Pumpkinseed						+	+
Bluegill		+	+		+		
Black crappie	+						+
Walleye	+						+

soft silt and clay bottoms in the deeper regions. All of the substrates exist in almost continuous strata along the entire length of the lake (Harman, 1972).

Molluscs were absent in the channery, the probable reason being that stone movement caused by wave action would tend to crush anything living there. The next lower zone was inhabited by *Lymnaea emarginata,* which occurred on silt-covered substrates with a low angle of repose. *Helisoma anceps,* a species less sensitive to steep slopes, also occurred in this zone. *Helisoma anceps* has the largest range, and most likely the highest density of any mollusc in the lake. It was found throughout the length of the lake but was scarce in areas of dense vegetation (Harman, 1972).

Viviparus georgianus is the only other species that grazes on a silty bottom. This organism occupied a different microhabitat than *L. emarginata* and *H. anceps* in that it burrowed more deeply into the silt that covers the zone. All three species inhabited the north end of the lake, but *V. georgianus* was not found at the southern end. Harman (1972) interprets this as an indication that the species had only recently invaded the lake via its outlet, and was then in the process of moving southward. *Viviparus georgianus* was common in areas of the Seneca River fed by the Skaneateles Outlet.

Two immature specimens of *Helisoma trivolvis* were found on *Najas flexilis,* but adults were absent (Harman, 1972). The species is commonly found on vegetation when young, but moves to silty substrates at maturity. *Amnicola limnosa* and *Gyraulus parvus* were the only gastropods normally found on vegetation in the lake. *Amnicola limnosa* was usually associated with beds of *Potamogeton. Gyraulus parvus* was found on *Potamogeton* at the northern end of the lake, and in an area of mixed species to the south.

Campeloma decisa, a burrowing snail, usually occured in soft sand and organic substrates. It was abundant under rotting logs, and in or near sandy bottomed areas (Harman, 1972). Dead specimens of *H. campanulatam,* *Valvata tricarinata,* and *A. lustrica* were obtained in deep samples, but no living populations were observed (Harman, 1972). *Strophitus undulatus,* *Anondonta cataracta,* and *Elliptio complanata,* members of the Unionidae that require soft but firm substrates, were distributed on the sand and clay bottoms in the southern two-thirds of the lake (Harman, 1972).

Harman (1972) attributes the occurrence of relatively few species of molluscs in Skaneateles to the fact that the various substrates on which they live form horizontal bands running the length of the lake with relatively few interruptions such as beds of rooted macrophytes. This permits a given species to have access to the whole of a specific environment. Thus, species with similar requirements would be severe competitors for a given resource. As a result, those species that are better adapted to a biotope, or the first to colonize it, would tend to dominate that substrate type throughout the entire lake.

Eaton (1928) provided a brief general description of the bottom fauna of the Finger Lakes in the Oswego River drainage basin. Samples were collected during the biological survey of 1927. During the survey, 632 samples were taken with an Eckman dredge, supplemented by 212 scoop dredge samples where the Eckman dredge did not work. Several hundred shallow water samples were collected near shore with a Needham dredge.

The shallow waters of all the lakes contained a fair supply of snails (principally *Physa, Limnea, Amnicola, Valvata,* and *Planorbis*), bivalves (*Sphaerium* with less than 1% *Pisidium*), and insect larvae. The snail *Hydracarina* was distributed from 0.3 to 23 m (1–75 ft), and was found on occasions at depths up to 69 m (225 ft). The insect larvae ranged to depths of 15 m (50 ft), and were primarily mayflies (*Hexagenia, Heptagenia, Ephemera,* and *Caenis*), caddis flies (mostly *Molanna, Leptocerus, Helicopsyche, Phryganea, Triaenodes,* and *Mystacides*), and midges.

In the deeper waters of all the lakes there was a plentiful supply of small crustaceans (*Pontoporeia hoyi*) and chironomid larvae (*Chironomus,* with about 1% *Tanypus, Palpomyia,* and *Protenthes*), a fair supply of small worms (Oligochaetes) and bivalves (*Sphaerium*). There was a fairly plentiful supply of *Mysis* in most of the lakes. *Pontoporeia filicornis* was collected in Keuka, Cayuga, and Skaneateles Lakes. Large bivalves were poorly distributed.

The abundance of *Pontoporeia* and *Chironomus* in the deep waters was thought to provide a plentiful supply of food for lake trout and whitefish. The shallow water forms were not present in sufficient abundance to support a large shallow water fish community. The *Mysis* and crayfish populations could not be quantified.

Trophic State

The eight Finger Lakes being considered range roughly from oligotrophic (Skaneateles) to eutrophic (Honeoye). The primary factors affecting trophic state are thought to be mean depth, i.e., shallow lakes tend to be more productive than deeper ones (Rawson, 1955; Larkin, 1964), and nutrient input as influenced by such factors as domestic wastes, drainage basin size, and land use patterns (Oglesby and Schaffner, 1975).

In an effort to better define the relationship between nutrients and lake productivity, Vollenweider (1968) introduced the concept of specific loading to a lake, relating this in a general way to lake trophic state as a function of mean depth. A later modification was introduced by dividing mean depth by mean hydraulic retention time (Vollenweider and Dillon, 1974; Vollenweider, 1975). Nutrient loading was calculated as being the amount added to a unit surface area of the lake in a year's time.

Estimates of nutrient input can be obtained in several ways. Budgets can be constructed by annually measuring the contributions from various sources, e.g., streams, precipitation, and sewage, but such an approach requires the expenditure of great time and effort if the lake is of any size. A somewhat simpler method involves calculating nutrient input based on such factors as drainage basin area, land use, population, domestic waste treatment methods, and annual precipitation. This method necessitates a rather detailed description of the drainage basin, plus estimates of the nutrient contribution from each component.

Specific phosphorus loadings to the eight Finger Lakes are presented in Table 39. They range from 0.23 gm P m^{-2} yr^{-1} (Skaneateles) to 0.97 gm P m^{-2} yr^{-1} (Owasco). It can be seen that there is a rather poor agreement between estimates for a given lake by different researchers. This is thought to be due to the method of estimation rather than natural variability (Schaffner and Oglesby, 1976). When these data along with additional values for selected New York lakes (Schaffner and Oglesby, 1976) are plotted on Vollenweider's (1976) trophic state graph (Fig. 41), the majority of the lakes fall in the "eutrophic" category. Two exceptions from the eight Finger Lakes are Skaneateles, which is "mesotrophic," and Keuka, ranked as being both "eutrophic" and "oligotrophic." Field data have already been presented which indicate that most of the Finger Lakes are assigned too high a degree of eutrophy by this method of positioning them on the trophic continuum.

Oglesby and Schaffner (1975) modified Vollenweider's approach in several ways in their study of a group of New York lakes. Total phosphorus

TABLE 39

Specific Phosphorus Loadings (L_{SP}) to the Various Lakes

Lake	L_{SP} (gm TP m^{-2} yr^{-1})
Hemlock	0.43[a]
Canadice	0.32[a], 0.36[b]
Honeoye	0.38[a], 0.83[b]
Keuka	0.45[a], 0.10[c]
Seneca	0.64[a], 0.38[d]
Owasco	0.97[a]
Skaneateles	0.23[a]
Otisco	0.55[a]

[a] Oglesby and Schaffner (1975).
[b] Stewart and Markello (1974).
[c] Anonymous (1974b).
[d] Anonymous (1974d).

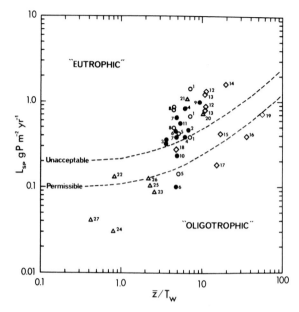

Fig. 41. Trophic state classification of the Finger Lakes as a function of specific phosphorus loading (L_{SP}), mean depth (z) and mean hydraulic retention time (T_W). Adapted from a figure in Vollenweider (1976). Additional lakes are shown for comparative purposes. Filled circles are the Finger lakes being considered in this publication; open circles are the rest of the Finger Lakes (Schaffner and Oglesby, 1976). Open diamonds are selected New York lakes (Schaffner and Oglesby, 1976). Open triangles are selected North American lakes (Vollenweider and Dillon, 1974). Specific lakes are identified in the tabulation on p. 455. Units: L_{SP} —g total P m^{-2} yr^{-1}; z-meters; T_W -years.

(TP) loading was replaced by biologically available phosphorus (BAP) loading. The latter is a composite of soluble phosphorus and that which may be desorbed from suspended stream sediments. The rationale for using this fraction and its definition are considered in some detail in their publication. Surface area loading was replaced by a volumetric loading factor, i.e., loading to the mean summer mixed volume of the lake. This modification was introduced to adjust loading for the effects of lake depth. This procedure gives a high positive weighting to shallowness for thermally unstratified lakes and minimizes the influence of depth for those that do not mix completely during the warmer months. Finally, loading was related to summer chlorophyll, summer transparency, and winter total phosphorus in order to express lake response in terms of quantifiable descriptors of water quality.

The primary result in using BAP rather than TP is to reduce the relative importance of the agricultural contribution of phosphorus to a lake (Oglesby and Schaffner, 1975). For example, it can be seen in Table 40 that

Fig. 41. Map Symbols

Number	Lake
1	Conesus
2	Hemlock
3	Canadice
4	Honeoye
5	Canandaigua
6	Keuka
7	Seneca
8	Cayuga
9	Owasco
10	Skaneateles
11	Otisco
12	Oneida
13	Canandarago
14	Saratoga
15	Lower St. Regis
16	Schroon
17	Sacandaga Reservoir
18	Chautauqua
19	Carry Falls Reservoir
20	Ontario
21	Erie
22	Michigan
23	Huron
24	Superior
25	Lake 227, Experimental Lakes Area, Ontario
26	Lake 239, Experimental Lakes Area, Ontario
27	Tahoe

the calculated agricultural contribution was 54% of the TP to Otisco Lake, but only 30% of the BAP. When all eight lakes are considered, agricultural inputs of TP range from 18 (Honeoye) to 54%, whereas BAP inputs range from 8 to 30%. Forested land contributes from 5 (Seneca) to 37% (Canadice) of both forms of phosphorus. All of the phosphorus from domestic waste treatment systems was considered to be biologically available, and contributed 28 (Otisco) to 60% (Seneca) of the TP, and 43 to 73% of the BAP. Phosphorus inputs from the other components were relatively minor.

The input of BAP to the summer mixed volume of the 13 New York lakes produced the mixed volume loading parameter Λ'_{mp}. This was in turn related to several trophic state indexes rather than to the subjective oligotrophic–eutrophic continuum. Winter total phosphorus correlated quite well with phosphorus loading. The former parameter has been suggested as an index of phytoplankton activity the following summer (Sawyer,

TABLE 40

Estimated Contributions of TP and BAP from Various Sources in the Lake Drainage Basins[a]

Lake	Type of P	Active agriculture		Inactive agriculture		Forest		Residential		Precipitation	
		kg yr⁻¹	%	kg yr⁻¹	%	kg yr⁻¹	%	kg yr⁻¹	%	kg yr⁻¹	%
Hemlock	TP	1066	34	198	6	735	24	73	2	142	5
	BAP	432	21			521	25	73	4	142	7
Canadice	TP	199	24	33	4	301	37	19	2	51	6
	BAP	78	14			213	37	19	3	51	9
Hoenoye	TP	400	15	83	3	944	35	165	6	138	5
	BAP	168	8			670	32	165	8	138	7
Keuka	TP	5817	26	1391	6	2327	11	1383	7	929	4
	BAP	1746	12			1651	11	1383	9	929	6
Seneca	TP	22,065	21	3857	4	5575	5	3714	4	3466	3
	BAP	8804	10			3955	5	3714	4	3466	4
Owasco	TP	10,013	39	1643	6	1923	7	747	3	528	2
	BAP	3935	22			1364	8	747	4	528	3
Skaneateles	TP	2932	35	566	7	646	8	531	6	709	8
	BAP	1200	20			458	8	531	9	709	8
Otisco	TP	1907	46	343	8	393	9	207	5	150	12
	BAP	767	30			279	11	207	8	150	6

456

Lake	Type of P	Sewered population kg yr⁻¹	%	Unsewered population kg yr⁻¹	%	Upstream lake kg yr⁻¹	%	Total input kg yr⁻¹	Specific P loading (gm m⁻² yr⁻¹) L_{sp}	L'_{sp}	Mixed depth P loading (mg m⁻³ yr⁻¹) Λ_{mp}	Λ'_{mp}
Hemlock	TP			890	29			3104	0.43		51	
	BAP			890	43			2058		0.29		34
Canadice	TP			218	27			821	0.32		41	
	BAP			218	38			579		0.22		29
Hoenoye	TP			957	36			2687	0.38		78	
	BAP			957	46			2098		0.30		61
Keuka	TP			9099	43			20,946	0.45		45	
	BAP			9099	61			14,808		0.32		32
Seneca	TP	42,411	40	20,650	20	4189	4	105,927	0.60		52	
	BAP	42,411	49	20,650	24	2962	3	85,962		0.49		43
Owasco	TP	3202	12	7898	30			25,954	0.97		87	
	BAP	3202	18	7898	45			17,674		0.66		59
Skaneateles	TP			2948	35			8332	0.23		22	
	BAP			2948	50			5846		0.16		15
Otisco	TP			1152	28			4152	0.55		55	
	BAP			1152	45			2555		0.34		34

[a] Total inputs are expressed as specific loading and mixed depth loading.

457

1947; Sakamoto, 1966; Dillon and Rigler, 1974). Loading can be directly related to summer phytoplankton standing crop, as expressed by the chlorophyll concentration, and the latter to water transparency (Table 41). As might be expected, there is also a strong correlation between winter total phosphorus and transparency.

In summary, it was possible to calculate the phosphorus input to the eight Finger Lakes by using various drainage basin parameters. The loading estimates so obtained were in turn related to winter total phosphorus, summer chlorophyll, and Secchi disc transparency. The interdependency of each was demonstrated. Such an approach allows the assessment of the impact of a number of watershed components on a lake in terms of the whole system, and provides insight as to how the modification of individual loading units may affect lake quality.

Nitrogen input data for the Finger Lakes are less common, but Dr. Jay Bloomfield, New York State Department of Environmental Conservation (personal communication), has provided the authors with a figure showing the surface area loadings to the eight lakes (Fig. 42). Owasco receives the greatest input (13 gm N m^{-2} yr^{-1}), Hemlock, Canadice, Honeoye, and Otisco somewhat less (~ 8 gm N m^{-2} yr^{-1}), and Keuka, Seneca, and Skaneateles the least (< 6 gm N m^{-2} yr^{-1}). Most of the nitrogen input is calculated as coming from nonpoint sources.

A number of attempts have been made since the early 1950's to develop simple lake trophic state indexes (TSI's) that could be related to such factors as phytoplankton standing crop, quantity of benthos, and fish production. Lake mean depth (Rawson, 1952, 1955, 1960; Northcote and Larkin, 1956; Hayes, 1957; Sakamoto, 1966) and edaphic influences, as represented by total dissolved solids (TDS), conductivity, and alkalinity (Rawson, 1951, 1960; Northcote and Larkin, 1956; Hayes and Anthony, 1964) were

TABLE 41

Correlation Matrix and Regression Equations for Λ'_{mp}, TP_w, CHL_s, and SDT_s[a]

	Λ'_{mp}	TP_w	CHL_s	SDT_s
Λ'_{mp}	—	0.91	0.85	**[b]
TP_w		—	0.91	0.90
CHL_s			—	0.92
SDT_s				—

[a] TP_w, 0.34 Λ'_{mp} + 0.99; CHL_s, 0.57 TP_w − 2.90; CHL_s, 0.23 Λ'_{mp} − 3.36; LOG_{10} SDT, 0.96 − 0.61 LOG_{10} CHL_s; LOG_{10} SDT, 1.36 − 0.76 LOG_{10} TP_w.

[b] ** Regression not calculated.

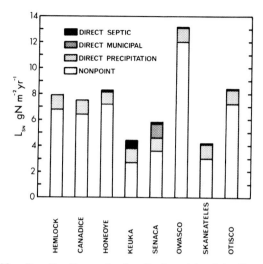

Fig. 42.   Specific nitrogen loading (L_{SN}) to eight of the Finger Lakes.

examined initially, followed by the subsequent development of the more complex morphoedaphic index (MEI) of Ryder (1961, 1965). Numerous attempts have been made to apply the MEI with varying degrees of success, e.g., Harvey and Fry (1973), Henderson *et al.* (1973), and Michalski *et al.* (1973). More recently, Carlson (1975) presented a system for trophic classification of lakes using a scale of 0–100 based on single parameter indexes known to be closely related to changes in trophic state, i.e., Secchi disc transparency, chlorophyll *a* concentration, and total phosphorus levels.

Schaffner and Oglesby (1976) have applied a variety of the above indexes to a selected group of 27 New York lakes which included the Finger Lakes. The results for the eight being considered in this monograph are summarized in Table 42, and a brief description of each index can be found in Table 43.

Carlson's Indexes.   The summer Secchi disk transparency of the 27 lakes considered by Schaffner and Oglesby (1976) varied from 1.2 to 9.3 m. The resulting TSI(SD)'s (supportive data) had a range of 28 (George) to 57 (Lower St. Regis). The eight Finger Lakes were in the 30's, 40's, and 50's (Table 42). Skaneateles had the lowest index (30) with a three-year range of 30–36. Keuka had a low index value in 1973 (32), but had been considerably less transparent the previous year with a TSI(SD) of 47. Seneca Lake was the least transparent of the eight during 1971–1973 with a range of 42–52, but when the indexes for the three-year period are averaged, Hemlock, Seneca, and Owasco are about the same, being 43, 46, and 44, respectively.

TABLE 42

Trophic State Indexes (TSI) and Supportive Data

Lake	Year	SDT (m)	TSI (SD)	Chl (mg m^{-3})	TSI (Ch)	TP (mg m^{-3})	TSI (TP)	TDS (mg liter^{-1})	MEI (TDS)	Specific conductance (μ S cm^{-1} at 25°C)	MEI (Cond.)
Hemlock	1971	4.3	39	5.4	47	10.6	34	136	10		
	1972	2.7	46	7.6	50	9.2	32				
	1973	3.0	44	5.0	46	10.2	33				
Canadice	1973	5.2	36	2.0	37			76	5	115	7
Honeoye	1973	3.0	44	25.7	62	19.0	42	119	24	166	34
Keuka	1972[a]	2.4	47	8.0	51	13.0	37	165	5	241[b]	8
	1973	7.0	32	1.8	36	14.2	38				
Seneca	1965[c]	3.1	44	4.8	46	22.0	44				
	1972[d]	1.8	52	13.0	56	15.0	39	276	3	790[b]	9
	1973[e]	3.6	42	8.6	52	10.2	33			769	9
Owasco	1965							160	6	263	9
	1971	3.8	41	6.6	49	17.5	41	167	6	280[b]	10
	1972	3.1	44	5.0	46	9.9	33			262	9
	1973	2.5	47	4.9	46						
Skaneateles	1965[e]							142	3	224	5
	1971	7.9	30	1.1	32	28.5	48	144	3	275	6
	1972	6.6	33	2.6	40	6.1	26				
	1973	5.2	36	1.3	33						
Otisco	1965[e]							194	19	287	28
	1973	5.2	36	1.8	36	9.6	32	183	18	293	29

[a] Anonymous (1974b).
[b] Anonymous (1975b).
[c] Anonymous (1966).
[d] Anonymous (1974d).
[e] Shampine (1973).

TABLE 43

Description of Trophic State Index

Carlson's trophic state index (TSI):
  Summer Secchi disc transparency

  TSI (SD) = 10(6 − ln SD/ln 2)
    SD = Secchi disc transparency in meters

  Summer chlorophyll *a* (surface)

  TSI (Ch) = 10[6 − (2.04 − 0.68 ln Ch)/ln 2]
    Ch = summer surface chlorophyll *a* in mg m^{-3}

  Summer total phosphorus (surface)

  TSI (TP) = 10[6 − ln(64.9/TP)/ln 2]
    TP = summer surface total phosphorus in mg m^{-3}

Morphoedaphic index:

  $$MEI = \frac{TDS \ (or \ cond.)}{\bar{z}}$$

    TDS = mean total dissolved solids in mg liter^{-1}
    Cond. = mean specific conductance μ S cm^{-2}
    $\bar{z}$ = mean depth of lake in meters

Specific phosphorus loading:

  L_{sp} = SP/A$_1$
    SP = annual total phosphorus input to a lake in gm P yr^{-1}
    A$_1$ = surface area of lake in m^2

Chlorophyll concentrations in the New York lakes ranged from 1.1 to 25.7 mg m^{-3}. TSI(Ch) values were distributed like the TSI(SD)'s (Schaffner and Oglesby, 1976). Of the eight Finger Lakes, Skaneateles had the lowest index (32) and Honeoye the highest at 62 (Table 42). Skaneateles varied from 32 in 1971 to 40 in 1972, then down to 33 in 1973. Seneca Lake had an index of 46 in 1965 and 56 in 1972. Keuka was 36 in 1973 after having been 51 in 1972.

The summer total phosphorus concentrations in the New York lakes varied from 6.1 mg m^{-3} in Skaneateles (1973) to 68.4 mg m^{-3} in Oneida (1973). The highest concentration measured in any of the eight Finger Lakes was observed in Skaneateles (1972). TSI(TP) had the greatest span when all of the New York Lakes are considered (Schaffner and Oglesby, 1976). It ranged from a low of 20 in Schroon Lake to 61 in Oneida. About 75% of the values were in the 30's and 40's. With the exception of Skaneateles' (1973) index of 26, all of the eight Finger Lakes fell into those two categories (Table 43). Skaneateles had the greatest range for an individual lake, with values of 48 and 26 on successive years.

As a group, the New York Lakes considered by Schaffner and Oglesby (1976) fell into the mid range of Carlson's (1975) trophic index scale, but when the indexes were compared with each other, there was a lack of any consistently strong correlation, with the correlation coefficients ranging from 0.41 to 0.71. The reason for this lack of correlation is not known, but it is thought to be due at least in part to the fact that all of the data that were used were not truly comparable, having been obtained from a variety of sources. The data also seem to indicate that year-to-year variations of 10 units or more are to be expected for a given lake.

Morphoedaphic Index.   The morphoedaphic index (MEI) was originally developed as an empirical tool to predict fish yield in unexploited Canadian Lakes (Ryder, 1965). It was later proposed (Ryder *et al.*, 1974) that its application might be more universal. This assumption was supported to some extent by the success of Welcomme (1976) in applying modifications of the MEI to African rivers. Although fish production data are rare for the Finger Lakes, it can be assumed that higher levels of production do reflect the primary processes going on within the lake, and that the MEI's of the various lakes would be related to these.

With this in mind, MEI's were calculated using TDS and conductivity. For the eight lakes MEI's ranged from 3 (Seneca and Skaneateles) to 24 when TDS was used, and from 5 (Skaneateles) to 34 (Honeoye) with conductivity (Table 42). Schaffner and Oglesby (1976) reported a range of 3–32 and 4–44 with TDS and conductivity, respectively, in the selected group of New York lakes that they looked at (Table 42).

Interpretation of the results is difficult for a number of reasons. As it was originally derived, the MEI did not take into account the impact of human populations. The MEI's for lakes receiving large nutrient inputs above those of geochemical origin should obviously be adjusted to allow for these. Methods to permit such modifications are not currently available. Oglesby (in press) has proposed a modification in the MEI to adjust for large differences in mean depth, and the reader is referred to his work for a more detailed consideration of the application of the MEI.

CONCLUSIONS AND RECOMMENDATIONS

The high intrinsic value of the Finger Lakes is illustrated by the following overview (Oglesby, 1974). "[They] constitute a dominant landscape feature and a major natural resource of upstate New York. Their basins contain a population of some 191,000 (1970 census). With 419 miles [674 km] of shoreline and 210 square miles [544 km²] of surface area they provide an extensive, diverse and beautiful recreational environment. Four of the Finger Lakes serve as vital water supplies for [major] cities outside their basins, and the larger lakes produce important ameliorating effects on the climates of the land surrounding them." Additionally, to the scientist, they represent one of the world's best laboratories for studies in comparative limnology and, to the local citizenry and industry, an economical site for the disposal of wastes.

Looking toward the future it seems inevitable that all of the current uses of the Finger Lakes will intensify. Conflicts between various users have occurred in the past but have generally been resolved in a reasonable way on a case-by-case basis. It is highly unlikely that such a procedure will provide satisfactory solutions to the more complex and larger scale problems envisaged for the future.

It appears to us that a management strategy for the Finger Lakes, and the means to implement it, are vitally needed. In order to develop such a strategy the following are required: (1) base line information concerning the resource, (2) an understanding of primary cause and effect relations sufficient to furnish reasonable predictability, (3) a clearly defined, but not inflexible, set of objectives, and (4) a synthesis of the first three elements into a plan. We believe that the prerequisites for developing a strategy are met sufficiently well in the case of the Finger Lakes to insure that a viable management plan could now be formulated.

In a previous publication (Oglesby, 1974) an attempt was made to provide management guidelines for the Finger Lakes. The following represents a modification of these and is provided to indicate some of the specific considerations which a management plan would have to address and the state of knowledge regarding each:

1. Summer concentrations of phytoplankton are primarily controlled by inputs of soluble phosphorus, modified by lake depth and volume. Regression models expressing these relations have been developed.

2. Sources of soluble phosphorus to the Finger Lakes have been quantitatively estimated. Human wastes dominate in most cases. Decreases in summer phytoplankton standing crop may be feasibly brought about by controlling phosphorus input in the form of municipal sewage discharge where this is an appreciable fraction of the phosphorus loading. Agriculture is not a major contributor of soluble phosphorus to the Finger Lakes, and control of this and all other diffuse sources poses difficult technical and economic questions (Casler and Jacobs, 1975).

3. Relatively strong data bases have been established for Skaneateles, Owasco, Cayuga, Hemlock, and Conesus Lakes. Others have been the subject of preliminary studies only. A regular program of surveillance should

be instituted for all of the Finger Lakes. In addition to providing the base line data needed for informed management, such a program would generate an ideal test of the regression models relating phytoplankton to soluble phosphorus loading.* An efficient surveillance program could be readily designed from knowledge gained in our previous studies.

4. Problems of excessive growths of rooted aquatic plants occur in some of the lakes. Repetitive cosmetic actions to remove plants in selective areas by mechanical or chemical means are the only techniques currently feasible for alleviating such nuisance growths. Technical advice is needed to assist lake associations and local governments in designing and implementing control programs. Institutional arrangements for organizing and funding such projects should be developed.

5. Both direct and indirect manipulation of planktonic and rooted flora will affect the quantity and perhaps quality of fish production. In general, decreases in primary productivity will result in a lowering of yields to fishermen (Oglesby, in press). For this reason it is imperative that management be carried out in an ecological framework rather than separated into the traditional categories of public health, water quality, and fisheries.

6. The regulation of lake level poses a substantial problem in the case of some of the Finger Lakes. Maximizing storage for water supply, maintaining unused storage capacity for downstream flood control, protection of shoreline from ice damage through winter lowering of water levels, and abstraction for water supply are regulatory measures frequently in conflict. The problem of damage from high water levels is exacerbated for some of the lakes by inadequate hydraulic capacity of the drainage network that carries their water to Lake Ontario. Comprehensive planning and regulation are needed to optimize the system.

Once information and understanding such as that illustrated in the above six observations are synthesized into a management plan, mechanisms for implementation have to be developed. An in-depth discussion of this is beyond the scope of the chapter, but, briefly, we believe that the value of the Finger Lakes is such that a considerable increase in resources for their proper management is justified. For example, a group of aquatic scientists should be permanently assigned to provide surveillance, develop approaches for dealing with specific problems, carry out informational programs, and

---

* A unique opportunity is provided by the New York State ban on phosphate builders in household laundry detergents. The effects of this on average summer phytoplankton concentrations is readily predicted from our method of calculating loading (Oglesby and Schaffner, 1975). If predictions were to be borne out by observations, this would provide a verification of the models.

provide user services such as wind and wave height information for boaters. Acting under the guidance of such a group, individual managers should be appointed to carry out more detailed programs for each of the larger lakes.

In summary, three points should be emphasized. First, knowledge of the Finger Lakes and an understanding of functional processes within them is sufficient to permit the formulation of rational management strategies. Second, the Finger Lakes are extremely valuable resources to the people of New York State, and a quantum increase in effort is both justified and needed to insure their future protection and optimal utilization. Finally, the Finger Lakes constitute a natural ecological and geographical unit and should be managed as such.

REFERENCES

Abraham, W. J. (1976a). "An Evaluation of the Hemlock Lake Trout Fishery," Mimeo. rep. New York State Department of Environmental Conservation, Albany, New York.

Abraham, W. J. (1976b). "An Evaluation of the Lake Trout Stocking Policy on Canadice Lake," Mimeo. rep. New York State Department of Environmental Conservation, Albany, New York.

Anonymous. (1932). "Fifteenth Census of the United States: 1930. Population," Vol. III, Part 2. US Govt. Printing Office, Washington, D.C.

Anonymous. (1966). "A List of Common and Scientific Names of Fishes from the United States and Canada," 2nd ed. Spec. Publ. No. 2. Am. Fish. Soc.

Anonymous. (1968). "Keuka Lake: Fisheries Investigations 1963–1968." New York State Conservation Department Bureau of Fish Management, Region 2, Olean, New York.

Anonymous. (1970a). "Census of Population. General Population Characteristics," Final Rep. PC (1)-B34, New York. US Govt. Printing Office, Washington, D.C.

Anonymous. (1970b). "Census of Housing. Block Statistics for Selected Areas of New York State," Final Rep. HC (3)-163. US Govt. Printing Office, Washington, D.C.

Anonymous. (1972). Allocthonous Pollutants in Seneca Lake—Summer 1972. Mimeo. Rep. National Science Foundation Student Originated Studies Program. Hobart and William Smith Colleges, Geneva, New York.

Anonymous. (1974a). Quality of public water supplies of New York. November 1970–April 1972. Open-file report. *U.S. Geol. Surv.*

Anonymous. (1974b). "Report on Keuka Lake," Working Pap. No. 160. U.S. Environmental Protection Agency—National Eutrophication Survey, Corvalis, Oregon.

Anonymous. (1974c). "Report on Owasco Lake," Working Pap. No. 163, U.S. Environmental Protection Agency—National Eutrophication Survey, Corvalis, Oregon.

Anonymous. (1974d). "Report on Seneca Lake," Working Pap. No. 170. U.S. Environmental Protection Agency—National Eutrophication Survey.

Anonymous. (1975a). "Water Quality," Appendix 7 to the Report of the Great Lakes Basin Framework Study. Public Information Office, Great Lakes Basin Commission, Ann Arbor, Michigan.

Anonymous. (1975b). Quality of public water supplies of New York. May 1972–May 1973. Open-file report. *U.S. Geol. Surv.*

Baston, L., Jr., and Ross, B. (1975). "The Distribution of Aquatic Weeds in the Finger Lakes of New York State and Recommendations for Their Control," Public Service Legislative Studies Program. New York State Assembly.

Berg, C. O. (1966). Middle Atlantic states. "Limnology in North America" (D. G. Frey, ed.), pp. 191–237. Univ. Wisconsin Press, Madison.

Birge, E. A., and Juday, C. (1914). A limnological study of the Finger Lakes of New York. Doc. No. 791. *Bull. U.S. Bur. Fish.* **32**, 525–609.

Birge, E. A., and Juday, C. (1921). Further limnological observations on the Finger Lakes of New York. Doc. No. 905. *Bull. U.S. Bur. Fish.* **37**, 210–252.

Bouldin, D. R., Johnson, A. H., and Lauer, D. A. (1975). The influence of human activity on the export of phosphorus and nitrate from Fall Creek. *In* "Nitrogen and Phosphorus: Food Production, Waste and the Environment" (K. H. Porter, ed.), pp. 59–120, Ann Arbor Sci. Publ., Ann Arbor, Michigan.

Carlson, R. E. (1975). "A Trophic State Index for Lakes," Contrib. No. 141. Limnol. Res. Cent., University of Minnesota, Minneapolis.

Carter, D. B. (1966). Climate. *In* "Geography of New York State" (J. H. Thompson, ed.), Syracuse Univ. Press, Syracuse, New York.

Casler, G. L., and Jacobs, J. J. (1975). Economic analysis of reducing phosphorus losses from agricultural production. *In* "Nitrogen and Phosphorus: Food Production, Waste and the Environment" (K. S. Porter, ed.), pp. 167–215. Ann Arbor Sci. Publ., Ann Arbor, Michigan.

Chamberlain, H. D. (1975). A comparative study of the zooplankton communities of Skaneateles, Owasco, Hemlock, and Conesus Lakes. Ph.D. Thesis, Cornell University, Ithaca, New York.

Child, D., Oglesby, R. T., and Raymond, L. S., Jr. (1971). Land Use Data for the Finger Lakes Region of New York State. Publ. No. 33. Cornell Univ. Water Res. Mar. Sci. Cent., Ithaca, New York.

Chiotti, T. (1976). "Compilation and Updating of Survey and Inventory Data," Appendix 201 to Federal Aid Proj. (F-34-R) Report. New York State Department of Environmental Conservation, Ithaca, New York.

Claassen, P. W. (1928). Biological studies of polluted waters in the Oswego watershed. *In* "A Biological Survey of the Oswego River Systems," Suppl. 17th Annu. Rep., 1927, Chapter VI, pp. 133–139. New York State Conservation Department, Albany, New York.

Cline, M. G., and Arnold, R. W. (1970). Working draft soil association maps for New York. Unpublished.

Collison, R. C., and Mensching, J. E. (1932). Lysimeter investigations II. Composition of rainwater at Geneva, N. Y., for a 10-year period. *N. Y., Agric. Exp. Stn., Geneva, Tech. Bull.* **1009**, 1–78.

Conklin, H. E., and Linton, R. E. (1969). "Economic Viability of Farm Areas" (a map). Cornell University, Ithaca, New York.

Crain, L. J. (1974). "Ground-Water Resources of the Western Oswego River Basin, New York," Basin Planning Rep. ORB-5. New York State Department of Environmental Conservation, Albany, New York.

Crain, L. J. (1975). "Chemical Quality of Ground Water in the Western Oswego River Basin, New York," Basin Planning Rep. ORB-3. New York State Department of Environmental Conservation, Albany, New York.

Dadswell, M. J. (1974). Distribution, ecology, and postglacial dispersal of certain crustaceans and fishes in eastern North America. *Natl. Mus. Can., Publ. Zool.* No. 11, pp. 1–110.

Davis, C. C. (1964). Evidence for the eutrophication of Lake Erie from the phytoplankton records. *Limnol. Oceanogr.* **9**, 275–283.

DeLaubenfels, D. J. (1966). Vegetation. *In* "Geography of New York State" (J. H. Thompson, ed.), pp. 90–103. Syracuse Univ. Press, Syracuse, New York.

Dethier, B. E. (1966). Precipitation in New York State. *N.Y. Agric. Exp. Stn. Ithaca, Bull.* **1009**, 1–78.

Dethier, B. E., and Pack, A. B. (1963a). "Climatological Summary," RURBAN Climate Ser. No. 1, Ithaca, New York. New York State College of Agriculture, Ithaca, New York.

Dethier, B. E., and Pack, A. B. (1963b). "Climatological Summary," RURBAN Climate Ser. No. 3, Geneva, New York. New York State College of Agriculture, Ithaca, New York.

Dillon, P. J., and Rigler, F. H. (1974). The phosphorus-chlorophyll relationship in lakes. *Limnol. Oceanogr.* **19**, 767–773.

Dudley, W. R. (1886). "The Cayuga Flora," Vol. II. Bull., Cornell University, Ithaca, New York.

Eaton, E. H. (1928). The Finger Lakes fish problem. *In* "A Biological Survey of the Oswego River System," Suppl. 17th Annu. Rep., 1927, Chapter II, pp. 40–66. New York State Conservation Department, Albany, New York.

Fisher, D. W., Gambell, A. W., Likens, G. E., and Bormann, F. H. (1968). Atmospheric contributions to water quality of streams in the Hubbard Brook Experimental Forest, New Hampshire. *Water Resour. Res.* **4**, 1115–1126.

Gorham, E. (1958). Atmospheric pollution by hydrochloric acid. *Q. J. R. Meteorol. Soc.* **84**, 274–276.

Greeley, J. R. (1927). Fishes of the Genesee region with annotated list. *In* "A Biological Survey of the Genesee River System," Suppl. 16th Annu. Rep., 1926, Chapter IV, pp. 47–66. New York State Department of Conservation, Albany, New York.

Greeley, J. R. (1928). Fishes of the Oswego watershed. *In* "A Biological Survey of the Oswego River System," Suppl. 17th Annu. Rep., 1927, Chapter IV, pp. 84–107. New York State Conservation Department, Albany, New York.

Greeson, P. E., and Williams, G. E. (1970). Characteristics of New York lakes. Part 1B. Gazeteer of lakes, ponds and reservoirs by drainage basins. *U.S., Geol. Surv., Bull.* **68B**, 1–122.

Hall, D. J., and Waterman, G. C. (1967). Zooplankton of the Finger Lakes. *Limnol. Oceanogr.* **12**, 542–544.

Harman, W. N. (1972). Benthic substrates: Their effect on fresh-water Mollusca. *Ecology* **53**, 271–277.

Harman, W. N., and Berg, C. O. (1970). Fresh-water Mollusca of the Finger Lakes region of New York. *Ohio J. Sci.* **70**, 146–150.

Harvey, H. H., and Fry, F. E. J. (1973). Sport fish index. *In* "The Approach, Theory, Methodology and Application of a Lakeshore Capacity Model," Environ. Sci. Publ. EG-10, pp. 139–190. University of Toronto.

Hayes, F. R. (1957). On the variation in bottom fauna and fish yield in relation to trophic level and lake dimensions. *Trans. Am. Fish. Soc.* **14**, 1–32.

Hayes, F. R., and Anthony, E. H. (1964). Productive capacity of North American lakes as related to the quantity and the trophic level of fish, the lake dimensions and the water chemistry. *Trans. Am. Fish. Soc.* **93**, 53–57.

Henderson, J. F., Ryder, R. A., and Kudhongania, A. W. (1973). Assessing fishery potentials of lakes and reservoirs. *J. Fish. Res. Board Can.* **30**, 2000–2009.

Holland, R. E., and Beeton, A. M. (1972). Significance to eutrophication of spatial differences in nutrients and diatoms in Lake Michigan. *Limnol. Oceanogr.* **17**, 88–96.

Hutchinson, G. E. (1967). "A Treatise on Limnology," Vol. 2. Wiley, New York.

Johnson, N. M., Likens, G. E., Bormann, F. H., Fisher, D. W., and Pierce, R. S. (1969). A working model for the variation in streamwater chemistry at the Hubbard Brook Experimental Forest, New Hampshire. *Water Resour. Res.* **5**, 1353–1363.

Juday, C. (1925). *Senecella calanoides,* a recently described fresh-water copepod. *Proc. U.S. Natl. Mus.* **66**, 1–6.

Kling, H. J., and Holmgren, S. K. (1972). Species composition and seasonal distribution in the Experimental Lakes Area, northwestern Ontario. *Fish. Res. Board Can., Tech. Rep.* **337**, 1–56.

Knox, C. E., and Nordenson, T. J. (1955). Average annual runoff and precipitation in the New England-New York area. *U.S., Geol. Surv., Hydrol. Inv. Atlas* **HA-7**, 1–6.

Larkin, P. A. (1964). Canadian lakes. *Verh. Int. Ver. Theor. Angew. Limnol.* **15**, 76–90.

Likens, G. E. (1972). "The Chemistry of Precipitation in the Central Finger Lakes Region," Publ. No. 50. Cornell Univ. Water Resour. and Mar. Sci. Cent., Ithaca, New York.

Likens, G. E. (1974a). "The Runoff of Water and Nutrients from Watersheds to Cayuga Lake, New York," Publ. No. 81. Cornell Univ. Water Resour. and Mar. Sci. Cent., Ithaca, New York.

Likens, G. E. (1974b). "Water and Nutrient Budgets for Cayuga Lake, New York," Publ. No. 82. Cornell Univ. Water Resour. and Mar. Sci. Cent., Ithaca, New York.

Likens, G. E., and Bormann, F. H. (1972). Nutrient cycling. *In* "Ecosystems, Structure and Function" (J. Weins, ed.), pp. 25–67. Oregon State Univ. Press, Corvallis.

Likens, G. F., Bormann, F. H., Johnson, N. M., Fisher, D. W., and Pierce, R. S. (1970). Effects of forest cutting and herbicide treatment on nutrient budgets in the Hubbard Brook watershed-ecosystem. *Ecol. Monogr.* **40**, 23–47.

MacIntire, W. H., and Young, J. B. (1923). Sulfur, calcium, magnesium and potassium content and reaction of rainfall in different points in Tennessee. *Soil Sci.* **15**, 205–227.

Maylath, R. E. (1973). Provisional report on Seneca Lake chlorides. New York State Department of Environmental Conservation. Unpublished report.

Michalski, M. F. P., Johnson, M. G., and Veal, D. M. (1973). "Muskoka Lakes Water Quality Evaluation," Rep. No. 3. Ontario Ministry of the Environment.

Mills, E. L. (1975). Phytoplankton composition and comparative limnology of four Finger Lakes, with emphasis on lake typology. Ph.D. Thesis, Cornell University, Ithaca, New York.

Moore, E. (1928). Introduction. *In* "A Biological Survey of the Oswego River System," Suppl. 17th Annu. Rep., 1927, pp. 9–16. New York State Conservation Department, Albany, New York.

Muenscher, W. C. (1928). Plankton studies of Cayuga, Seneca and Oneida Lakes. *In* "A Biological Survey of the Oswego River System," Suppl. 17th Annu. Rep., 1927, Chapter VII, pp. 140–157. New York State Conservation Department, Albany, New York.

Northcote, T. G., and Larkin, P. A. (1956). Indices of productivity in British Columbia lakes. *J. Fish. Res. Board Can.* **13**, 515–540.

Oglesby, R. T. (1974). "Limnological Guidance for Finger Lakes Management," Publ. No. 89. Cornell Univ. Water Resour. and Mar. Sci. Cent., Ithaca, New York.

Oglesby, R. T. (1976). "The Limnology of Cayuga Lake, New York," Chapter 1, this volume.

Oglesby, R. T. (In press). "Community composition and yield of fish in relation to lake phytoplankton standing crop and production and to morphoedaphic factors," *J. Fish. Res. Board Can.*

Oglesby, R. T., and Schaffner, W. R. (1975). The response of lakes to phosphorus. *In* Nitrogen and Phosphorus: Food Production, Waste and the Environment" (K. S. Porter, ed.), Chapter 2, pp. 25–57. Ann Arbor Sci. Publ., Ann Arbor, Michigan.

Oglesby, R. T., Hamilton, L. S., Mills, E. L., and Willing, P. (1973). "Owasco Lake and its

Watershed," Publ. No. 70. Cornell Univ. Water Resour. and Mar. Sci. Cent., Ithaca, New York.

Oglesby, R. T., Schaffner, W. R., and Mills, E. L. (1975). "Nitrogen, Phosphorus and Eutrophication in the Finger Lakes," Publ. No. 94. Cornell Univ. Water Resour. and Mar. Sci. Cent., Ithaca, New York.

Oglesby, R. T., Vogel, A., Peverly, J. H., and Johnson, R. (1976). Changes in submerged plants at the south end of Cayuga Lake following tropical storm Agnes. *Hydrobiologia* **48**, 251–255.

Olsen, G. W., Witty, J. E., and Marshall, R. L. (1969). Soils and their use in the five-county area around Syracuse. *Cornell Misc. Bull.* **80.**

Parker, C. E. (1975). Finger Lakes fishing. *Conservationist* **29**, 18–47.

Patalas, K. (1972). Crustacean plankton and the eutrophication of St. Lawrence Great Lakes. *J. Fish. Res. Board Can.* **29**, 1451–1462.

Pearson, F. J., Jr., and Fisher, D. W. (1971). Chemical composition of atmospheric precipitation in the northeastern United States. *U.S., Geol. Surv., Water-Supply Pap.* **1535-P**, 1–23.

Raffieux, P. (1671–1672). The Jesuit reflections and other documents. **56**, 48–52.

Rawson, D. S. (1951). The total mineral content of lake waters. *Ecology* **32**, 669–972.

Rawson, D. S. (1952). Mean depth and fish production of large lakes. *Ecology* **33**, 513–521.

Rawson, D. S. (1955). Morphometry as a dominant factor in the productivity of large lakes. *Verh. Int. Ver. Theor. Angew. Limnol.* **12**, 164–175.

Rawson, D. S. (1960). A limnological comparison of twelve large lakes in northern Saskatchewan. *Limnol. Oceanogr.* **5**, 195–211.

Rayback, R. J. (1966). The Indian. *In* "Geography of New York State" (J. H. Thompson, ed.), pp. 113–120. Syracuse Univ. Press, New York, New York.

Rickard, L. V., and Fisher, D. W. (1970). "Geological Map of New York. Finger Lakes Sheet." New York State Museum and Science Service, Albany, New York.

Ryder, R. A. (1961). Fisheries management in northern Ontario. *Ont. Fish. Wildl. Rev.* **1**, 13–19.

Ryder, R. A. (1965). A method for estimating the potential fish production of north-temperate lakes. *Trans. Am. Fish. Soc.* **94**, 214–218.

Ryder, R. A., Kerr, S. R., Loftus, K. H., and Regier, H. A. (1974). The morphoedaphic index, a fish yield estimator—review and evaluation. *J. Fish. Res. Board Can.* **31**, 663–688.

Sakamoto, M. (1966). Primary production by phytoplankton community in some Japanese lakes and its dependence on lake depth. *Arch. Hydrobiol.* **62**, 1–28.

Sawyer, C. N. (1947). Fertilization of lakes by agriculture and urban drainage. *J. N. Engl. Waterworks Assoc.* **61**, 109–127.

Schaffner, W. R., and Oglesby, R. T. (1976). "A Review of Trophic State Indices for Selected New York Lakes," Mimeo. Rep. U.S. Environmental Protection Agency, Washington, D.C.

Schindler, D. W., and Nighswander, J. E. (1970). Nutrient supply and primary production in Clear Lake, eastern Ontario. *J. Fish. Res. Board Can.* **27**, 2009–2036.

Scott, W. B., and Crossman, E. J. (1973). Freshwater fishes of Canada. *Bull., Fish. Res. Board Can.* **184**, 1–966.

Shampine, W. J. (1973). "Chemical Quality of the Surface Water in the Eastern Oswego River Basin, New York," Basin Planning Rep. ORB-6. New York State Department of Environmental Conservation, Albany, New York.

Stewart, K. M., and Markello, S. J. (1974). Seasonal variations in concentrations of nitrate and total phosphorus, and calculated nutrient loading for six lakes in western New York. *Hydrobiologa* **44**, 61–89.

Stout, N. J. (1958). Atlas of forestry in New York. *State Univ. Coll. For. Syracuse Univ., Bull.* **41.**

Thornthwaite, C. W. (1948). An approach toward a rational classification of climate. *Geol. Rev.* **38,** 55–94.

Train, R. E., Cahn, R., and MacDonald, G. J. (1970). "Environmental Quality. The First Annual Report of the Council on Environmental Quality." US Govt. Printing Office, Washington, D.C.

Vollenweider, R. A. (1968). Scientific fundamentals of the eutrophication of lakes and flowing waters, with particular reference to phosphorus and nitrogen as factors in eutrophication. *OECD Tech. Rep.* **DAS/CS1/68.27,** 1–159.

Vollenweider, R. A. (1975). Input-output models with special reference to the phosphorus loading concept in limnology. *Schweiz. Z. Hydrol.* **37,** 53–83.

Vollenweider, R. A. (1976). Advances in defining critical loading levels for phosphorus in lake eutrophication. *Mem. Ist. Ital. Idrobiol. Dott. Marco de March:* **33,** 53–83.

Vollenweider, R. A., and Dillon, P. J. (1974). The application of the phosphorus loading concept to eutrophication research. *Natl. Res. Counc. Can., Publ.* **13690,** 1–42.

Vollenweider, R. A., and Nauwerck, A. (1961). Some observations on the C-14 method for measuring primary production. *Verh. Int. Ver. Theor. Angew. Limnol.* **14,** 134–139.

von Engeln, O. D. (1961). "The Finger Lakes Region: Its Origin and Nature." Cornell Univ. Press, Ithaca, New York.

Wagner, F. B. (1928). Chemical investigation of the Oswego watershed. *In* "A Biological Survey of the Oswego River System," Suppl. 17th Annu. Rep., 1927, Chapter V, pp. 108–132. New York State Conservation Department, Albany, New York.

Welcomme, R. L. (1976). Some general and theoretical considerations on the fish yield of African rivers. *J. Fish Biol.* **8,** 351–364.

Whitehead, H. C., and Feth, J. H. (1964). Chemical composition of rain, dry fallout, and bulk precipitation at Menlo Park, California, 1957–59. *J. Geophys. Res.* **69,** 3319–3333.

Wright, T. D. (1969). Plant nutrients. *In* "Ecology of Cayuga Lake and the Proposed Bell Station (Nuclear Powered)" (R. T. Oglesby and D. J. Allee, eds.), Publ. No. 27, pp. 197–214. Cornell Univ. Water Resour. and Mar. Sci. Cent., Ithaca, New York.

Index